Sustainable Landscaping

Sustainable Landscaping
Principles and Practices

Second Edition

Marietta Loehrlein

CRC Press is an imprint of the
Taylor & Francis Group, an **informa** business

Second edition published 2021
by CRC Press
6000 Broken Sound Parkway NW, Suite 300, Boca Raton, FL 33487-2742

and by CRC Press
2 Park Square, Milton Park, Abingdon, Oxon, OX14 4RN

© 2021 Taylor & Francis Group, LLC

First edition published by CRC Press 2013

CRC Press is an imprint of Taylor & Francis Group, LLC

Reasonable efforts have been made to publish reliable data and information, but the author and publisher cannot assume responsibility for the validity of all materials or the consequences of their use. The authors and publishers have attempted to trace the copyright holders of all material reproduced in this publication and apologize to copyright holders if permission to publish in this form has not been obtained. If any copyright material has not been acknowledged please write and let us know so we may rectify in any future reprint.

Except as permitted under U.S. Copyright Law, no part of this book may be reprinted, reproduced, transmitted, or utilized in any form by any electronic, mechanical, or other means, now known or hereafter invented, including photocopying, microfilming, and recording, or in any information storage or retrieval system, without written permission from the publishers.

For permission to photocopy or use material electronically from this work, access www.copyright.com or contact the Copyright Clearance Center, Inc. (CCC), 222 Rosewood Drive, Danvers, MA 01923, 978-750-8400. For works that are not available on CCC please contact mpkbookspermissions@tandf.co.uk

Trademark notice: Product or corporate names may be trademarks or registered trademarks, and are used only for identification and explanation without intent to infringe.

Library of Congress Cataloging-in-Publication Data
Library of Congress Control Number: 2020937412

ISBN: 9780367250898 (hbk)
ISBN: 9780367559755 (pbk)
ISBN: 9780429285974 (ebk)

Typeset in Times
by codeMantra

Contents

Preface xv
Author xvii

Chapter 1 Sustainable Landscaping 1

Terms to Know 1
What Is Sustainable Landscaping? 1
History and Background 2
Sustainable Sites Initiative 2
Environmental Issues and Landscaping 5
    Global Climate Change 6
        Carbon Emissions 8
    Air Quality 8
    Water Issues 9
    Pesticide Use and Toxicity 9
    The Waste Stream 10
Non-Renewable and Renewable Resources 10
The Role of the Landscaping Industry in Sustainability ... 11
Sustainability Audit 13
    How to Implement the Sustainability Audit 13
Summary 13
Review Questions 15
Enrichment Activities 15
Further Reading 15

Chapter 2 Sustainability in the Plantscape 17

Terms to Know 17
Introduction 17
Environment Aspects of Plants 18
    Carbon Sequestration 18
    Oxygen Release 19
    Cooling Effect of Plants 19
    Structural Effects of Plants 20
Turfgrass 20
    Drought-tolerant Turf 20
    Salt-Tolerant Turf 21
    Weeds in Turf 21
    Aeration 21
    Mowing 21
Woody Plants 22
    Pruning 22

	Invasive Plants	23
	Ecology and Plants	26
	Ecoregions of the United States	27
	Forests and Woodlands	27
	Prairies	29
	Meadows	30
	Riparian Zone Habitat	32
	Desert Ecosystems	34
	Ecological Landscaping	37
	Native and Non-Native Plant Selection	37
	Attracting Wildlife	38
	Native Pollinators	38
	Birds and Mammals	39
	Fire-wise Landscaping	45
	Programs for Habitat Development	45
	The Backyard Habitat	45
	The Golf Course Habitat	45
	Summary	47
	Review Questions	48
	Enrichment Activities	48
	Further Reading	48
Chapter 3	The Sun and the Sustainable Landscape	51
	Terms to Know	51
	Introduction	52
	Studies Related to the Effects of the Sun on the Landscape	52
	Solar Energy	54
	Solar Heat Gain	56
	Reflecting and Absorbing Light	56
	Thermal Emissivity	56
	Heat Capacity	57
	Solar Reflectance Index	58
	Heat Loss and Heat Transfer	59
	Urban Heat Island	59
	Landscaping Practices to Mitigate the Urban Heat Island Effect	61
	Cooling Effect of Plants	62
	Shading Air Conditioners	65
	Green Roofs	65
	Cooling Paved Surfaces	65
	Other Considerations	67
	Structure Orientation	67
	Sun Exposure	67
	Reducing Heat Gain in Summer	67
	Increasing Heat Gain in Winter	68
	Optimizing Solar Incidence for Warmth in Winter	68

Contents

	Summary	68
	Review Questions	69
	Enrichment Activities	69
	Further Reading	69

Chapter 4 The Wind and Energy Conservation 73

- Terms to Know 73
- Introduction 73
- Trapping Cold Air on a Slope 75
- Planting for Insulative Properties 75
- The Cooling Effects of Wind 75
- Windbreaks to Reduce Heat Loss 77
  - Energy Usage in Winter 78
  - Designing the Windbreak 79
    - Height 79
    - Shape and Size 80
    - Plants for a Windbreak 80
- Wind in the Urban Landscape 81
  - The Wind and Urban Pollution 82
- Summary 83
- Review Questions 84
- Activities 84
- Further Reading 84

Chapter 5 Water Issues 87

- Terms to Know 87
- Introduction 88
- The Water Cycle 88
- Water Sources 88
- Potable Water 90
- Polluted Water 90
  - Excessive Nutrients in Water 90
  - Pesticides in Water 92
    - Human and Environmental Effects of Pesticides in Water 93
    - Contributions from Urban Areas 93
    - Insecticides 95
    - Herbicides 96
    - Other Pollutants 97
- Preventing and Treating Contaminated Water 97
- Reducing Use of Pesticides in the Landscape 97
- Bioremediation and Phytoremediation 98
- Wetlands and Constructed Wetlands 98
  - The Nature, Function, and Value of Wetlands 100
  - Components of a Wetland 100

	Constructed Wetlands 100
	Constructed Wetlands Design 101
	Siting a Constructed Wetland 103
	Natural Components of a Constructed Wetland 103
	Water in a Wetland 103
	Substrates in a Wetland 104
	Plants in a Wetland 104
	Built Components of a Constructed Wetland 104
	Summary 105
	Review Questions 106
	Activities 106
	Further Reading 106
Chapter 6	Water Conservation 109
	Terms to Know 109
	Introduction 109
	Precipitation 110
	Drought and Water Shortage 110
	Plant Water Requirements 111
	Rainwater Collection Systems 112
	Above-Ground Water Storage 112
	Below-Ground Water Storage 113
	Calculating Rainfall Amounts 114
	Irrigation and Water-Use Efficiency 114
	Drought-Tolerant Plants 116
	Water-Wise Gardening 116
	Mulch 119
	How to Apply Mulch 119
	Effectiveness of Different Mulches 120
	Gray-Water Use 121
	Timing of Gray-Water Usage 122
	Problems to Avoid 123
	Gray-Water Delivery 123
	Plant Safety Concerns 123
	Summary 125
	Review Questions 126
	Enrichment Activities 127
	References 127
Chapter 7	Managing Excess Water in the Landscape 131
	Terms to Know 131
	Introduction 131
	Stormwater Runoff 133
	The Urban Water Cycle 135

Contents

Solutions to Excess Water in the Landscape 135
    Drainage 136
    Landscape Swales 136
    Rain Gardens 136
        Siting a Rain Garden 137
        Calculating the Area Required for a Rain Garden 137
        How to Build a Rain Garden 138
        Plants for Rain Gardens 138
    Rainwater Collection 139
    Permeable Pavement Materials 140
    Green Roofs 141
        Green Roofs and Media Depth 143
        Green Roof Design 143
        Plants for Green Roofs 144
Green Walls 144
Summary 145
Review Questions 146
Enrichment Activities 146
Further Reading 146

Chapter 8   Soil Health 149

Terms to Know 149
Introduction 149
Soils and Construction Activities 151
Brownfields 152
Soil Testing 153
    Physical Properties of Soil 154
        Soil Texture 154
        Soil Structure 156
    Chemical Properties of Soil 158
        Soil pH 158
        Cation Exchange Capacity 158
        Salinity and Deicers 159
Addressing Problems with Soil Chemistry 160
Soil Organic Matter 161
    Living Organisms in Soil 161
        Earthworms 161
        Nematodes 162
        Decomposers 162
        Symbionts 162
Essential Plant Nutrients 164
Improving Soil Health for Landscaping 164
    Correcting Compaction 165
    Preserving and Replacing Topsoil 165
    Reducing Subsoil at the Surface 165

Providing Adequate Soil Quantity for Root Growth 166
Amending with Organic Matter 166
Composting 166
Using Mulch 166
Sustainable Fertilization 167
Summary 167
Review Questions 167
Enrichment Activities 168
Further Reading 168

Chapter 9 Sustainable Fertilization 171

Terms to Know 171
Introduction 171
Fertilizer 172
Toxic Fertilizers 172
Plant Fertilizer Requirements 173
Nitrogen 173
Phosphorus and Potassium 173
Calcium and Magnesium 174
Sulfur 174
Iron 175
Forms of Fertilizers 175
Fertilizer Sources 175
Mineral Fertilizers 176
Nitrogen 176
Phosphorus 176
Potassium 178
Effect of Inorganic Fertilizers on Soil Health 178
Contamination of the Environment 178
Reducing Nutrient Runoff and Leaching 178
Organic Fertilizers 179
Animal Manure 179
Municipal Solid Waste 182
Compost Solutions 184
Other Organic Fertilizers 185
Green Manure and Inter-Planting 185
Summary 186
Review Questions 187
Enrichment Activities 187
Further Reading 187

Chapter 10 Improving Landscape Soils with Organic Matter 189

Terms to Know 189
Introduction 189

Contents

- Organic Matter in the Landscape 190
- Fate of Organic Matter 191
- Organic Matter and Soil Health 191
- Types of Organic Matter for Landscaped Areas 192
  - Organic Amendments 192
  - Mulch 192
    - Mulch Materials 194
    - How to Apply Mulch 195
    - Problems with Mulch 196
  - Organic Soil Amendments 196
    - Peat Moss 197
    - Compost 197
      - How to Build a Compost Bin System 197
      - What to Add to the Compost 197
      - Carbon and Nitrogen: Finding the Right Balance 198
      - Proper Conditions 198
      - Moisture 199
      - Turning 199
      - Compost Solutions 199
  - Grass Clippings 199
- Summary 199
- Review Questions 200
- Enrichment Activities 201
- Further Reading 201

Chapter 11 Pesticides in the Landscape 203

- Terms to Know 203
- Introduction 204
- Pesticide Use in the Landscape 205
- Types of Pesticides 208
  - Synthetic Pesticides 208
  - Naturally Occurring Pesticides 209
- Pesticide Regulation 209
  - FIFRA 210
  - FFDCA 210
    - The EPA and DDT 210
  - Restricted Use Pesticides 211
- Human Health Hazards 211
  - Epidemiology 212
  - Acute and Chronic Effects 214
  - Carcinogens 214
  - Teratogenic Effects 215
  - Endocrine Disrupters 215
- Cholinesterase Inhibitors 216
- Other Health Effects 218

Environmental Hazards..218
Pesticide Handling...219
    Safety Issues...219
    Storage and Disposal..220
Summary...221
Review Questions..222
Enrichment Activities...222
Further Reading...223

Chapter 12   Integrated Pest Management...225

Terms to Know..225
Introduction...226
Avoidance..227
Cultural Practices...227
    Sanitation...228
    Proper Irrigation/Watering..228
    Soil Health and Compaction...229
Genetically Improved Plants...229
Treatment...230
    Determining Pest or Disease Presence...........................231
        Sticky Traps...232
        Pheromone Traps...233
        Branch Beating..234
Phenology and Degree Days...234
Action Thresholds..236
Economic Thresholds..236
Aesthetic Injury Level...237
Alternative Pest Controls..237
Biological Controls..237
    Botanicals...239
    Non-toxic Pesticides..239
    Insect Growth Regulators...240
Summary..241
Review Questions...241
Activities..242
Further Reading...242

Chapter 13   Energy: Sources and Uses..243

Terms to Know..243
Introduction...244
Energy Sources..245
Scope of the Problem...245
    Non-renewable Resource..245
    Expense...245

Contents xiii

       Emissions 245
       Other Pressures 247
   Government Support for Renewable Energy 247
   Energy for Electricity 248
       Generation of Electricity 248
            Non-renewable Fossil Fuel Energy 249
            Renewable Energy 252
       Fuel for Tools, Equipment, and Transportation 254
            Non-renewable 254
            Renewable 256
            Electric Vehicles 257
            Hydrogen Power 257
   Energy-Efficient Lighting 258
   Summary 258
   Review Questions 259
   Activities 259
   References 260

Chapter 14 Tools and Equipment 261

   Terms to know 261
   Introduction 261
   Power Tools used in the Landscape 262
   Types of Engines 263
   Sustainability Issues Concerning Landscape Tools
   and Equipment 264
       Air Pollution from Landscape Tools and Equipment 264
       Fugitive Dust and Particulate Matter 266
       Noise from Landscape Tools and Equipment 266
       Solutions 267
            Reducing Emissions 268
            Other Technological Advances 268
            Reducing Noise 269
            Sidewalk Vacuum 269
            Reducing use of Power Tools 269
            Transportation Efficiencies 269
            Landscape Design 270
            Landscape Maintenance 271
   Summary 271
   Review Questions 272
   Activities 272
   Further Reading 273

Chapter 15 Sustainable Landscape Materials and Products 275

   Key Terms 275

Introduction 275
Landscape Construction Materials.. 276
Recycled Materials for Landscape Products 278
Life Cycle Assessment 278
Waste Management 279
Recycled Materials 280
    Plastic Lumber 280
    Rubber 280
    Crumb Rubber 281
    Concrete and Asphalt 282
    Glass 282
Renewable Resources 283
    Sustainably Harvested Lumber 283
Salvaged Materials 285
    Urban Wood 285
Local Materials 285
Summary 286
Review Questions 287
Activities 287
References 288

Appendix A: Sustainability Audit 289
Appendix B: Important Websites Used as Resources in this Book 297
Index 299

Preface

WHAT DOES "SUSTAINABLE" MEAN?

The idea of sustainability refers to the impacts of human activity on the earth and its environment. The commonly used definition originated in the United Nations Brundtland Report, also called *Our Common Future*, which states that sustainable development is "development that meets the needs of the present without compromising the ability of future generations to meet their own needs".

Landscaping brings beauty to our surroundings. It can provide the soothing, healthful effects of the experience of nature. Such effects are numerous and well-documented, and contribute to both our physical and emotional well-being. However, in the quest for beauty and even perfection in nature, the landscaping industry uses some practices that are harmful to our environment.

Sustainable practices are those practices which attempt to minimize or eliminate harm to our environment. Examples include water conservation, preventing erosion and soil degradation, minimizing air, soil, and water pollution, and reducing greenhouse gas emissions.

In the context of this book, I use the term *sustainable* to refer to solutions to the problems of environmental harm caused by humans in the process of constructing, implementing, and managing our residential and commercial landscapes.

HOW DOES IT WORK?

Beginning with the design process, a landscape audit is done. In an existing landscape to be renovated, the audit itemizes an inventory on the site and examines unsustainable practices, with a goal of replacing those with sustainable practices. In a new development, sustainable practices are integrated into the overall design process. Appendix A provides a case study using the audit on an existing landscape and discusses the results of the recommended changes.

Chapter 1 provides an overview and context of landscape sustainability. Each chapter that follows provides a detailed look at plant usage, sun, wind, water, soil and fertilizer, pesticides, energy usage, tools and equipment, and landscape materials.

I have drawn from many sources to provide a comprehensive view of all aspects of the practice of landscaping. At the end of each chapter, I have provided a "Further Reading" list that includes sources that were used within the chapter. I encourage the reader to refer to those sources for more in-depth coverage of the topics. The reader may also refer to Appendix B for a succinct list of important websites that provided information for this book.

Undisputedly, the aim of landscaping is to provide aesthetically pleasing surroundings in a built environment. The goal of this book is to provide a means of accomplishing that while doing the least harm to the environment. An important component of a successful outcome may very well include a shift in the concept of what is visually pleasing. For example, a manicured lawn may be replaced by a planting of native plants. This could look like a grouping of trees, shrubs, or a wildflower planting. Such a planting has the benefits of providing habitat for birds and pollinators as well as creating a sense of space with its own character and charm.

Another example would be replacing individual trees surrounded by mulch islands with a grouping of trees with a larger area of mulch tying the group of trees together. This eliminates lawn, which has different horticultural requirements than the trees, and turns the area into more of a replica of a small woodland. Benefits include better tree health and longer-lived trees, reduced gas emissions from mowing the lawn, and potentially reduced water and fertilizer use on that turf area. The case studies in each chapter provide examples of projects creating aesthetically pleasing areas while addressing environmental concerns with beneficial results.

This book is a call to everyone involved in the landscape industry to imagine a new aesthetic, an aesthetic that places an emphasis on the environmental outcomes and then considers how best to work within that context. I believe that this approach can bring forth many creative, effective, and beautiful solutions.

Author

Dr. Marietta Loehrlein is a Professor Emeritus of Horticulture and Landscaping at Western Illinois University in Macomb, Illinois, USA.

While there, she developed a new course, Sustainable Landscaping, for which there were no textbooks, so she undertook the project of writing a textbook that would be immediately pertinent to the topic. She didn't want to simply address the "how-to" of sustainable landscaping, but also to examine the related issues, such as energy sources, landscape tools, equipment, and materials, and soil- and water-related environmental issues.

Her half-acre backyard is a showcase of sustainable landscaping practices: she has reduced what had been an all-turf lawn area by planting a small prairie, a small woodland, and many species of native trees and shrubs. The stream that runs through the property supports a riparian community that facilitates spring bird migration. The native garden areas support a large number of wildlife species, some of which are never seen, as they are nocturnal and/or live in subterranean habitats. However, ground-dwelling bees and at least a dozen species of songbirds are regularly seen. It is both a pollinator-friendly area and a certified wildlife habitat by the National Wildlife Federation.

Dr. Loehrlein previously published *Home Horticulture: Principles and Practices* (Cengage). She is an Evansville, Indiana, native, earned her college degrees at the University of Arizona (B.S., M.S.) and the Pennsylvania State University (Ph.D.), and was a research associate for Sun World International in a fruit tree breeding program in Central California. She holds a patent on the regal pelargonium "Camelot".

1 Sustainable Landscaping

OBJECTIVES

- Understand the history of current sustainability issues
- Identify the major issues of sustainable landscaping
- Name sustainable landscape practices
- Discuss the Sustainable Sites Initiative (SITES™) and its role in landscaping
- Compare and contrast the sustainable landscape audit with the conventional site analysis

TERMS TO KNOW

Anthropogenic climate change
Ecosystem services
Global climate change
Greenhouse gases
LEED
Non-renewable resources
Renewable resources
Sustainable Sites Initiative
Sustainable landscaping

WHAT IS SUSTAINABLE LANDSCAPING?

In this text, **sustainable landscaping** refers to landscape practices that preserve our planet and our environment without depleting and damaging our air, water, and soil. In other words, it promotes practices that support and nurture all life forms and their habitats. Sustainable landscape practices address the issues of **renewable** and **non-renewable resources**; emission of **greenhouse gases** that contribute to **global climate change**; and air, water, and soil quality. Global climate change refers to a change in the average global temperature. **Anthropogenic** climate change is due to human activities, such as release of carbon dioxide or other **greenhouse gases**.

Sustainable landscape practices address a number of different issues. They include the following:

- Proper plant selection and placement
- Using plants for energy efficiency
- Creating and preserving wildlife habitat
- Designing for energy efficiency
- Landscaping for wildfire safety
- Bioremediation, phytoremediation, and constructed wetlands

- Water conservation and aquifer replenishment
- Stormwater management
- Protecting, building, and maintaining healthy soil
- Responsible fertilizing
- Reducing or eliminating pesticide usage
- Reducing emission of **greenhouse gases** in landscape practices
- Protecting air and water quality
- The use of **renewable** and **non-renewable resources** in landscape materials.

HISTORY AND BACKGROUND

Sustainable landscaping does not have a single line of development that can be traced back through time. There are many individuals, groups of private individuals, governmental entities, and others who have developed ideas, engaged in and promoted practices, and educated others about the issues that comprise sustainability (Table 1.1). Doxon (1991, 1996) and Fretz et al. (1993) explored the idea of sustainability and how it applies to horticulture and landscaping. These papers are some of the earliest published in the academic horticulture literature on the topic.

There are various approaches to sustainability in landscaping practices. Over the years, there have been differing parameters included in what constituted sustainability. There are also, by necessity, regional differences. The term *sustainability* has not always been used to discuss the concepts that this book addresses. Some of the other terms that apply to certain aspects of the umbrella term *sustainable landscaping* include organic practices, ecological design, energy-efficient landscapes, regenerative design, and water-wise landscaping. There are some who prefer to use words other than "sustainable" because they feel it has been overused, or may have negative connotations. Nevertheless, sustainability is a widely used term, and has proven to be flexible enough to incorporate many different issues and practices. Table 1.2 shows some of the developments occurring in sustainable landscaping at state and regional levels.

SUSTAINABLE SITES INITIATIVE

In 1998, The US Green Building Council (USGBC) developed the Leadership in Energy and Environmental Design (LEED®) rating system for building construction. Points are assigned for addressing various environmental issues related to building construction. Energy efficiency, water management, reducing carbon dioxide emissions, improved indoor environmental quality, and stewardship of resources are some of the major areas addressed in this points-based system.

One component of LEED certification addressed landscaping issues. However, this became more fully developed when a steering committee led by the American Society of Landscape Architects (ASLA) was formed. The outcome of that organization is the **Sustainable Sites Initiative** (SITES™). The ASLA worked in conjunction with the Lady Bird Johnson Wildflower Center and the United States Botanic Garden, and together they released a preliminary report in 2007. The goal of the SITES™ is to foster a transformation in land development and management practices.

TABLE 1.1
Events Related to Sustainable Landscaping Practices

Year	Event	Comments
1947	Federal Insecticide, Fungicide, and Rodenticide Act (FIFRA) passed	Federal law regulating the use and labeling of pesticides
1948	The Federal Water Pollution Control Act passed	The first major US law to address water pollution
1949	*Sand County Almanac* by Aldo Leopold published	Author proposes "The Land Ethic" idea
1962	*Silent Spring* by Rachel Carson published	About the dangers of pesticides on human health and the environment
1969	National Environmental Policy Act passed (signed on January 1, 1970)	An Act to establish a national policy for the environment
1970	Rabb and Guthrie	Coined the term "Integrated Pest Management"
1970	The First Earth Day	Marked the beginning of the "environmental movement"
1970	The Clean Air Act signed	Established air quality standards to protect the public
1972	DDT banned in the United States	High-profile case of pesticide toxicity and persistence in the environment brought to the public's attention by Rachel Carson
1977	The Clean Water Act passed	Major amendments to the Water Pollution Act of 1948
1978	Permaculture developed by Bill Mollison	Provided a systems approach to sustainable agriculture
1984	Life Cycle Assessment	Provided a framework for evaluating environmental impact of products and materials
1987	*Our Common Future* (Brundtland Report for the United Nations) published	Released report with definition of "sustainability"
1987	Audubon International founded	Fosters environmental awareness and practices in communities and golf courses through certification programs
1992	The first Earth Summit was held in Rio de Janiero, Brazil	Laid international groundwork for sustainable practices in many facets of human activity
1993	Forest Stewardship Council founded	Developed standards for sustainably harvested lumber
1998	US Green Building Council establishes LEED (Leadership in Energy and Environmental Design)	Provides guidelines and points for energy efficient buildings
2002	"Xeriscaping" coined by Ken Ball at the Denver Water Department	Developed a structural framework for water conservation in landscaping
2005	American Society of Landscape Architects begins discussions about the Sustainable Sites Initiative	Developed a system to reward sustainable landscaping practices

TABLE 1.2
Select Developments That Promote a Broad Range of Sustainable Landscaping Practices

Arizona	In 2010, The Arizona Landscape Contractors Association officially adopted the standards in the book *Sustainable Landscape Management Standards for Landscape Care in the Desert Southwest* by Janet Waibel
California	Rob Maday, landscape architect, developed website resource for sustainability http://www.landscaperesource.com/site-features#feature1
Colorado	Front Range Sustainable Landscaping Best Management Practices http://frslc.wetpaint.com/page/Sustainable+Landscape+Principles+for+Colorado
Georgia	Georgia Department of Natural Resources Sustainability Division http://www.gasustainability.org/
Illinois	Chicago Climate Action Plan to address urban heat island issues – has landscaping components http://www.chicagoclimateaction.org/pages/adaptation/11.php
Regional	EPA provides information specific to various regions http://www.epa.gov/region1/topics/waste/greenscapes.html

Guidelines and performance benchmarks of the Initiative were published in 2008 and can be viewed at their website: http://www.sustainablesites.org. The guiding principles of a sustainable site can be seen in Table 1.3.

Three pilot projects were certified in 2011. They are Novus Headquarters Campus in St. Charles, Missouri; The Green at College Park at The University of Texas at Arlington in Arlington, Texas; and Woodland Discovery Playground at Shelby Farms Park in Memphis, Tennessee. Details on these projects and subsequent projects may be viewed at the Sustainable Sites website.

The main areas covered by the Sustainable Sites Initiative are as follows:

1. Site selection
2. Pre-design assessment and planning
3. Site design – ecological components
4. Site design – human health components
5. Site design – materials selection
6. Construction
7. Operations and maintenance.

Prerequisites and credits are benchmarks that are defined within each topic area. Prerequisites are those benchmarks that must be met in order for a site to qualify as a sustainable site. For each prerequisite or credit benchmark, twelve **ecosystem services** are acknowledged (see Table 1.4). One or more ecosystem services may be addressed by a particular benchmark. Ecosystem services are defined as a good or service provided by a natural element, such as plants, air, or water, that is directly or indirectly beneficial to humans. The US Green Building Council will incorporate the benchmarks into future versions of the LEED® ratings system.

TABLE 1.3
Guiding Principles of a Sustainable Site

Do no harm
Apply the precautionary principle
Design with nature and culture
Use a decision-making hierarchy of preservation, conservation, and regeneration
Provide regenerative systems as intergenerational equity
Support a living process
Use a systems-thinking approach
Use a collaborative and ethical approach
Maintain integrity in leadership and research
Foster environmental stewardship

TABLE 1.4
Ecosystem Services Used in the Sustainable Sites Initiative

1. Global Climate Regulation
2. Local Climate Regulation
3. Air and Water Cleansing
4. Water Supply and Retention
5. Erosion and Sediment Control
6. Hazard Mitigation
7. Pollination
8. Habitat Functions
9. Waste Decomposition and Treatment
10. Human Health and Well-Being Benefits
11. Food and Renewable Non-food Products
12. Cultural Benefits

ENVIRONMENTAL ISSUES AND LANDSCAPING

The environmental issues that are driving many of the changes towards sustainability in landscaping include global climate change, air and water quality, water management, and concerns about the waste stream. Some figures provided by the Environmental Protection Agency (EPA) are as follows:

- Gasoline-powered landscape equipment (mowers, trimmers, blowers, chainsaws) account for over 5 percent of our urban air pollution.
- Residential application of pesticides is typically at a rate ten times that of farmers per acre; it has many unintended results.
- A lawn has <10 percent of the water absorption capacity of a natural woodland – a reason for suburban flooding.
- Yard trimmings (mostly grass clippings) comprise 13.1 percent of municipal solid waste collected and most end up in landfills (Figure 1.1).

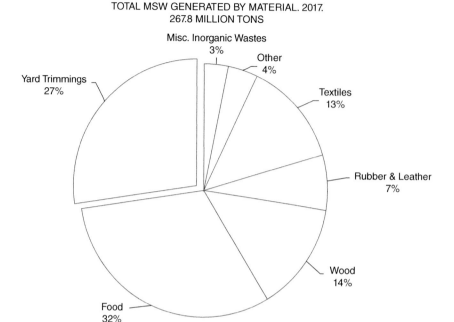

FIGURE 1.1 Total MSW generated, by material, 2017. US EPA.

Global Climate Change

Global climate change is sometimes used interchangeably with global warming because of the accumulation of gases in the earth's atmosphere that are leading to an overall increase in the average annual temperature globally. Due to this warming, multiple changes in climate are resulting, with varied effects that include warming in some areas, cooling in others, increased flooding due to bigger and more numerous rainfalls in some areas, and deeper droughts in other areas. The polar icecaps are expected to melt more than usual in the annual cycle, causing the ocean water levels to rise and moisture in the air to increase.

One of the measurable effects of a warmer climate is the rising temperature of the world's oceans (Figure 1.2). They are now warmer now than they have ever been in the last 50 years. The surface layer of the ocean, which has grown much warmer since the late 1800s, is heating up at a rate of 0.2°F (0.11°C) per decade. As the oceans warm up, weather patterns are affected, leading to more powerful tropical storms. Warmer oceans also have an impact on sea life, such as corals and fish. Furthermore, warmer oceans are one of the main causes of rising sea levels. According to the EPA, the amount of summer ice in the Arctic Ocean in recent years was the smallest it has ever been in the forty or so years since scientists started measuring the area covered by ice. Arctic ice is also getting thinner. In 2012, the global ocean surface temperature was 0.81°F (0.45°C) above average and the year ranked as the ninth warmest such period on record. The year 2018 was the fourth

Sustainable Landscaping

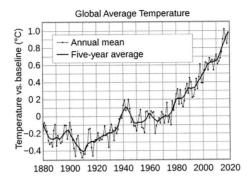

FIGURE 1.2 Average global sea surface temperature, 1880–2015. NASA Goddard Institute for Space Studies – https://data.giss.nasa.gov/gistemp/graphs_v4/.

warmest year in the 139-year-old records of the National Oceanic and Atmospheric Administration (NOAA). Table 1.5 shows global land and ocean temperatures, and indicates that record highs have been experienced globally.

Some scientists think that increased effects of climate forces known as El Nino and La Nina have intensified as part of the overall phenomena of global warming. The gases that contribute to global climate change include carbon dioxide, methane, nitrous oxide, and ozone (Figure 1.3).

For many years, *global climate change* was referred to as the *greenhouse effect*, due to the nature of the warming that occurs. The gases mentioned above are sometimes called *greenhouse gases*. The problem is that they trap infrared energy within the earth's atmosphere, which is felt as heat. Carbon dioxide is considered the main culprit in global climate change. The carbon dioxide concentration of the air is increasing largely because of the use of fossil fuels. Compared to a pre-industrial (1880s) atmospheric concentration of around 270 parts per million (ppm), the average concentration has increased to close to 400 ppm in 2019. Methane is a more important greenhouse gas than carbon dioxide, but is present at smaller concentrations than carbon dioxide.

TABLE 1.5
The Average September Temperature Across the World's Land and Ocean Surfaces

September	Anomaly °C	Anomaly °F	Rank (Out of 140 Years)	Records Year(s)	°C	°F
Land	+1.42 ± 0.26	+2.56 ± 0.47	1st Warmest	Warmest: 2019	+1.42	+2.56
			140th Coolest	Coolest: 1884	−0.78	−1.40
Ocean	+0.78 ± 0.14	+1.4 ± 0.25	2nd Warmest	Warmest: 2015	+0.84	+1.51
			139th Coolest	Coolest: 1903, 1904, 1908	−0.47	−0.85
Land and Ocean	+0.95 ± 0.15	+1.71 ± 0.27	1st Warmest	Warmest: 2015, 2019	+0.95	+1.71
			140th Coolest	Coolest: 1912	−0.53	−0.95

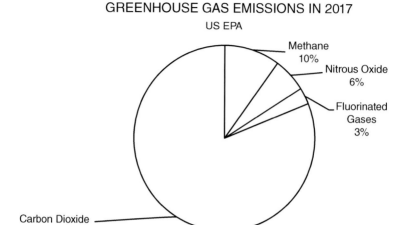

FIGURE 1.3 US greenhouse gas emissions by gas.

Carbon Emissions

In landscaping, human activity generates carbon emissions through burning of fossil fuels, such as are used for automobiles and heating of homes and water. Fossil fuels are those fuels that are derived from deposition of plant and animal matter that have not fully decomposed, usually because of anaerobic conditions. Such fuels include coal, natural gas, and petroleum. These fuels all contain very high levels of carbon.

The amount of carbon dioxide emitted through human activities is calculated into a figure called a **carbon footprint**. Carbon footprint calculators are available at various websites (see http://www.carbonfootprint.com/calculator.aspx for an example). The annual carbon footprint for an individual is directly related to the fuel efficiency of the car he or she drives, the number of miles driven each year, the energy efficiency of their home, and other energy-related activities. In the United States, the average carbon footprint for an individual is calculated to be around 16–20 tons of carbon dioxide per year.

AIR QUALITY

Air pollution results from the emission of particulates in the air that can cause asthma and other respiratory problems for humans. The emission of greenhouse gases could be considered another type of air pollution. The major greenhouse gas emitted during landscape practices is carbon dioxide.

Gasoline-powered engines are responsible for much of the air pollution generated by the landscape industry. Another source of air pollution comes from burning of landscape waste, particularly leaves and woody plant material. The smoke from this burning exacerbates breathing problems in people suffering from respiratory ailments such as emphysema, bronchitis, asthma, and allergies. Smoke from burning fires also releases harmful greenhouse gases, including nitrous oxides, carbon monoxide, volatile organic compounds, and particulate matter (PM). Volatile organic compounds contribute to the formation of smog. PM creates haze. The problems associated with so-called backyard burning are familiar to many municipalities, who have often responded by banning it or by limiting the types of materials that can be burned. The EPA has developed ambient air quality trends for PM. Under the Clean Air Act, EPA sets and reviews national air quality standards for PM. Cities in the southwest United States are in particular danger of unhealthy levels of PM due to regional climate factors.

WATER ISSUES

There are three categories of water issues with respect to landscaping. They are water quality, water conservation, and water management.

Water quality is affected by landscape practices such as pesticide and fertilizer use, which often are carried as runoff into local streams and rivers. Urban stormwater also carries pet and yard waste, motor oil, anti-freeze, household hazardous wastes, paint, and street litter. Landscapers and homeowners can reduce the amounts of herbicide, fertilizer, and pesticides by ensuring that they only apply necessary amounts, by following label directions, by using the least toxic chemicals possible, and by using alternative, less harmful methods when available.

Water conservation is an issue in the arid regions of the country that cover much of the west and southwest. It is also an issue in urban areas where population growth is straining the water supply. Expansion of municipal water treatment facilities is a very expensive undertaking. Many communities are looking for methods of replenishing groundwater on-site as a means of reducing costs. Water-conserving landscape designs can be part of the solution when low rainfall or water availability is a problem.

On-site stormwater management is another aspect of water issues in the landscape. Besides replenishing groundwater, on-site management of stormwater reduces demands on waste water treatment facilities. Some of the means for managing stormwater are permeable paving, green roofs, and rain gardens. The latter also remove excess nutrients from the water. Both green roofs and rain gardens help reduce carbon dioxide and provide oxygen. Green roofs can also reduce energy requirements of buildings.

PESTICIDE USE AND TOXICITY

Pesticides include herbicides for killing weeds as well as insecticides and miticides for plant pests. Fungicides are designed to control fungal infections in plants and sometimes in the soil. Pollution from both runoff of pesticides from landscaping and air-borne pollutants that end up in water bodies can be problematic.

Homeowners use tenfold the amount of pesticides per acre than farmers, according to the EPA. Pesticides are often not stored properly, nor are they always used according to label directions. Excess pesticide material is disposed of improperly,

FIGURE 1.4 Yard waste can be recycled into landscape mulch. (Photo credit: Marietta Loehrlein.)

and its disposal is not regulated very closely, if at all. This results in a large amount of pesticides in wastewater and in trash that ends up in landfills. Detectable amounts of pesticides have been found in 5–10 percent of wells tested. All of this result in pesticide poisoning that could be avoided, as well as groundwater contamination and stormwater pollution that adversely affect wildlife as well as human life.

THE WASTE STREAM

Approximately 36 percent of landscape waste is organic matter, about 50 percent of which is comprised of grass clippings. Most organic materials can be shredded (Figure 1.4) or composted to provide valuable nutrients to landscapes and gardens. Some municipalities also offer recycling services, arrange for special pick-up days for yard wastes, provide special drop-off locations, or a combination of these. In addition to reducing air pollution, precious landfill space is saved by not mixing yard wastes with other trash.

NON-RENEWABLE AND RENEWABLE RESOURCES

Non-renewable resources are defined as those resources that cannot be regenerated. In other words, there is a finite amount of them, and they will eventually no longer be available. Non-renewable resources are also those resources for which natural regeneration takes so long that they will not be available in any reasonable time span. Examples of non-renewable resources are oil, gas, and minerals. **Renewable resources** include energy that is produced from wind and water, or biological products that can be grown repeatedly each year, such as bio-fuels. Some bio-diesel is made from used cooking oils and other sources that are constantly being generated (Figure 1.5).

Sustainable Landscaping

FIGURE 1.5 Sunflower oil can be used for fuel and is a renewable resource. (Photo credit: Marietta Loehrlein.)

THE ROLE OF THE LANDSCAPING INDUSTRY IN SUSTAINABILITY

Those who work in the landscape industry and related professions have an important role to play in helping to reduce the pollution caused by power equipment, insecticides and pesticides, overuse of fertilizers, and other practices. Furthermore, there are other issues related to human population growth (material resources, loss of wildlife habitat to human activities and land development, water use and waste water treatment) that can be addressed by the landscape industry.

All of the aforementioned issues are being addressed by landscape professionals (see Case Studies 1.1 and 1.2). New technologies and products are becoming available to address these challenges. Local and regional ideas and practices are emerging in response to the many developments that are occurring. It is important for the professional to be informed and involved, and most of all to be concerned and educated about the many options that present themselves. As much as possible this book will present products and ideas that are currently available and being used or practiced.

Case Study 1.1: Environmental Business Award Winner: Pacific Landscape Company

Reference: Lawn & Landscape. December 30, 2011. http://www.lawnandlandscape.com/ll121611-environmental-business-winners.aspx. Viewed December 30, 2011.

Pacific Landscape Company (http://www.pacscape.com/) is located in Hillsboro, Oregon. The environmentally friendly practices they have implemented include installed solar panel to reduce energy consumption by 95 percent, bioswale to reduce runoff, a rain garden for the same purpose, using hybrid cars as company vehicles, a 0 percent waste initiative, organic fertilizers, and reduction in use of pesticides. They also rely on weather-based irrigation. Altogether, the company estimates a savings of at least 55 million gallons of water annually. This translates into a significant savings and return on their investment in 2 years.

Another change the company has made is replacing all of its hand-held two-cycle equipment. By doing so, they have cut emissions by 80 percent. The president of Pacific Landscape Company, Bob Grover, has cited customer requests as a motivating factor in providing and maintaining landscapes using sustainable practices. As a result of their efforts, they were named one of Portland's Best Green Companies by *Oregon Business Magazine*, received a Lawn & Landscape 2011 Environmental Business Award, and have been certified with an Ecological Business Certification by a local organization. Details about their sustainability practices can be viewed at their website.

Case Study 1.2: Environmental Business Award Winner: Ruppert Landscape Company

Reference: Lawn & Landscape. December 30, 2011. http://www.lawnandlandscape.com/ll121611-environmental-business-winners.aspx. Viewed December 30, 2011.

Ruppert Landscape Company (http://www.ruppertlandscape.com/about/leadership.html) is located in Laytonsville, Maryland. They were recognized for two projects that transformed under-utilized spaces into usable areas that incorporated sustainable landscaping practices. In one case, green roof gardens were installed at a cancer center. During construction, no mechanical equipment was used in order to minimize disruption to nearby patients. Both extensive and intensive green roofs were installed over a 20,000 ft^2 area.

In another project, a parking lot, trash collection area, and numerous walkways were transformed into a plaza. The space had been dimly lit and uninviting, but became a gathering place for students, faculty, and staff. It also serves as an outdoor classroom and for athletic activities. Sustainable landscape practices included installing an underground cistern, a rain barrel, rain gardens, a bioswale, pervious paving, and native plants. The goal of collecting rainwater for irrigation, maintenance, and other amenities resulted in 100 percent success. In addition to water savings, runoff to the Potomac River basin was eliminated.

In addition to these projects, the company also uses hybrid cars and solar panels providing energy at their offices, and the company is housed in a LEED-certified building. In addition to the Lawn & Landscape awards, the company has also received the Landscape Contractors Association (LCA) Excellence in Landscape competition, The Professional Landcare Network's (PLANET) 42nd Annual Environmental Improvement Awards Program, and the Associated Builders and Contractors Metro Washington Chapter's Excellence in Construction Awards Program.

SUSTAINABILITY AUDIT

A **sustainability audit** serves as a detailed gathering of information that resembles the more traditional site analysis that is conducted prior to designing a new landscape project. The purpose of an audit is to make an inventory of the site with particular regard to sustainability issues. Appendix A explains in more detail the components of a sustainability audit. There is a blank audit form provided in Appendix A. In addition to the form, a drawing of the property should be made, indicating key features of the site. This is similar to a conventional site analysis or a functional diagram. The diagram of the property can be as simple as a sketch drawn approximately to scale or as complex as a base map drawn to scale or photocopy of original blueprints of the house or other building(s) showing boundary lines of the whole property.

Notes may be made on the diagram indicating the direction of the prevailing wind, shaded areas, and "hot spots" in the landscape. However, the sustainability audit differs from a conventional site analysis in that particular attention will be paid to use of plants for energy conservation, reducing water runoff from the site, and other sustainability issues.

How to Implement the Sustainability Audit

An audit consists of a walk-through of the site accompanied with note-taking on the problems that are identified throughout the process. Following the audit, a thorough review of the problems and potential solutions should be conducted and documented.

At the completion of the sustainability audit, recommendations can be made to the client. In addition to plant usage for energy efficiency, other sustainable practices should be implemented. Some examples are additions of a compost pile, use of recycled materials for new projects, or replacing large turf areas with trees, shrubs, or flower beds. The process is the same as conventional landscape design, but the design phase will incorporate more sustainable solutions.

SUMMARY

Sustainability refers to activities that do not harm our planet or our environment and that do not poison our air, water, and soil. It promotes practices that are self-sustaining, and sustain our lives on this planet, as well as protect other life forms and their habitats.

Sustainable practices address issues that include using appropriate plants, including use of native plants in landscape designs; landscape designing for energy efficiency; managing water and drought in the landscape; reducing and eliminating contamination of air, water, and soil; sustainable fertilizing and managing soil health; and renewable and non-renewable resources.

Sustainable landscaping does not have a single line of historical development, but rather is a convergence of various issues and practices. Regional differences exist with regard to sustainability issues.

The Sustainable Sites Initiative is a points-based system that was designed to help in transforming land development and management practices. The program covers the spectrum of landscape activities from site selection to construction to maintenance.

There are several environmental issues driving changes to landscape practices. Global climate change is an important phenomenon exhibiting complex effects on climate, weather, and living organisms around the world. The major contributors to global climate change are the so-called greenhouse gases, which include carbon dioxide, methane, nitrous oxide, and ozone. Carbon emissions from burning fossil fuels play a major role in global climate change. The amount of emissions an individual contributes can be calculated using online sources.

Air pollution from PM and gases in the air contributes to respiratory problems for humans. Gasoline-powered engines are the cause of much of the air pollution generated by landscape practices. Burning of landscape waste also contributes to air pollution problems. Smoke exacerbates respiratory problems and releases harmful greenhouse gases, volatile organic compounds, and PM. Volatile organic compounds contribute to the formation of smog. PM creates haze. Many municipalities have banned burning or have limited the types of materials that can be burned.

Landscaping issues in relation to water are of three types: water quality, water conservation, and water management. Water quality is adversely affected by chemicals used in landscaped areas. Urban stormwater carries pollutants such as herbicides, fertilizers, pesticides, pet and yard waste, motor oil, anti-freeze, household hazardous wastes, paint, and street litter. Water conservation is an issue both in arid areas of the country and in developed areas where potable water is expensive to provide. It is also an issue during periods of drought. Water management concerns stormwater runoff and replenishment of groundwater as well as minimizing the demands on water treatment facilities.

Pesticide use in landscaping is an important environmental issue. Homeowners use tenfold the amount of pesticides per acre than farmers. When excess pesticide is disposed of improperly, it leads to a large amount of pesticides in wastewater and landfills. Improper use and storage of pesticides leads to pesticide poisoning and water pollution.

The waste stream is affected by landscape waste, as well as kitchen waste. Both of these materials can be valuable resources when composted and applied as a soil amendment.

Non-renewable resources are those materials that cannot be regenerated in a reasonable time span or that exist in a finite amount. Some non-renewable resources are oil, gas, and minerals. Renewable resources are not depleted, such as solar, wind, and hydroelectric resources, or come from sources that can be grown anew each year, such as bio-fuels.

Landscape professions can help reduce air and water pollution by changing their practices. New technologies and products are becoming increasingly available. Legal issues and regulations are providing incentives for practitioners to comply with more environmentally healthy practices.

The first step in implementing sustainable landscaping solutions is to conduct a sustainability audit. In addition to a checklist, a diagram of the site is used. Once the audit is complete, the design process proceeds as usual for conventional landscaping, except sustainable solutions are used.

REVIEW QUESTIONS

1. Name three federal laws that provided a foundation for sustainable landscaping practices, including the years they were passed into law.
2. Discuss the term *sustainability* in relation to landscape practices.
3. What is LEED? What is the Sustainable Sites Initiative™?
4. What are ecosystem services?
5. What is a carbon footprint and why does it matter?
6. In what ways do landscape practices contribute to global climate change?
7. What are the major water issues relating to landscaping?
8. Why is pesticide usage considered a problem in the landscaping industry?
9. What are landscape professionals doing to address the problems associated with sustainability?
10. How does a sustainable landscape audit differ from a conventional site analysis?

ENRICHMENT ACTIVITIES

1. Conduct a sustainability audit on a real site. Do a walk-through and record observations using the form provided in Appendix A. Make a diagram of the property beforehand to make notes on during the audit. Meet with clients to discuss what improvements can be made to make the site more sustainable.
2. Determine your carbon footprint using a carbon calculator on the internet.
3. Interview a local landscape company that is implementing sustainable landscaping practices. Find out what motivated them to implement such practices, what is rewarding about using those practices, and what they would advise to a student who wants to do the same.
4. Investigate the sustainable landscape issues in your town, state, or region and compile a reference or resource list. Identify landscape practices that present solutions to the issues on your list.
5. Attend a landscape industry field day or trade show in your area and identify the sustainable landscape products available. Discuss the history of product development for at least three different sustainability-related products with the sales person.

FURTHER READING

American Society of Landscape Architects (ASLA). Sustainable design. http://www.asla.org/sustainabledesign.aspx. Viewed January 1, 2012.
Bisgrove, R. and P. Hadley. 2002. Gardening in the global greenhouse: the impact of climate change on Gardens in the UK. Technical report. UKCIP, Oxford.
Doxon, L.E. 1991. Sustainable horticulture. *HortScience* 26(12): 1454–1455.

Doxon, L.E. 1996. Landscape sustainability: environmental, human, and financial factors. *HortTechnology* 6(4): 362–365.

Fretz, T.A., D.R. Keeney, and S.B. Sterrett. 1993. Sustainability: defining the new paradigm. *HortTechnology* 3(2): 118–126.

Hertsgaard, M. 2011. *Hot: Living through the Next Fifty Years on Earth.* Houghton-Mifflin-Harcourt, Boston, MA, 339 pp.

Hulme, M. and G.J. Jenkins. 2002. Climate change: scenarios for the United Kingdom: the UKCIP02 scientific report. Tyndall Centre for Climate Change Research, School of Environmental Sciences, University of East Anglia, Norwich, UK.

Leahy, S. 2011. Climate change could be worsening effects of El Nino, La Nina. http://ipsnews.net/news.asp?idnews=54087. Retrieved May 10, 2011.

Leopold, A. 1949. *A Sand County Almanac.* Oxford University Press, New York.

National Oceanic and Atmospheric Administration. Answers to La Nina frequently asked questions. https://www.climate.gov/enso. Retrieved July 5, 2019.

Sustainable Sites Initiative. http://www.sustainablesites.org/. Retrieved June 30, 2011.

Templeton, S.R., D. Zilberman, and S.J. Yoo. 1998. An economic perspective on outdoor residential pesticide use. *Environ. Sci. Tech.* 32(17): 416A–423A.

The Greenhouse Effect. http://www.ucar.edu/learn/1_3_1.htm. Retrieved December 29, 2010.

Wikipedia. Carbon cycle. http://en.wikipedia.org/wiki/Carbon_cycle. Viewed February 7, 2008.

2 Sustainability in the Plantscape

OBJECTIVES

- Understand the connection between the environment and plant placement and use
- Relate carbon usage by plants to carbon emissions
- Identify ecosystem services provided by plants
- Be able to describe and discuss what an ecosystem is
- Understand the functions and interactions of ecosystems
- Distinguish between the benefits and disadvantages of native and non-native plants
- Understand the role of wildlife in the landscape
- Name some programs that have been designed to encourage wildlife habitat development in residential or urban areas
- Discuss the relationship between landscape practices and wildfire safety

TERMS TO KNOW

Anthropogenic
Biome
Carbon footprint
Carbon sequestration
Ecosystem
Ecosystem services
Evapotranspiration
Humus
Pre-emergent
Understory
Xeriscaping
Xerophyte

INTRODUCTION

There are many ways to look at the "plantscape", or plants in the landscape and sustainability. The aesthetic aspects of the plantscape include considerations for human enjoyment and well-being. The environmental aspects of plants include **carbon sequestration**, the associated release of oxygen to the atmosphere, and the cooling effect of evapotranspiration. Plants are an integral part of an ecosystem, which is a dynamic complex of plant, animal, and microorganism communities and their non-living environment interacting as a functional unit. The ecological aspects include plant-soil interactions and the food and habitat plants provide for insects, birds, and other wildlife.

Plants can provide shade, trap cold air on a slope, create dead-air space near a wall, and funnel the wind or cause it to lift over structures and areas. Turfgrass also serves a functional role in the landscape. It is useful as a groundcover that can support recreational activities and light traffic, cool the area around a structure, all the while providing a pleasing color and texture.

This chapter examines the environmental and biological aspects of plants in the landscape, and looks at special considerations for plant maintenance levels. Trees with problematic growth habits and invasive plants will be discussed in this context. Sustainable turf management practices are also addressed in this chapter.

ENVIRONMENT ASPECTS OF PLANTS

Plants provide many benefits to the environment. In addition to the fundamental physiological aspect of removing carbon dioxide from the atmosphere (known as carbon sequestration) and release of oxygen to the atmosphere that occurs during photosynthesis, plants also provide services such as erosion control and removal of pollutants from the soil and air.

CARBON SEQUESTRATION

Plants take up carbon dioxide from the air in the process of photosynthesis and incorporate it into plant tissues and molecular structures. Thus, through the process known as carbon sequestration, carbon is removed from the atmosphere and stored for varying amounts of time in plant leaves, roots, and stems, including wood. This carbon may begin to be recycled back to the environment through decomposition when plant matter is removed or shed from the plant.

Plant tissues are decomposed by many different organisms. Organisms eventually return the carbon back to the environment as carbon dioxide. Many studies are currently underway to examine how plants sequester and store carbon due to the interest in using carbon sequestration as a strategy to help reduce levels of atmospheric carbon dioxide. In order to reduce excess levels of carbon dioxide that contribute to global climate change, the carbon would have to be stored for long periods of time.

Three naturally occurring options for storing carbon dioxide are to store it in plants, in soil, or in the ocean. The carbon cycle (Figure 2.1) illustrates the flux of carbon on a global scale. Approximately 610 gigatons of carbon are stored in vegetation, with another 1,500 gigatons stored in soils. By comparison, there are about 750 gigatons in the atmosphere and around 39,000 gigatons in the oceans. Cement production and the burning of fossil fuels release around 400 gigatons of carbon dioxide to the environment.

According to a study conducted by Puoyat et al. (2006), urban soils in residential areas often contain relatively high levels of soil organic carbon. Some cities, such as Chicago, Illinois, and Oakland, California, have higher amounts of soil organic carbon than it was prior to urban development. Urban trees also store carbon.

Sustainability in the Plantscape

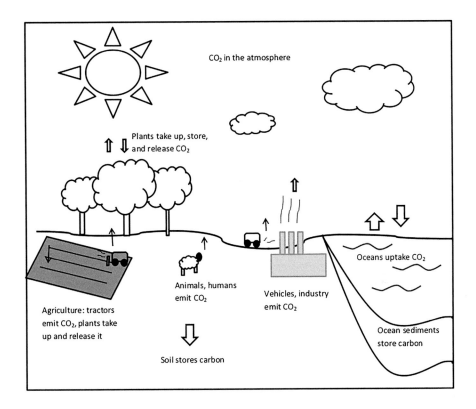

FIGURE 2.1 The carbon cycle diagram illustrates how carbon is cycled into and out of various components of the natural and constructed, or "built", environment. (Illustration by author.)

Nowack and Crane (2002) estimated that urban trees in the lower 48 states store 700 million tons of carbon, at a value of $14.3 trillion in 2002 dollars.

OXYGEN RELEASE

In photosynthesis, as plants take up carbon dioxide, they also release oxygen. The rate of photosynthesis is affected by temperature, sunlight, and levels of carbon dioxide. The amount of oxygen released from plants in the landscape can be significant. Whereas tree species and the growing environment can affect the amount of oxygen plants produce in a given year, estimates for a mature tree are that it can provide enough oxygen for two people annually. Similarly, a 2,500 ft² area of turf produces enough oxygen for four people.

COOLING EFFECT OF PLANTS

Plants take up water through their roots in a process known as absorption, and they lose water through their leaves in a process known as transpiration. Temperature, sunlight, and wind can affect the transpiration rate. When plants transpire, they cool the surrounding air. This creates a microclimate around the plants that, along with providing shade, has a cooling effect. This cooling effect of plants can be experienced by walking on grass barefoot in the summertime. The ground and grass are noticeably cooler than surrounding surfaces, including bare ground. According to Beard and Johns, when bermudagrass (*Cynodon dactylon*) is 100°F, surrounding asphalt is 140°F and artificial turf is 162°F. Even though this study was done in 1985, temperature studies of artificial turf have been conducted in numerous sites across the country, and even with more modern turf systems, the artificial turf remains significantly hotter than natural turf.

Plants also cool the surrounding area by providing shade. The cooling effects of shade are discussed in more detail in Chapter 3.

STRUCTURAL EFFECTS OF PLANTS

Plants can affect their surrounding environment by blocking the wind, channeling the wind, blocking the sun, and creating dead air space. All of these effects can aid in energy savings on heating or cooling houses or other structures. By planting shrubbery near a house or building, dead air space can be created between the shrubs and the house. This dead air space can help reduce heat loss from the house in winter. Shading of air conditioning units also may have a beneficial effect if done correctly.

This topic and wind screens designed to reduce energy use are discussed in more detail in Chapter 4. They are of particular use in areas with cold winters and open spaces. Wind screens can be designed to slow the wind speed around buildings, lift the wind over the desired area, and may also serve as living snow fences.

TURFGRASS

Turfgrass is often cited as an unsustainable plant due to the input requirements for its maintenance, especially when aesthetic perfection is the goal. This is due to the need for irrigation, fertilization, chemical pest control, and the use of fuels and the subsequent emissions of greenhouse gases, to maintain a uniformly colored and textured surface. With residential application of pesticides typically twenty times that of farmers, alternatives should be developed and utilized. Most turfgrass species are not drought-tolerant. They will turn brown during dormancy, whether it is induced by drought or temperatures that are too cold or too hot.

Turfgrass does have its benefits, however. As its leaves lose moisture through transpiration, it cools the surrounding air. It prevents erosion, and can help build soil structure over time. It aids in carbon sequestration and oxygen production, provides a place for recreation, and is aesthetically pleasing.

DROUGHT-TOLERANT TURF

There are two species of turfgrass that are particularly drought-tolerant: tall fescue (*Festuca arundinacea*) and buffalo grass (*Buchloe dactyloides*). Tall fescue is a cool-season grass that also does well in the intermediate zone. Buffalo grass is native to North America from central Montana east to Minnesota and south to eastern coastal

Sustainability in the Plantscape

Louisiana, and westward to eastern Arizona and northern Mexico. It has undergone hybridization improvement to develop it into a more aesthetically pleasing turf-type grass. Unimproved forms turn somewhat brownish in winter. Buffalo grass does not do well in shady areas, nor does it tolerate heavy foot traffic. In less than ideal situations, it thins out and weeds move in. Both of these grasses can be grown at 3 in. before requiring mowing.

There are some types of fine fescue (*Festuca rubra*) that are sold as "low mow" or "no mow" turfgrass. Whereas all of the fine fescues are somewhat drought-tolerant, they do not tolerate heat very well, and grow best in partial shade.

SALT-TOLERANT TURF

Paspalum grass (*Paspalum vaginatum*) has been under development as a salt-tolerant golf course grass in coastal areas, where it may replace bermudagrass. Whereas management of Paspalum grass can be as demanding as other golf course turf species, it has the added benefit of tolerating higher salinity levels in water. Golf courses that are exposed to flooding, hurricanes, or brackish water may benefit from using this grass. Paspalum may also respond better to use of effluent or other water that has high salt content or in locations with saline soils.

WEEDS IN TURF

Weeds in turf can be symptoms of other problems. For example, the presence of clover in turf is an indication of low nitrogen. Some proponents encourage the inclusion of clover in turf for that reason. Plantain (*Plantago* spp.) species are prevalent in areas that have been compacted. Thus, the presence of plantain suggests the need to aerate the lawn. Nutsedge (*Cyperus* spp.) grows in wet areas. Wet areas should be evaluated for the cause of wetness (compaction, poor drainage pattern in the lawn) and corrected. Dandelions (*Taraxacum officinale*), violets (*Viola* spp.), crabgrass (*Digitaria* spp.), and creeping Charlie (*Glechoma hederacea*) can be pervasive and difficult to control. However, they tend to be less of a problem in areas where a healthy stand of turf is growing. Other weeds come in where the ground has been disturbed and nothing else is being cultivated. Lambsquarters (*Chenopodium* spp.), velvetleaf (*Abutilon theophrasti*), Johnsongrass (*Sorghum halepense*), and many other undesirable plants are opportunistic in this way.

AERATION

Lawn aeration is a cultural practice that can alleviate compaction and the weeds and diseases that it encourages. Lawn areas become compacted due to regular foot traffic and even rain fall. Aeration once a year is usually recommended in areas that experience routine use.

MOWING

Raise mowing height during dormant periods for turf. In the northern half of the United States, this is summer during droughty periods, in the south it is in winter.

WOODY PLANTS

Woody plants can be high maintenance when they have poor structural form, are not trained or pruned properly, or when they are invasive. Poor structural growth is prevalent in some tree species. Narrow crotch angles (Figure 2.2), co-dominant leaders, multiple branches at a node (Figure 2.3), and rubbing or crossing branches are some examples of poor structural growth of trees. Eventually, such problems can lead to splitting off of major limbs, or entire trunks, and pest or disease infestation due to exposure of the inner layers of the tree to the elements. Thus, these plants should be avoided or removed and replaced with better-suited trees when possible.

Pruning

Poor pruning practices, such as tree topping, stubbing, and stripping bark during pruning, can all lead to the early death of a tree and should be avoided. Proper pruning requires training and knowledge of tree management practices. Larger trees require the work of a certified arborist.

A medium to low maintenance shrub will not produce a lot of new stems or suckers every year. Suckers, water-sprouts, and leggy growth are examples of undesirable growth. Evergreen shrubs such as yews and boxwoods that require hedging to maintain a desirable shape are also high-maintenance plants that should be avoided. Many types of roses also require regular pruning to maintain a desirable size. They are also prone to numerous pests and diseases. Even landscape-type roses can be considered high maintenance.

FIGURE 2.2 Narrow crotch angles are weak structural components on a tree. (Photo by author.)

Sustainability in the Plantscape

FIGURE 2.3 Flowering pear with multiple branches at one point results in a weakened trunk. (Photo by author.)

INVASIVE PLANTS

The USDA National Invasive Species Information Center defines invasive plants as "introduced species that can thrive in areas beyond their natural range of dispersal. These plants are characteristically adaptable, aggressive, and have a high reproductive capacity. Their vigor combined with a lack of natural enemies often leads to outbreak populations". (https://www.invasivespeciesinfo.gov/) (Table 2.1). There are various ways an invasive plant may be introduced to a property. Birds may leave seeds from invasive species in their droppings and squirrels may plant them there.

TABLE 2.1
Selected List of Invasive Landscape Plants

Name	Botanical Name	Comments
Amur Honeysuckle	*Lonicera maackii*	Soil conservation service introduction; seeds readily, birds spread; Manchuria, Korea
Autumn-Olive	*Elaeagnus umbellata*	Soil conservation service introduction; birds spread; salt-tolerant
Boxelder	*Acer negundo*	Aggressive growth in adverse conditions; Native
Eastern Cottonwood	*Populus deltoides*	Numerous seeds germinate readily; fragile; moves into abandoned land; fast growth

(Continued)

TABLE 2.1 (*Continued*)
Selected List of Invasive Landscape Plants

Name	Botanical Name	Comments
English ivy	*Hedera helix*	Vines can cover tree trunks and limbs; grows in dense shade
European alder	*Alnus glutinosa*	Escaped from cultivation; fixes atmospheric nitrogen
Green ash (red ash)	*Fraxinus pensylvanica*	Fast growth; seeds germinate readily
Thornless honey locust	*Gelditsia triacanthos* var. inermis	Fast growth; native
Japanese honeysuckle	*Lonicera japonica*	Vine, noxious weed; Asian origin
Japanese wisteria	*Wisteria floribunda*	Aggressive growth requires aggressive pruning to keep under control; Japan
Mimosa, silk tree	*Albizia julibrissin*	Seeds prolific, germinate readily; brittle wood; shallow roots raise sidewalks, patios; Non-native
Multiflora rose	*Rosa multiflora*	Used as rose rootstock; escaped from cultivation. Invades fields, woodlands; Japan and Korea
Oriental bittersweet	*Celastrus orbiculatus*	Twining vine; more vigorous than American bittersweet, which is also overly vigorous
Purple loosestrife	*Lythrum salicaria*	Noxious weed or similarly designated in 32 states
Russian mulberry	*Morus alba*	Introduced for silkworm production; birds distribute seeds; fruit is messy
Russian-olive	*Elaeagnus angustifolia*	Roadsides; salt-tolerant
Siberian elm	*Ulmus pumila*	Fast growth, brittle wood. Siberia, Manchuria, Korea, northern China
Silver maple	*Acer saccharinum*	Moist soils, woodlands, streambanks; fast growth; many seeds germinate easily. Native
Tree of Heaven	*Ailanthus altissima*	Leaves and stems smelly; seeds prolifically, germinates readily, suckers; Non-native
Virginia creeper	*Parthenocissus quinquefolia*	Vine that creeps along ground, can root at every leaf node, attach itself to vertical supports; fast growth; tolerates a wide range of conditions; Native
Weeping willow	*Salix alba* "Tristis"	Seeds germinate readily; fast growth, weak wood; invades septic and sewer systems. Naturalized in United States – native to Europe, Siberia, and Asia

Some plants produce excessive amounts of seed and some spread through rhizomes. Invasive plants can spread into natural areas and crowd out native plants. Purple loosestrife, English ivy, silver maple, amur maple, tree of heaven, and Virginia creeper are all examples of invasive landscape plants. Landscape professionals should avoid the use of invasive plants. Regional, climatic, and environmental differences do occur, so that species that are not invasive in one location may be quite invasive elsewhere. It is necessary to familiarize oneself with invasive and otherwise undesirable plants in one's own region.

Case Study: English Ivy

Reference: Invasive.org. Center for Invasive Species and Ecosystem Health. http://www.invasive.org/browse/subinfo.cfm?sub=3027. Viewed December 29, 2011.

Plant Conservation Alliance's Alien Plant Working Group. http://www.nps.gov/plants/alien/fact/hehe1.html. Viewed December 29, 2011.

Whereas it is important to substitute invasive plants in the landscape with those that have proven to be non-invasive, eradicating existing plants can be a daunting task. English ivy (*Hedera helix*) is an evergreen vine with attractive glossy dark green three- to five-lobed leaves. It is widely used as a vine and groundcover, as well as decoratively in hanging baskets and even in floral arrangements.

Some eighteen states have reported English ivy to be invasive in natural areas. It grows well in diverse conditions, from full sun to full shade, dry to moderately moist (but not wet) soil with a slightly acidic (6.5) pH. It is this low-maintenance quality of the plant combined with its ability to replace turf in difficult areas such as shaded spots and slopes that have made it attractive as a landscape plant. In addition to vigorous growth, mature plants produce berries that are spread by birds.

Removal of English ivy can be effected by manual, mechanical, and chemical control methods. Vines that are climbing up trees are especially important to control due to their ability to shade out the tree's ability to photosynthesize. Glyphosate and triclopyr are the two most recommended herbicides. Both work by translocating throughout the plant (systemic), necessitating application to only one part of it. This works especially well for vines climbing up trees, or spreading throughout an area that is otherwise not accessible. Herbicide applications may be made any time of year, since this plant is evergreen, as long as temperatures exceed 55°F–60°F.

Case Study: Tampa Bay, Florida – Brazilian Pepper Trees

Reference: http://www.tampabaywatch.org/index.cfm?fuseaction=content.home&pageID=26. Viewed December 29, 2011.

Florida Exotic Pest Plant Council. Publications. http://www.fleppc.org/publications.htm. Viewed December 29, 2011.

Every region has its own set of conditions that allows a plant to be invasive. In Tampa Bay, one of the invasive plants that is receiving a lot of attention is the Brazilian pepper tree, *Schinus terebinthifolius*, introduced in 1898. It is a problem because, among other things, it is displacing native mangrove trees which provide vital habitat for coastal wildlife. The rate at which it is spreading is only one of the problems with this plant, which is also allelopathic, meaning that it adversely affects growth of other plants. To date, it has taken over in excess of 700,000 ac of land in Florida.

Tampa Bay Watch is organized to coordinate the efforts of volunteers and community groups in removing exotic plants. Such oversight is necessary given the multi-pronged approach required: application of toxic herbicides, use of specialized equipment, proper identification of plant species, and training of professional staff.

Other names for Brazilian pepper are Florida holly, Christmas berry, and pepper tree. In addition to mangroves, rare species of plants have been displaced by this federally designated noxious weed, which include Beach Jacquemontia (*Jacquemontia reclinata* House) and Beach Star (*Remirea maritime* Aubl.). Fruits are eaten by wildlife and disseminated. Chemicals in the leaves, stems, flowers, and fruits are irritating to human skin and respiratory passages.

Florida Exotic Pest Plant Council (FLEPPC) maintains a webpage with informational brochures and publications. One such brochure suggests native alternatives for invasive species in the landscape. Some of the recommended alternatives for Brazilian pepper are Yaupon holly (*Ilex vomitoria*), persimmon (*Diospyros virginiana*), and pignut hickory (*Carya glabra*).

ECOLOGY AND PLANTS

In considering the effect of plants on the environment, plant communities and ecosystems provide important clues to more sustainable landscapes. The conventional manicured landscape is one kind of aesthetic that has been promoted as an ideal to strive towards. However, such landscapes require large expenditures of labor and money to maintain. The goal of designing and planting more ecologically balanced landscapes is a more sustainable approach. In keeping with the goal of sustainability, recreation or mimicking of an ecosystem usually means that the landscape will evolve over time. It does not mean that there will be no maintenance, however. It does mean that the maintenance requirements will differ from those in a conventional landscape setting.

The Sustainable Sites Initiative addresses the ecological aspects of plants during the site selection and site design process by establishing prerequisites and providing credits in specific ways. They include:

- Limit development on farmland
- Protect floodplain functions
- Conserve aquatic ecosystems
- Conserve habitats for threatened and endangered species
- Designate and communicate vegetation and soil protection zones

Sustainability in the Plantscape

- Conserve and use native plants
- Conserve and restore native plant communities
- Conserve healthy soils and appropriate vegetation
- Conserve special status vegetation
- Control and manage invasive plants
- Use appropriate plants
- Optimize biomass
- Protect and restore riparian and wetland buffers
- Use vegetation to minimize building energy use
- Understand and implement best practices in wildfire safety
- Reduce urban heat island effects

The following section describes and briefly discusses general ecosystem types that are found in the United States. This discussion is, by necessity, only introductory, but should serve as a guide. The Further Reading section at the end of the chapter provides references with much greater detail for individual locales.

ECOREGIONS OF THE UNITED STATES

Regional ecosystems that have been identified in the United States are categorized by moisture and temperature. Four categories provided by the US Forest Service are: Polar Domain, Humid Temperate Domain, Dry Domain, and Humid Tropical Domain. Each Domain has one or more Divisions, and within each Division are one or more Provinces (https://www.fs.fed.us/land/ecosysmgmt/colorimagemap/ecoreg1_provinces.html). Province types include Arctic Tundra, Meadow, Eastern Broadleaf Forest, Ouachita Mixed Forest, Prairie Parkland, California Dry Steppe, Central Plains Steppe and Shrub, Intermountain Semi-desert and Desert, and many others. Figure 2.4 identifies the major ecoregions in the United States.

Forests and Woodlands

Forest and woodland landscapes are dominated by shade trees and understory plants that can survive in the shaded environment. Understory plants may be smaller shade-loving trees, shrubs, vines, and herbaceous plants. Eastern woodlands are highly regarded for their fall foliage and spring-blooming trees. Woodland wildflowers are numerous and may be interspersed with ferns for foliage interest in late spring through the end of summer. There are a number of different forest and woodland ecosystems in the United States. They vary in their make-up according to the climate, soil, and hydrology.

Temperate Coniferous Forests

Temperate coniferous forests occur along the western coastal area and in the southeastern United States. The dominant plants found in the higher-rainfall areas of the western coniferous forests include Douglas Fir (*Pseudotsuga menziesii*), red cedar (*Thuja plicata*), redwood (*Sequoia sempervirens*), and sitka spruce (*Picea sitchensis*). In lower rainfall areas of the west, ponderosa pine (*Pinus ponderosa*), Engelmann spruce (*Picea engelmannii*), and lodgepole pine (*Pinus contorta*) are the dominant

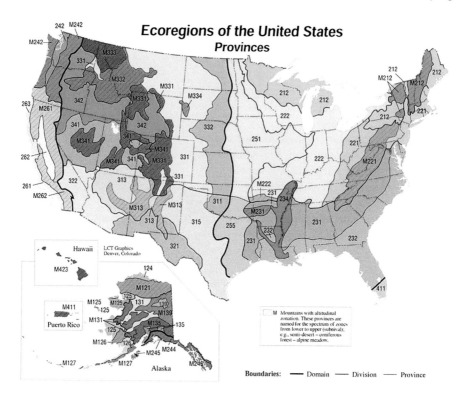

FIGURE 2.4 Ecoregions of the United States. (Source: R. G. Bailey. USDA Forest Service.)

plants. Understory plants, or those lower growing plants that occur under the canopy of the taller trees, include shrubs, ferns, liverworts, and lichens. Few flowering herbaceous plants are able to thrive in the low light levels of the coniferous forest. The soils tend to be nutrient poor, as little organic matter is returned to the soil from the decomposition of needles. Table 2.2 lists woody and herbaceous plants that can be found in Pacific Coast forests.

In the southeast, the coniferous forest covers much of the coastal plain of Louisiana, Mississippi, Alabama, Georgia, and Florida. The dominant plants include longleaf pine (*Pinus palustris*), pitch pine (*Pinus rigida*), and slash pine (*Pinus elliotti*). The soils in the southeast forests tend to be sandier and nutrient poor. The dominant understory plant is wiregrass (*Aristida stricta*), with over 3,000 other native herbaceous and shrub species. Table 2.3 lists some woody and herbaceous plants that can be found in southeastern forests.

The natural ecosystem of this region has been disturbed almost beyond recognition, and has been replaced by mixed hardwoods. Suppression of fire, establishment of pitch pine plantations, urban development, and agricultural practices have contributed to this change. Wildlife native to this area include many species of reptiles and amphibians, butterflies, mammals, and birds, including the endangered red-cockaded woodpecker.

TABLE 2.2
Plants Found in Pacific Coast Forests

Common Name	Botanical Name
Woody	
Douglas fir	*Pseudostuga menziesii*
Mountain dogwood	*Cornus nuttallii*
Ninebark	*Physocarpus malvaceous*
Oregon ash	*Fraxinus latifolia*
Redwood	*Sequoia sempervirens*
Rustyleaf, mock azalea	*Menziesia ferruginea*
Sitka spruce	*Picea sitchensis*
West coast rhododendron	*Rhododendron macrophyllum*
Western hemlock	*Tsuga heterophylla*
Western red cedar	*Thuja plicata*
Herbaceous	
Adder's tongue, yellow fawn lily	*Erythronium citrinum*
Alpine heuchera	*Heuchera glabra*
Alpine shooting star	*Dodecathon alpinum*
Bluebells, fringed lungwort	*Mertensia ciliata*
Dwarf lewisia	*Lewisia pymaea*
Llyall's star tulip	*Calochortus lyallii*

Temperate Deciduous Forests

The temperate deciduous forests once filled most of the eastern half of the United States. Urban development and agricultural activities have contributed to significant changes in the original ecosystem. In its undisturbed state, various tree species have been dominant, depending on local conditions. Some species of the dominant trees include basswood (*Tilia* spp.), beech (*Fagus* spp.), cottonwood (*Populus* spp.), elm (*Ulmus* spp.), hickory (*Carya* spp.), magnolia (*Magnolia* spp.), maple (*Acer* spp.), oak (*Quercus* spp.), and willow (*Salix* spp.). Understory plants include a diverse array of shrubs, ferns, mosses, and many herbaceous species of plants. See Table 2.4 for a list of some of the native herbaceous plants of deciduous forests. The soils are rich in nutrients, with an annual addition of organic matter that quickly decomposes into rich humus, consisting of sticky compounds that help build soil structure.

Prairies

In the United States, the prairie ecosystem is temperate grassland. The dominant prairie plant in the United States, particularly in areas of moderate to high rainfall, is bluestem (*Andropogon* species). In areas of lower rainfall, buffalograss (*Buchloe dactyloides*) is the dominant species. Other grasses are also found in the prairie biome. For a list of plants found in prairies, see Table 2.5. Prairie soils are typically very fertile, having been built up over many years with annual additions of organic matter. Many prairie plants have deep root systems (Figure 2.5), which contribute to their drought tolerance.

TABLE 2.3
Some Native Plants of the Southeastern United States

Woody	Species
American holly	*Ilex opaca*
Flowering dogwood	*Cornus florida*
Laurel oak	*Quercus hemisphaerica*
Live oak	*Quercus virginiana*
Longleaf pine	*Pinus palustris*
Marshall's hawthorn	*Crataegus marshallii*
Pawpaw	*Asimina triloba*
Pitch pine	*Pinus rigida*
Red buckeye	*Aesculus pavia*
Slash pine	*Pinus elliottii*
Southern magnolia	*Magnolia grandiflora*
Yaupon Holly	*Ilex vomitoria*
Herbaceous	**Species**
Birthwort	*Aristolochia tomentosa*
Early black-cohosh	*Cimicifuga racemosa*
Eastern spring-beauty	*Claytonia virginica*
Goat's-beard	*Aruncus dioicus*
Heart-leaf	*Asarum virginica*
Indian pink	*Spigelia marilandica*
Leather flower	*Clematis viorna*
Purple lobelia	*Lobelia elongate*
Swamp rose mallow	*Hibiscus moscheutos*
Thermopsis	*Thermopsis caroliniana*
Wild Ginger	*Asarum canadense*
Wild passionflower	*Passiflora incaarnata*

To maintain a prairie, woody species may have to be actively eradicated. In natural systems, this was accomplished by burning. This method is still used. An alternative to fire is mowing once or twice each year, after flowers have been allowed to bloom, cross-pollinate, and set seed, allowing native pollinators and other insects to benefit from the food provided by the prairie plants.

Meadows

A meadow is similar to a prairie in that it consists of herbaceous, rather than woody, plants. Different types of meadows have been identified, such as wet meadows, alpine meadows, and saline meadows. Wet meadows occur in poorly drained areas and may resemble marshes or grasslands. Some are found in higher mountainous areas, others are in low-lying areas, or may even be found in agricultural lands. They are often wet for many months, but may have standing water for only a portion of that time. Alpine meadows are found at higher elevations. They are generally basin-shaped, and may

TABLE 2.4
Native Herbaceous Species of Eastern US Deciduous Woodland

Common Name	Botanical Name
Canopy	
American basswood	*Tilia Americana*
Burr oak	*Quercus macrocarpa*
Chestnut oak	*Quercus prinus*
Chinkapin oak	*Quercus muhlenbergii*
Eastern arborvitae	*Thuja occidentalis*
Eastern hop-hornbeam	*Ostrya virginiana*
Eastern red cedar	*Juniperus virginiana*
Eastern white pine	*Pinus strobus*
Northern red oak	*Quercus rubra*
Northern white oak	*Quercus alba*
Pignut hickory	*Carya glabra*
Shag-bark hickory	*Carya ovata*
Sugar maple	*Acer saccharum*
White ash	*Fraxinus americana*
Woody Understory	
Alder-leaf buckthorn	*Rhamnus alnifolia*
American bladdernut	*Staphylea trifolia*
Bear oak	*Quercus ilicifolia*
Black huckleberry	*Gaylussacia baccata*
Black raspberry	*Rubus occidentalis*
Common juniper	*Juniperis communis*
Common red raspberry	*Rubus idaeus*
Downy arrowwood	*Viburnum rufinesquianum*
Dwarf chinkapin oak	*Quercus prinoides*
Eastern poison ivy	*Toxicodendron radicans*
Eastern prickly gooseberry	*Ribes cynosbati*
Eastern teaberry	*Gaultheria procumbens*
Late lowbush blueberry	*Vaccinium angustifolium*
Limber honeysuckle	*Lonicera dioica*
Mountain maple	*Acer spicatum*
Round-leaf dogwood	*Cornus rugosa*
Sheep-laurel	*Kalmia angustifolia*
Smooth sumac	*Rhus glabra*
Stiff dogwood	*Cornus foemina*
Striped maple	*Acer pensylvanicum*
Toothache tree	*Zanthoxylum americanum*

(Continued)

TABLE 2.4 (*Continued*)
Native Herbaceous Species of Eastern US Deciduous Woodland

Common Name	Botanical Name
Herbaceous Understory	
Bigleaf aster	*Aster macrophyllus*
Blue cohosh	*Caulophyllum thalictroides*
Bristle-leaf sedge	*Carex eburnean*
Brittle bladderfern	*Cystoperis fragilis*
Broadleaf sedge	*Carex platyphylla*
Bulblet bladderfern	*Cystopteris bulbifera*
Canadian wildginger	*Asarum canadense*
Common lady fern	*Athyrium filix-femina*
Eastern bottlebrush grass	*Elymus hystrix*
Herbrobert	*Geranium robertianum*
Maidenhair spleenwort	*Asplenium trichomanes*
Marginal wood fern	*Dryopteris marginalis*
Northern maidenhair	*Adiantum pedatum*
Pennsylvania sedge	*Carex pensylvanica*
Rattlesnake fern	*Botrychium virginianum*
Rattlesnake-weed	*Hieracium venosum*
Savin-leaf ground-pine	*Lycopodium sabinifolium*
Silver false spleenwort	*Deparia acrostichoides*
Virginia strawberry	*Fragaria virginiana*
Walking fern	*Asplenium rhizophyllum*
Wavy hair grass	*Deschampsia flexuosa*
White baneberry, doll's eyes	*Actaea pachypoda*
White woodaster	*Aster divaricatus*
Woman's tobacco	*Antennaria plantaginifolia*

be very wet throughout much of the year. Alpine plants can be found there. Saline meadows occur at the mouths of rivers where they enter the ocean, and the build-up of minerals over many miles of draining the land has occurred. Salt-tolerant plants thrive in these meadows.

Riparian Zone Habitat

A riparian habitat is land adjacent to rivers, streams, ponds, and lakes. Plants that thrive in very moist conditions do well in riparian habitats. Plants along a bank or shore help to reduce the erosion effects of moving water. They also provide habitat for wildlife, which could include shore birds, dragonflies, muskrats, and others. Since rivers and other bodies of water usually pass through or are adjacent to property owned by more than one person, maintaining plantings along them creates a wildlife

TABLE 2.5
Herbaceous Flowering Plants of the Prairies

Common Name	Botanical Name	Soil Moisture	Sun Exposure	Bloom Time (Months)
Bloodroot	*Sanguinaria canadensis*	Moderate	Partial	April to May
Blue wild indigo	*Baptisia australis*	Mod. Moist	Full to partial	May to July
Butterfly weed	*Asclepias tuberosa*	Moist to dry	Full to partial	June to August
Canadian milk vetch	*Astragallus candensis*	Moderately moist to mod. Dry	Full to partial	June to August
Cardinal Flower	*Lobelia cardinalis*	Moist	Full to shade	July to October
Culver's root	*Veronicastrum virginicum*	Moderate	Full to partial	June to August
Dotted St. John's wort	*Hypericum punctatum*	Moderate to dry	Full to partial	June to September
Dutchman's breeches	*Dicentra cucullaria*	Wet to moderate	Shade	April to May
Early sunflower	*Heliopsis helianthoides*	Moderate	Full to partial	June to September
False aster	*Boltonia asteroides*	Mod. Moist to wet	Full to partial	August to October
Foxglove beardtongue	*Penstemon digitalis*	Moderate to dry	Full	June to July
Golden Alexanders	*Zizia aurea*	Moderate	Full to partial	April to June
Hairy woodmint	*Blephilia hirsuta*	Moist to mod. Moist	Shade to partial	June to September
Indian paintbrush	*Castilleja coccinea*	Moderately moist to moderately dry	Full to partial	April to September
Missouri evening primrose	*Oenethera macrocarpa*	Dry	Full to part	May to June
Narrow-leaved coneflower	*Echinacea angustifolia*	Dry	Full	June to July
New England aster	*Aster nova-angliae*	Moist to wet	Full to partial	August to October
Obedient plant	*Physostegia virginiana*	Moderate to moist	Full to partial	August to September
Prairie coreopsis	*Coreopsis palmate*	Moderate to dry	Full to partial	June to August
Prairie spiderwort	*Tradescantia bracteata*	Dry	Full	May to July
Purple prairie clover	*Petalostemum (Dalea) purpureum*	Moderate to dry	Full to partial	June to September

(Continued)

TABLE 2.5 (Continued)
Herbaceous Flowering Plants of the Prairies

Common Name	Botanical Name	Soil Moisture	Sun Exposure	Bloom Time (Months)
Rattlesnake Master	*Eryngium yuccifolium*	Moderate	Full	July to September
Scarlet gaura	*Gaura coccinea*	Dry	Full	May to August
Showy goldenrod	*Solidago speciosa*	Moderate to dry	Full to partial	August to October
Showy tick trefoil	*Desmodium canadense*	Moderate	Full to partial	July to August
Sky blue aster	*Aster azureus*	Moist to dry	Full to partial	August to October
Small fringed gentian	*Gentiana procera*	Wet	Full	September to October
Solomon's plume	*Smilacina racemosa*	Moist to dry	Sun or shade	April to June
Southern wild hyacinth	*Camassia angusta*	Moderately moist to mod. dry	Full to partial	May to July
Swamp marigold	*Bidens aristosa mutica*	Mod. moist to wet	Full to partial	August to October
Swamp milkweed	*Asclepias incarnata*	Wet to moist	Full	June to August
Tall bellflower	*Campanula americana*	Moderately moist to mod. dry	Shade to partial	July to October
Tall ironweed	*Vernonia altissima*	Moderate	Full	August to October
White wild indigo	*Baptisia leucantha (alba)*	Mod. moist to dry	Full to partial	June to July
Yellow coneflower	*Ratibida pinnata*	Moderate to dry	Full to partial	July to September

corridor which is attractive to many animals and birds. Creating and maintaining diverse riparian plantings can help with excluding invasive species such as purple loosestrife and glossy buckthorn. Tables 2.6 and 2.7 list native trees and shrubs that may be found in riparian plant communities.

Desert Ecosystems

A desert habitat is an arid environment that receives <10 in. of rain per year. This is a particularly challenging environment in which to do landscaping. Plants that have evolved in a desert climate are drought-tolerant xerophytes. Many are also salt tolerant. There are numerous different deserts in the United States, some having plant species that evolved only in one place. An example of this is the Saguaro cactus, which only exists in the Sonoran Desert of Arizona. California, Colorado, Idaho, New Mexico, Oregon, Texas, Utah, Washington, and Wyoming also have desert ecosystems. Table 2.8 lists native desert plants.

Sustainability in the Plantscape

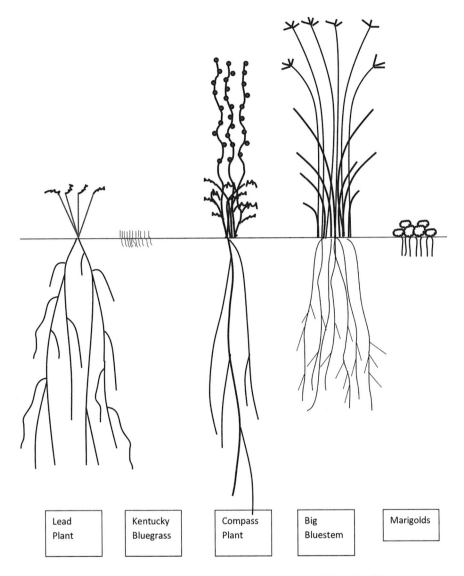

FIGURE 2.5 Root systems of various plants can grow to several feet (a half meter or more) long. By contrast, many turf grasses and annual bedding plants have roots of only a few inches (centimeters). (Illustration by author.)

Xeriscaping is a term that refers to arid-land gardening. It is also described as "water-wise" gardening. It was developed out of a need to reduce water consumption in arid regions but can also be used where water quality is poor. It is receiving more interest in areas where rainfall is adequate, but availability of treated water is limited. This is true in many areas of the midwest and southeast where droughty periods occur annually or periodically. Xeriscaping is discussed in more detail in Chapter 6.

TABLE 2.6
Native American Trees for Riparian Zones

Common Name	Botanical Name	Hardiness Zones
American elm	*Ulmus americana*	3–9
American sycamore	*Platanus occidentalis*	4–9
Basswood	*Tilia americana*	3b–8
Black walnut	*Juglans nigra*	4–9
Boxelder[a]	*Acer negundo*	3–9
Butternut	*Juglans cinerea*	3–7
Cottonwood[b]	*Populus deltoides*	3a–9
Green ash	*Fraxinus pensylvanica*	3–9
Honey locust	*Gleditsia triacanthos*	4–9
Ironwood, American hophornbeam	*Ostrya virginiana*	3b–9
Red maple	*Acer rubrum*	3b–9
River birch	*Betula nigra*	3b–9
Silver maple[b]	*Acer saccharinum*	3–9
Sugar maple	*Acer saccharum*	3–8
Swamp white oak	*Quercus bicolor*	4–8

[a] Can be invasive.
[b] Fast-growing and weak-wooded.

TABLE 2.7
Native American Shrubs for Riparian Zones

Common Name	Botanical Name	Hardiness Zones
Arrowwood viburnum	Viburnum dentatum	2(3)–8
Black chokeberry	Aronia melanocarpa	3–8(9)
Buttonbush	Cephalanthus occidentalis	5–11
Common witchhazel	Hamamelis virginiana	3b–8(9)
Cranberry bush viburnum	Viburnum trilobum	2–7
Eastern wahoo	Euonymus atropurpurea	4–9
Elderberry	Sambucus canadensis	3(4)–9
Gray dogwood	Cornus racemosa	3b–8
Red chokeberry	Aronia arbutifolia	4–9
Redosier dogwood	Cornus sericea	2–7
Spicebush	Lindera benzoin	4–9
Winterberry	Ilex verticillata	3–9

TABLE 2.8
Native Plants of the Southwest Desert

Common Name	Botanical Name
Trees	
Acacia	*Acacia* spp.
Mesquite	*Prosopis* spp.
Mimosa	*Mimosa* spp.
Palo verde	*Parkinsonia* spp. (*Cercidium* spp.)
Shrubs/Cactus	
Agave	*Agave* spp.
Barrel cactus	*Echinocactus* spp., *Ferocactus* spp.
Cholla	*Opuntia fulgida*
Creosote	*Larrea tridentate*
Jojoba	*Simmondsia chinensis*
Ocotillo	*Fouquieria splendens*
Organ pipe	*Stenocereus thurberi*
Prickly pear	*Opuntia* spp.
Queen of the night	*Cereus* spp.
Saguaro	*Carnegiea gigantea*
Yucca	*Yucca* spp.

ECOLOGICAL LANDSCAPING

Ecological design is an important component of sustainable landscaping. However, the differences between conventional landscaping and ecological landscaping are more than a matter of biology. There is the issue of aesthetics and what is conventionally perceived as a "beautiful" landscape. This conventional view holds that a manicured lawn and neatly trimmed shrubbery is part of an ideal landscape. Ecological systems would allow leaves to remain in the landscape and permitted to decompose and return their nutrients on site, as one example of aesthetic difference. It may be possible that attitudes about what is beautiful in a landscape will change as people become more educated about the environmental and ecological benefits that accompany sustainable landscaping practices.

NATIVE AND NON-NATIVE PLANT SELECTION

Among the different components of an ecosystem, plant selection is one over which the individual has a great deal of control. Wildlife habitat can be developed through the selection and placement of plants. Native pollinators, animals, and even soil-dwelling organisms can benefit through various practices designed appropriately and implemented properly. For example, micro-organisms can be aided through practices which contribute to healthy soil, including composting yard waste and using natural mulches such as wood bark or chips. If there is a natural water feature, such as a pond, lake, river, or stream, niches of habitat can be developed that mimic naturally occurring sites. The use of native plants that have evolved in these environments play an important role, whereas removal of weedy or invasive plants helps with establishment of desirable ones.

A study published by Lerman et al. (2012) supported the notion that native plants benefit native wildlife. The group studied residential landscape types in Phoenix, Arizona, and found that landscapes with native plants provided more abundant resources for birds and had a more diverse range of bird species than those with fewer native plants.

The use of native plants in landscaping is promoted as one method of minimizing inputs such as irrigation and fertilizer. When used appropriately, native plants can serve as low-maintenance, and thus sustainable components in the landscape. However, non-native plants that are well-adapted to a particular site may also be an important component of sustainable landscaping. Either way, weedy, invasive, or otherwise undesirable plants should be avoided.

Whether only native plants are used or they are used in conjunction with non-native plants, the goal of sustainable landscaping is to establish and maintain the dynamic interaction between the plants that are used; the animals they attract, shelter, feed, or otherwise benefit; the microorganism communities associated with them; and the surrounding environment.

ATTRACTING WILDLIFE

Encouraging wildlife is an important component of many ecosystem enhancement programs. Wildlife includes larger, more visible animals such as foxes and deer, smaller animals, such as rabbits, voles, opossums, frogs, toads, and even includes insects. All of these, in combination with plant species and microorganisms, are needed for a healthy ecosystem.

NATIVE POLLINATORS

Pollinators play an essential, key role in nature and in agriculture. Through pollination, seeds and fruits develop, providing food and feed for many animals. Insects and other pollinators benefit directly from pollinating flowers by obtaining nectar and pollen. Bees are the most important pollinators in general, and honey bees are the most important pollinators for commercial crops. Seventy-five percent of flowering plants rely on bees for pollination, whereas honey bees represent at least $3 billion in production of agricultural and horticultural products. Other pollinators that play an important role are butterflies, moths, and flies, as well as animals, such as bats and birds.

There are 4,000 species of bees in North America alone. Most of the native bees do not live in colonies as honey bees do, but are solitary. Bees require nesting sites, and flowering plants for food. Many have coevolved with native plants, but introduced species of garden plants may also serve as a food source for them. Bees nest in various types of habitats, and the ideal habitat depends on the species of bee. Some bees live in tunnels or holes in woody plants, whereas others nest in tunnels in the ground, and still others use cavities such as rodent burrows or small spaces

Sustainability in the Plantscape

in constructed areas, such as outbuildings. Nesting sites for native bees can be created by providing areas of undisturbed ground, or leaving dead or decaying trees in place. Careful observation of a site may reveal the presence of ground-nesting bees in mounds of soil, compost piles, unattended hillsides, or other places.

Some native bees and butterflies have declining populations for no known reason. The Xerces Society for Invertebrate Population (www.xerces.org) reports that "three species of bees, the rusty-patched, the yellowbanded, and western bumblebee, have dropped over the past decade. A fourth species, Franklins' bumblebee (Oregon and northern California) has only been seen once in the past several years." The rusty patched bumble bee is an important pollinator of prairie wildflowers as well as numerous crops including cranberries, blueberries, apples, and alfalfa. Neonicotinoid insecticide has been identified as one of the possible causes of bee decline.

The Xerces Society has developed guidelines for numerous programs to encourage habitat for native pollinators along roadsides, in parks, and in golf courses. Guidelines for these programs can be viewed and downloaded at their website. The core principles they propose are: recognize the native pollinators and their habitat that are already on your site; adjust existing land management practices to avoid causing undue harm to the pollinators already present; and enhance, restore, or create habitat for native bees and butterflies (Tables 2.9 and 2.10).

In 2019, The Xerces Society reported a severe decline in western monarch butterfly populations. In their report "Record Low Number of Overwintering Monarch Butterflies in California" they revealed an all-time record low of monarchs at 213 sites, an 86 percent drop from the previous count done at Thanksgiving 2017, and a 99.4 percent decline from the numbers presented in the 1980s (Schultz et al. 2017). Thirty species of butterflies are currently listed as Endangered or Threatened on the US Fish and Wildlife Service Endangered Species List.

BIRDS AND MAMMALS

Birds and other wildlife are affected by the disturbance of land for development, as well as the alteration of plant communities in an area. Some birds and animals benefit from human habitation when they are able to adapt to conditions that include streets and traffic and the presence of humans in their habitat, and can take advantage of nesting places under eaves or on rooftops. Racoons, house wrens, nighthawks, and rabbits are commonly seen in many urban areas. Squirrels are seen more often in cities than farmlands. Animals that may not be seen, but do exist in and near urban areas include skunks, foxes, coyotes, deer, and hedgehogs. Frogs and snakes can survive in many urban habitats, as well.

Those wishing to encourage wildlife will be successful at some level if they can provide food and habitat. Even apartment balconies may be used to feed birds. Some ways to provide habitat include creating a brush pile in one area of the yard to encourage birds and animals that require some sort of protective cover and planting trees and shrubs to provide nesting areas for birds. Birdhouses are sometimes successful at attracting nesting birds. Food for birds includes most commonly seeds or fruit. Some birds prefer insects and will not visit feeders. Table 2.11 lists common bird food and the birds they are most likely to attract. Hummingbirds and orioles may

TABLE 2.9
Plants That Feed Butterfly Larvae

Butterfly	Larval Food Plant
Black swallowtail *Papilio polyxenes asterius*	Carrots *Daucus carota*
	Parsley *Petroselinum crispum*
	Dill *Anethum graveolens*
Cloudless sulphur *Phoebis sennae*	Sennas *Cassia* spp.
Dog face *Colias cesonia*	Leadplant *Amorpha canescens*,
	Clover *Trifolium* spp.
	Purple prairie clover *Petalostemon* spp.
Giant swallowtail *Papilio cresphopntes*	Prickly ash *Zanthoxylum americanum*
	Citrus trees *Citrus* spp.
Great Spangled Fritillary *Speyeria cybele*	Violets *Viola* spp.
Monarch *Danaus*	Milkweeds *Asclepias* spp.
Painted lady *Vanessa cardui*	Thistles *Circium* spp.
	Bachelor's buttons *Centaurea cyanus*
Pipevine swallowtail *Papilio philenor*	Pipevine *Aristolochia tomentosa*
Question mark *Polygonia interrogationis*	Elm *Ulmus* spp.
	Hackberry *Celtis* spp.
Red Admiral *Vanessa atalanta*	Nettles *Urtica* spp.
	Bachelor's buttons *Centaurea cyanus*
Red-spotted purple *Limentitis arthemis astyanax*	Willows *Salix* spp.
	Poplars, *Populus* spp.
Spicebush swallowtail *Papilio Troilus*	Spicebush *Lindera benzoin*
	Sassafras *Sassafras albidum*
Tiger swallowtail *Papilio glaucus*	Wild cherry, *Prunus* spp.
	Birch, *Betula* spp.
	Poplar, *Populus* spp.
	Ash, *Fraxinus* spp.
	Apple *Malus* spp.
	Tulip tree *Liriodendron tulipfera*
Viceroy *Limenitis archippus*	Pussy willow *Salix discolor*
	Plums, cherries *Prunus* spp.
Zebra swallowtail *Papilio marcelius*	Pawpaw *Asiminia triloba*

be attracted to nectar feeders. Birds will also feast on seed-producing flowers and grasses, and fruit from various trees and shrubs. Table 2.12 lists plants to feed birds.

Feeding wildlife does not necessarily mean allowing them to forage in your garden to eat as they please. Fencing in a vegetable garden prior to planting it may keep out unwanted visitors. However, many suburban developments have been built right in the middle of deer habitat, and the deer do not simply go away once the houses go up. In such areas, deer feeding on cultivated plants is a serious problem with few remedies. Table 2.13 lists plants that deer prefer not to eat. Other than planting such non-delicacies, plants and gardens may require fencing of various types, and even then it may not be possible to keep the deer away.

TABLE 2.10
Plants That Provide Nectar for Butterflies

Common Name	Botanical Name
Bee balm	*Monarda didyma*
Blazing stars	*Liatris* spp.
Buckeye	*Aesculus* spp.
Butterfly bush	*Buddleia davidii*
Butterfly milkweed	*Asclepias tuberosa*
Ironweed	*Vernonia* spp.
Lilac	*Syringa* spp.
Purple aster	*Aster* spp.
Redbud	*Cercis Canadensis*
Sedum	*Sedum alboroseum*
Spicebush	*Lindera benzoin*
Sweet Alyssum	*Alyssum* spp.
Verbena	*Verbena* spp.
Viburnums	*Viburnum* spp.
Wild bergamot	*Monarda fistulosa*
Zinnias	*Zinnia* spp.

TABLE 2.11
Foods That Attract Birds

Bird	Food
American goldfinch	Sunflower seeds (black-oil, black-striped, gray-striped)
Blue jay	Sunflower seeds (black-oil, black-striped, gray-striped)
Brown creeper	Suet, peanut butter
Cardinal	Safflower, sunflower seeds (black-oil, black-striped, gray-striped), millet (white), wheat
Carolina wren	Suet, pecan meats
Catbird	Oranges (halved)
Chickadee	Sunflower seeds (black-oil, black-striped, gray-striped)
Grackle	Sunflower seeds (black-oil, black-striped, gray-striped)
House sparrow	Millet (red, white), sunflower seeds (black-oil, black-striped, gray-striped)
Hummingbird	Sugar water
Nuthatch, white-breasted	Sunflower seeds (black-oil, black-striped, gray-striped)
Orioles	Oranges (halved), nectar (hummingbird feeders may attract)
Purple finch	Safflower, sunflower seeds (black-oil, black-striped, gray-striped)
Red-breasted nuthatch	Sunflower seeds (black-oil, black-striped, gray-striped), suet
Rufous-sided towhee	Millet, white proso, mixed seeds
Tufted titmouse	Sunflower seeds (black-oil, black-striped, gray-striped)
Woodpecker, downy	Suet, peanuts
Woodpecker, red-bellied	Suet, peanuts

TABLE 2.12
Plants That Attract Birds because of Their Food Value

Common Name	Botanical Name	Habitat	Bird Attracted
Ash	*Fraxinus* spp.	Woods, slopes, moist areas	Songbirds
Bald cypress	*Taxodium distichum*	Wet areas	Water fowl
Black chokeberry	*Aronia melanocarpa*	Moist, sandy woods	Songbirds, brown thrasher, meadowlark
Black oak	*Quercus velutina*	Dry woods	Grackles, jays, brown thrashers, woodpeckers
Buttonbush	*Cephalanthus occidentalis*	Streams, lakes, ponds, swamps	Waterfowl
Chestnut	*Castanea dentate*	Acidic upland	Woodpeckers
Elderberry	*Sambucus canadensis*	Moist soil, open woodlands	Songbirds
False indigo	*Amorpha fruticosa*	Moist soil rocky streams, alluvial soils	Songbirds
Gray dogwood	*Cornus racemosa*	Fencerows, roadsides, streambanks, prairies	Songbirds, bobwhite, catbird, thrushes
Hackberry	*Celtis occidentalis*	Most conditions	Mockingbirds, robins
Hazelnut	*Corylus americana*	Dry or moist woods	Songbirds, red-bellied woodpecker
Northern red oak	*Quercus rubra*	Well drained slopes	Grackles, jays, brown thrashers, woodpeckers
Persimmon	*Diospyros virginiana*	Dry woods, bottomland woods, field edges	Songbirds
Redbud	*Cercis canadensis*	Rich woods, ravines, fencerows	Songbirds
Red-twig dogwood	*Cornus sericea*	Marshes, moist fencerows, swamps, streambanks	Songbirds, bobwhite, catbird, thrushes, towhee, wood ducks
River birch	*Betula nigra*	Moist areas	Songbirds
Sassafras	*Sassafras albidum*	Roadsides, old fields, edges of woods	Songbirds
Serviceberry	*Amelanchier arborea*	Wooded hillsides, swamps, stream banks, wet woods	Baltimore oriole, bluebird, blue jay, brown thrasher, cardinal, catbird, flicker, junco
Sour gum	*Nyssa sylvatica*	Most conditions	Pileated woodpecker, robins
Spicebush	*Lindera benzoin*	Rich, moist woods, streambanks	Songbirds, flicker, kingbird, red-eyed vireo, woodthrush, veery
Sugar maple	*Acer saccharum*	Moist woods	Evening grosbeaks
Sweet gum	*Liquidambar styraciflua*	Moist areas	Goldfinch
Wild plum	*Prunus americana*	Woods, edges of streams, fencerows	Songbirds
Winterberry	*Ilex verticillata*	Wet woods, swamps, edges of streams, ponds	Songbirds, bluebird, brown thrasher, catbird, cedar waxwing, hermit thrush, purple finch, waterfowl

TABLE 2.13
Plants Deer Prefer Not to Eat

Common	Botanical Name	Zones
Trees		
Ash	*Fraxinus* spp.	3–9
Cedar	*Cedrus* spp.	6–9
Coast redwood	*Sequoia sempervirens*	7–9
Cypress	*Cupressus* spp.	7–9
Douglas fir	*Pseudotsuga menziesii*	4–6
False cypress	*Chamaecyparis* spp.	4–8
Fig	*Ficus carica*	7–9
Fir	*Abies* spp.	Varies
Hackberry	*Celtis occidentalis*	3–9
Hawthorn	*Crataegus* spp.	4–7
Japanese maple	*Acer palmatum*	5–8
Magnolia	*Magnolia* spp.	5–9
Maidenhair tree, ginkgo	*Ginkgo biloba*	4–8
Oak	*Quercus* spp.	3–10
Pine	*Pinus* spp.	2–7
Silk tree	*Albizia julibrissin*	6–9
Spruce	*Picea* spp.	2–7
Shrubs		
Barberry	*Berberis* spp.	3–8
Bottlebrush	*Callistemon* spp.	8–11
Boxwood	*Buxus* spp.	6–9
Butterfly bush	*Buddleia davidii*	5–9
California buckeye	*Aesculus californica*	7–8
Cotoneaster	*Cotoneaster* spp.	4–7
Currant	*Ribes* spp.	2–7
Eleagnus	*Eleagnus* spp.	2–8
Firethorn	*Pyracantha coccinea*	6–9
Flowering quince	*Chaenomeles* spp.	4–8
Glossy abelia	*Abelia X grandiflora*	6–9
Heath	*Erica* spp.	5–7
Heavenly bamboo	*Nandina domestica*	6–9
Holly	*Ilex* spp.	3–9
Japanese kerria	*Kerria japonica*	4–9
Juniper	*Juniperus* spp.	2–9
Lilac	*Syringa* spp.	3–7
Myrtle	*Myrtus communis*	9–10
Oleander	*Nerium oleander*	8–11
Oregon grape holly	*Mahonia aquifolium*	5–7
Scotch heather	*Calluna vulgaris*	4–6
St. Johns wort	*Hypericum* spp.	4–8
Sumac	*Rhus* spp.	3–9
Sweet box	*Sarcococca hookeriana*	6–8
Viburnum	*Viburnum* spp.	3–7

Skunks, groundhogs, rabbits, and other small mammals will find food on their own, often eating turfgrass or weeds, and special efforts are not required to attract them to an area. They will choose a habitable location that provides cover, food, and access to water.

Case Study: Portland, Oregon

Reference: Suutari, Amanda. USA – Oregon (Portland) – Sustainable City. In The EcoTipping Points Project: Models for Success in a Time of Crisis. http://www.ecotipping points.org/our-stories/indepth/usa-portland-sustainable-regional-planning.html. Viewed December 28, 2011.

Portland, Oregon, is a "Sustainable City", with long-term planning and a long record of actions towards sustainability. Their acts have not been isolated to landscaping practices alone, but incorporate comprehensive planning in all facets of city life. Among those practices, though, are some that are directly related to landscaping and horticulture. These include the Nature in Neighborhoods program and the Metropolitan Greenspaces Master Plan.

The Greenspaces Master Plan provides funding for local groups to organize neighborhood cleanups or restoration, education, or conservation; acquisition of open spaces within city limits, including 74 miles of stream and riparian frontage; assistance and incentives for homeowners to protect watersheds and wildlife habitat; maintaining up-to-date maps of parks and greenspaces in the region; and education on sustainable gardening practices.

The Nature in Neighborhoods program is an integral part of the Greenspaces Master Plan. It facilitates partnerships with localities to restore habitat and implement habitat-friendly development. Some of the areas it focuses on are connecting riparian corridors to upland regions, protecting wildlife habitat, and restoring degraded sites.

Case Study: Boston, Massachusetts Urban Wilds Program – Nira Rock

Reference: Gregory, Regina. USA – Massachusetts (Boston) – Nira Rock. In The EcoTipping Points Project: Models for Success in a Time of Crisis. http://www.ecotipping points.org/our-stories/region/usa-canada.html#nira. Viewed December 28, 2011.

Greening Communities by Design. Parks- Nira rock update. http://www.cogdesign.org/proj_2007_nirarock.html. Viewed December 28, 2011.

The Urban Wilds Initiative is administered by the Boston Parks and Recreation Department. Its main objective is to protect publicly owned land in the urban setting. In addition to managing the lands, some of the goals of the initiative are to promote their ecological integrity, protect the lands from development, and promote their use for recreation and education.

Sustainability in the Plantscape

To that end, Nira Rock, a former quarry, then a Depression-era jobs creation project, and later a favorite hang-out for teenagers, and eventually crack-cocaine addicts, was cleaned up and re-designed to make it more family-friendly. The Community Outreach Group for Landscape Design (COGDesign) developed a plan that would include improved access, enhanced trails, signage, circulation, improved native plantings, and community involvement. Designers Nina Shippen and Susan Opton organized the design and restoration of this 1.5 acre park in the Jamaica Plain neighborhood using numerous community and corporate volunteers. Photos of the finished work may be viewed at http://www.cogdesign.org.

Boston has more than thirty other Urban Wild sites, including Allendale Woods, Belle Isle, Condor Street Urban Wild, Dell Avenue Rock, Geneva Avenue Cliffs, Iroquois Woods, Puddingstone Garden, West Roxbury High School, and Willowwood Rock. More information is available at http://www.cityofboston.gov/parks/urbanwilds/.

FIRE-WISE LANDSCAPING

Numerous organizations provide fire safety practices and tips, including the National Interagency Fire Center. Proper plant management for preventing the spread of a fire include: spacing landscape vegetation so that fire cannot be carried to the structure or surrounding vegetation; removing branches from trees to height of 15 ft; and maintaining a fuel break around all structures.

Landscape construction materials should be considered for combustibility and flammability in areas that are particularly prone to wildfires.

PROGRAMS FOR HABITAT DEVELOPMENT

There are a number of programs that have been developed to provide incentives and a framework for designing and planting wildlife habitat in the managed landscape. Only two are considered herein, as they apply directly to two major landscape areas: residential and golf courses.

THE BACKYARD HABITAT

The National Wildlife Federation has developed guidelines for individual residences and certify their implementation in a program called the Backyard Habitat Program®. A certification form may be found at their website, www.nwf.org. To qualify as a backyard habitat, there are five criteria to meet: provide food for wildlife, a water source, a place for cover, a place for rearing young, and implement sustainable gardening practices.

THE GOLF COURSE HABITAT

The Audubon Society has developed several programs to encourage participation in ecosystem enhancement. These are: The Sustainable Communities Program, The

Green Neighborhoods Program, The Audubon Cooperative Sanctuary Program, and Audubon Partners for the Environment Program.

Hundreds of golf courses both domestically and internationally, have become certified in Audubon's habitat-enrichment program for golf courses called the Cooperative Sanctuary Program. After joining the program, golf courses evaluate their environmental resources and potential liabilities, and then develop an environmental plan. Audubon International provides a Site Assessment and Environmental Planning Form and also provides guidance. The six areas addressed in this program are:

1. Environmental Planning
2. Wildlife and Habitat Management
3. Chemical Use Reduction and Safety
4. Water Conservation
5. Water Quality Management
6. Outreach and Education

Participants then implement and document their environmental management practices in each of the areas to become eligible for designation as a Certified Audubon Cooperative Sanctuary.

Case Study: Bonita Bay Club East

Ref.: http://bonitabaygroup.com. Viewed December 27, 2011

Bonita Bay Club East is located near Naples, Florida, on the Gulf of Mexico. Bonita Bay golf courses, designed by Tom Fazio, are part of the Bonita Springs community. Two golf courses, The Cypress and The Sabal, have participated in the Audubon International Cooperative Sanctuary Program for many years. Re-certification must be attained every two years following the original certification.

Environmental efforts in numerous areas, including Wildlife and Habitat Conservation, Waste Management, and Environmental Planning, have not only resulted in certification in the Audubon program, but have also led to an Award for Excellence from the Urban Land Institute, as well as being designated the Environmental Leader in Golf by *Golf Digest*. With a total of 2,400 acres, more than half is dedicated to open space, which includes nature preserve areas and recreational areas.

Outreach and Education are important components of the Audubon Program. To meet this requirement, Bonita Bay has planted trees at a local Elementary School, hosts guided tours and provides educational programs for local schoolchildren, and planted a butterfly garden at a local city park. They also organize Earth Day activities.

The ongoing environmental efforts have included replacing turf that required irrigation with native grasses and other native vegetation. They were able to reduce irrigation and eliminate over thirty sprinkler heads altogether. Habitat enhancement and work to protect the native wildlife population are ongoing, as are efforts to limit pesticide use. Habitats include 900-acre cypress reserve, pine flatwoods, live oak, and cabbage palm hammocks. Ninety-eight percent of the landscape material used is native vegetation.

SUMMARY

There are many beneficial effects of plants in the landscape. These include carbon sequestration and the associated release of oxygen to the atmosphere in the process of photosynthesis; the cooling effect of evapotranspiration; structural effects; and the role plants play within an ecosystem. Plants may be drought-tolerant, tolerant of wet soils, or they may flourish to varying degrees in full sun to full shade. Plants provide ecosystem services through regulation of global and local climate, cleansing of air and water, erosion control, pollination, habitat functions, and plant-soil interactions.

One goal of sustainable landscaping is to establish and maintain a dynamic interaction between plants, animals, microorganisms, and the surrounding environment. Thus can wildlife habitat be developed through plant selection in the landscape. This includes the appropriate use of native plants in landscaping to provide low-maintenance, sustainable components in the landscape, as well as well-adapted, non-native plants. Turfgrass may be included in such a design, keeping in mind the maintenance requirements of each turfgrass species. Two drought-tolerant species of turfgrass are tall fescue and buffalograss. Turfgrass provides numerous benefits, as it cools surrounding air, prevents erosion, and builds soil structure. It aids in carbon sequestration, provides a place for recreation, and is aesthetically pleasing.

Woody plants with undesirable growth habits and poor structure should be avoided, as they require higher levels of maintenance. Invasive plants should also be avoided. If they do appear in a landscape, they should be eradicated as early as possible to prevent their establishment.

Ecosystem services account for the direct or indirect benefit to humans that are provided by processes involving the interaction of living and non-living elements, including vegetation and soil organisms, with water and air.

In the United States, numerous ecosystems have been described. These are complex interactions of plant, animal, and microorganism communities and their non-living environment. The main types of biomes that have been identified in North America are temperate coniferous forests, temperate deciduous forests, prairies or temperate grasslands, and deserts or xeric shrub lands. Understanding native ecosystems can serve as a guideline to sustainable landscape professionals in attempting to mimic or restore them.

Pollinators play an important role in ecosystems. Honey bees, native bees, birds, bats, and other animals contribute to pollination, which provides seeds, fruits, and nuts for wildlife. Wildlife can be encouraged to stay in areas where they find food and shelter and access to water.

Programs have been developed to help people design wildlife habitat in the managed landscape. Two notable ones are: The National Wildlife Federation Backyard Habitat Program® and The Audubon Cooperative Sanctuary Program.

REVIEW QUESTIONS

1. List the criteria for the NWF backyard habitat program.
2. What segment of the Green Industry is targeted by the Audubon Society Cooperative Sanctuary Program?
3. What is meant by carbon sequestration?
4. On a global scale, what is the location of the greatest amount of carbon storage?
5. What are the four main regional ecosystems identified in the United States?
6. What environmental condition favors understory plants?
7. What soil-building substance forms on the forest floor from the break-down of organic matter?
8. Which grass is the dominant species on the American prairie?
9. Riparian habitat is located near what?
10. What is the term used for drought-tolerant desert plants?

ENRICHMENT ACTIVITIES

Investigate the native ecosystem for your location using the USDA Forest Service map (http://www.fs.fed.us/rm/ecoregions/products/map-ecoregions-united-states/) or The EPA Ecoregions website (http://www.epa.gov/wed/pages/ecoregions/na_eco.htm). Research the native plant communities for your region.

Design a landscape using the NWF Backyard Habitat Certification program.

Make recommendations for a local golf course to meet the criteria for the Audubon Society's Cooperative Sanctuary Program.

Research native pollinators in your area and identify the endangered ones. Write an article about one of them.

Search for other habitat-development programs using the internet and make a presentation to your class on it.

FURTHER READING

American Society of Landscape Architects, Lady Bird Johnson Wildflower Center, United States Botanic Garden. 2008. The sustainable sites initiative guidelines and performance benchmarks (draft). Retrieved October, 2009.

Beard, J.B. and D. Johns. 1985. The comparative heat dissipation from three typical urban surfaces: asphalt, concrete, and a Bermuda grass turf. *Texas Agric. Exp. Station* 4329: 59–62.

Beard, J.B. and R.L. Green. 1994. The role of turfgrasses in environmental protection and their benefits to humans. *J. Environ. Qual.* 23(3): 452–460.

Biomes in Encyclopedia of Earth. http://www.eoearth.org/article/terrestrial_biome. Retrieved August, 2010.

Burns, D. (ed.) 2011. *The Xerces Society Guide to Attracting Native Pollinators.* Storey Publishing, North Adams, MA.

EPA. Heat Island effect. http://www.epa.gov/heatisland/ Retrieved June 25, 2010.

Harker, D., G. Libby, K. Harker, S. Evans, and M. Evans. 1999. *Landscape Restoration Handbook*, 2nd ed. Lewis Publishers, Boca Raton, FL, 865 pp.

Ingham, E.R. The soil food web in the soil biology primer. http://soils.usda.gov/sqi/concepts/soil_biology/soil_food_web.html. Retrieved June 29, 2010.

Kent, S., R. Heth, G. Morris, K. McConnaughay, and S. Morris. 2007. Carbon sequestration in urban turf soils. http://a-c-s.confex.com/a-c-s/2007am/techprogram/P35961.HTM. Viewed January 11 2009.

Lerman, S.B., P.S. Warren, H. Gan, and E. Shochat. 2012. Linking foraging decisions to residential yard bird composition. *PLoS One* 7(8): e43497. doi: 10.1371/journal.pone.0043497.

McClain, W. 1997. Prairie establishment and landscaping. Technical Publication #2. Division of Natural Heritage Illinois, Department of Natural Resources, Natural Heritage. Springfield, IL, 62 p.

Mugaas, R.J., M.L. Agnew, and N.E. Christians. 2008. The benefits of turfgrass. http://www.extension.umn.edu/distribution/horticulture/DG5726.html. Viewed August 12, 2008.

Newman, D.S., R.E. Warner, and P.C. Martin. 2003. Creating habitats and homes for Illinois wildlife. Illinois Department of Natural Resources and University Illinois. Metro Litho/Creative Drive, Oak Forest, IL, 212 p.

Nowack, D.J. and D.E. Crane. 2002. Carbon storage and sequestration by urban trees in the USA. *Environ. Pollut.* 116: 381–389. www.elsevier.com/locate/envpol.

Pollinator Partnership. http://pollinator.org/index.html. Viewed May 15, 2011.

Prairie Moon Nursery. 2010. *Catalog and Cultural Guide.* Prairie Moon Nursery, Winona, MN, 71 p.

Puoyat, R.V., I.D. Yesilonis, and D.J. Nowak. 2006. Carbon storage by urban soils in the United States. *J. Environ. Qual.* 35: 1566–1575.

Qian, Y.L. and R.F. Follett. 2002. Assessing long-term soil carbon sequestration in turfgrass systems using long-term soil testing data. *Agron. J.* 94: 930–935.

SaltScape Solutions. http://www.paspalumgrass.com/. Viewed January 25, 2013.

Schultz, C.B. 2017. Citizen science monitoring demonstrates dramatic declines of monarch butterflies in western North America. *Biol. Conserv.* doi: 10.1016/j.biocon.2017.08.019.

Shepherd, M. Making room for native pollinators: how to create habitat for pollinator insects on Golf Courses. The Xerces Society for Invertebrate Conservation. Retrieved June 26, 2010.

Shepherd, M., M. Vaughan, and S.H. Black. Pollinator-friendly parks. The Xerces Society for Invertebrate Conservation. Retrieved June 26, 2010.

Tangley, L. June/July 2009. "The Buzz on native pollinators" in *National Wildlife Magazine*, pp. 40–46.

The Xerces Society for Invertebrate Conservation. http://www.xerces.org/. Retrieved June 27, 2010.

U.S. Fish and Wildlife Service Endangered Species. https://www.fws.gov/endangered/. Retrieved April 19, 2019.

U.S. Forest Service. Ecosystem provinces map. https://www.fs.fed.us/land/ecosysmgmt/colorimagemap/ecoreg1_provinces.html. Retrieved April 19, 2019.

Wikipedia. Carbon cycle. http://en.wikipedia.org/wiki/Carbon_cycle. Retrieved June, 2008.

World Wildlife Fund. Southeastern conifer forests (NA0529). http://www.worldwildlife.org/wildworld/profiles/terrestrial/na/na0529_full.html. Retrieved July 6, 2010.

Xerces Society. Pollinators and roadsides: managing roadsides for bees and butterflies. http://www.xerces.org/wp-content/uploads/2010/05/roadside-guidelines_xerces-society1.pdf. The Xerces Society for Invertebrate Conservation. Retrieved June 26, 2010.

Zajicek, J.M. and J.L. Heilman. 1991. Transpiration by crape myrtle cultivars surrounded by mulch, soil, and turf grass surfaces. *HortScience* 26(9): 1207–1210.

3 The Sun and the Sustainable Landscape

OBJECTIVES

Upon completion of this chapter, the reader should be able to:

- Explain how solar radiation affects temperature of surfaces
- State the causes of the urban heat island
- Identify ways that plants affect the urban heat island
- Demonstrate how to ameliorate the effects of unwanted solar heat in the landscape
- Demonstrate how to maximize solar heating in the landscape in winter
- Implement landscape practices to reduce the heat island effect

TERMS TO KNOW

Albedo
Anthropogenic heat
Convection
Electromagnetic spectrum
Emissivity
Emittance
Evapotranspiration
Heat capacity
Heat gain
Microsurfacing
Radiative
Solar reflectance
Solar Reflectance Index (SRI)
Surface reflectivity
Thermal
Thermal conductivity
Thermal emittance
Transpiration
Urban heat island
Whitetopping

INTRODUCTION

Solar radiance and heat from the sun play an important role in buildings and the landscapes that surround them. A variety of materials are used in buildings and landscapes. Both natural materials, such as trees and turf, and "man-made" materials, such as lumber, glass, and steel are affected by solar radiation that contributes to the amount of heat that enters and remains on a site. **Heat gain** occurs when heat enters a building, usually by sunlight shining through a window, but also by being transmitted through materials, such as roofs and walls. Hardscaped areas of the landscape are also subject to heat from the sun, and subsequently can heat up to levels significantly higher than ambient temperatures, depending on their composition and their exposure to the sun. Shade provided by trees or structures can modify the amount of heat build-up on a driveway or other large paved areas by reducing the amount of sunlight striking the surface, or the amount of time it does so.

Taken together, the interactions that occur among the various materials have effects at two levels: the more immediate microclimate on-site and a collective effect that has come to be known as the **heat island** effect. Urban heat islands occur because of the materials that are used in the built environment, particularly buildings and pavement, and because of the removal of vegetation prior to development that is not replaced afterwards. **Anthropogenic heat** refers to manmade sources of heat that also contribute to the heat island effect. This chapter will examine the effects of solar radiation on materials and surfaces found in the landscape.

Building design plays a very important role, as does the way it is oriented on the lot. There is little the landscape designer can do to affect those decisions after the structures are in place; however, landscape architects may work with architects from the onset of a project to ensure such factors are taken into consideration. On a larger scale, urban planners should take these factors into consideration when deciding how to incorporate plantings and greenspaces into the urban setting.

STUDIES RELATED TO THE EFFECTS OF THE SUN ON THE LANDSCAPE

The effect of the sun in the landscape is a subject that is being increasingly studied in recent years. Some of the research conducted on this topic dates back several decades, but with more interest in global climate change, this area is receiving renewed attention. The US Green Building Council and Sustainable Sites Initiative™ have overlapping objectives in a shared goal of creating a more sustainable living environment. Solar radiation in the sustainable living environment affects surfaces, including roofs, pavements, and other hardscape materials. It also affects plants. However, surfaces interact differently than plants with solar radiation. This chapter will examine each of these areas.

The Sustainable Sites Initiative™ provides credits related to the topics addressed in this chapter as follows:

- Minimize building heating and cooling requirements with vegetation
- Reduce urban heat island effects

The Lawrence Berkeley National Laboratory is at the forefront of the research and dissemination of information related to solar radiation in the living environment. A number of websites are devoted to the various related topics. See the Further Reading section at the end of this chapter for websites and pertinent research. This chapter discusses the effects of solar radiation in the landscape, and examines the ways in which landscape design can be used to address the pre-existing problems at a site.

Case Study: Vegetation and the Urban Heat Island

Rose, L., H. Akbari, and H. Taha. 2003. Characterizing the fabric of the urban environment: A case study of greater Houston, Texas. Lawrence Berkeley National Laboratory. LBNL-51448;

Akbari, H. and S. Konopacki. 2002. Energy savings for heat island reduction strategies in Chicago and Houston (including updates for Baton Rouge, Sacramento, and Salt Lake City). LBNL-49638

Researchers at the Lawrence Berkeley National Laboratory conducted case studies in cooperation with the EPA Urban Heat island Pilot Project. Several cities were involved, including Houston. The summarized results for Houston are included here.

The objectives of the study were to quantify potential energy savings, economic benefits, and air quality improvements that could be realized if urban heat island mitigation strategies were implemented. The EPA strategies for mitigating the urban heat island can be viewed at their website and downloaded as a booklet (http://epa.gov/hiri/mitigation/index.htm).

The total land use area for Houston was estimated to be 3,430 km^2. Of that, ~56 percent was residential. The total impermeable areas were 740 km^2 for roofing (21 percent) and 1,000 km^2 for paved areas (29 percent). The combined areas for grass and tree cover comprised 1,320 km^2, or 39 percent of the total area. In Houston, trees primarily shaded streets, parking lots, grass, and sidewalks.

The authors concluded that in Houston, trees could potentially shade 20 percent of the roof area, 20 percent of the roads, 50 percent of the sidewalks, and 30 percent of the parking areas. By their estimates, a 12 percent increase in tree cover was possible. This amounts to an additional 8 million trees. However, other groups who studied the issue found that only about half of this increase, or an additional 4 million trees, was a realistic estimate.

Impermeable surfaces represented 50 percent of the total area of Houston. By increasing the albedo of parking areas, roads, roofs, and sidewalks, the ambient air temperature and the ozone levels could be reduced. Direct effects include planting trees near buildings to provide shade and use of high-albedo roofing materials. Indirect effects are urban reforestation with high albedo pavements and building surfaces. Combined strategies use both direct and indirect methods.

In a simulation assessing the savings using combined effects, the authors estimated that potential annual savings of $82 million could be realized, given current (2002) energy prices. In energy usage terms, an avoidance of 734 MW (megawatts) of peak power could be realized, and annual carbon emissions could be reduced by 170 ktC (kilotons of carbon). For just shade tree effects alone, a savings of $27.8 million could be realized as a result of a reduction in electricity usage of 421 GWh. Annual carbon emissions would be reduced by 58 ktC.

SOLAR ENERGY

When sunlight, or the sun's energy, enters the earth's atmosphere, it enters in many different wavelengths. Two of the notable effects it provides are light and warmth. Both phenomena are due to the wavelengths of the energy emitted by the sun: wavelengths which occur on a continuum known as the **electromagnetic spectrum** (Figure 3.1). Visible light, or light that humans detect, exists in the range of about 400 to about 700 nm, (1 nm = 10^{-7} m). Energy that is felt as heat from the sun falls in the **infrared** range (see Figure 3.1), which lies outside of the visible range. In addition to the range of wavelengths emitted by the sun, there is a range of intensities at different wavelengths (Figure 3.2).

When light strikes a surface, it may be re-radiated, but having lost some of its energy, the wavelengths are longer than the original ones. Infrared wavelengths, then, are somewhat longer than those visible to the human eye, but they can be felt as heat. They can also be detected using **thermal imaging** equipment. **Thermography** uses thermal imaging equipment and can indicate the temperature of objects that are radiating energy in the infrared range (Figure 3.3). Night vision goggles use this technology to allow people to "see" in the dark. Energy audits on buildings are conducted using thermal imaging cameras that can indicate whether insulation is present in the walls of a house, for example.

Solar radiation in the living environment affects surfaces and plants. Surfaces in the built environment include roofs, pavement, decking, and other landscape materials. In the landscape, plants often exist singly, but may also occur in groups, or as a monoculture, such as large turf areas.

Gamma rays, x-rays, ultraviolet light, visible light, infrared, microwaves, and radio waves all fall on this spectrum.

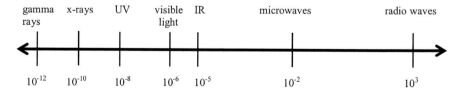

FIGURE 3.1 Electromagnetic spectrum of energy.

The Sun and the Sustainable Landscape

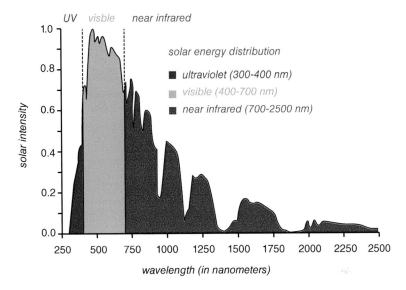

FIGURE 3.2 Solar energy intensity varies over wavelengths from about 250 to 2,500 nm. (Drawing courtesy EPA.)

FIGURE 3.3 Infrared photo of asphalt in sun and shade. Bar on right side of photograph indicates heat. White area in center of photograph corresponds to the hottest temperature in the picture, 116°F (46.7°C). Coolest area is in shade, at 87°F (30.5°C). (Photo by the author.)

Solar Heat Gain

Solar **heat gain** is the increase in temperature that results from the sun striking a surface. The surfaces which are struck by sunlight either re-radiate the energy or absorb it and then transmit it to surrounding air or objects.

In solar heat gain, the type of surface that sunlight first encounters is the most important factor. The properties of the surface material determine whether it will retain heat from the sun, and for how long, as well as whether it will be hot to the touch or not. The primary properties that affect materials and heat are the amount of light and heat that are reflected or absorbed by the material.

Reflecting and Absorbing Light

Light colors reflect more light and absorb less heat than dark colored surfaces. **Albedo** is a measure of how much solar radiance is reflected from a surface as a percentage of the amount received, and it can vary with the angle of incoming radiation.

The terms **surface reflectivity** of the sun's radiation and **solar reflectance** are used interchangeably with the term *albedo*. Together, these terms describe **radiative** properties. Albedo is expressed as a number between 0 and 1, or as a percentage between 0 and 100. If a surface has an albedo value of zero it does not have any solar radiance. A black surface is an example of a surface having zero albedo. A white surface, such as clean snow, has an albedo of one (100 percent). There is another aspect of albedo, however: the value assigned to a surface is adjusted according to the wavelength of radiation reflected (see Figure 3.1). Usually, more than a single wavelength is reflected, so the number given refers to an average across the electromagnetic spectrum. Thus, light and heat may be reflected from the surface, giving a different value than if light alone was reflected. Figure 3.4 shows various materials and their albedo values.

For hardscaping materials used in the landscape, albedo varies, but it also can change over time. For example, new asphalt has an albedo of around 5–10 percent and weathered asphalt has an albedo of 10–15 percent. This illustrates the effect of weathering on asphalt: it tends to lighten up over time. On the other hand, concrete tends to darken with age. New white Portland cement concrete has an albedo of 70–80 percent and the same cement weathered has an albedo of 40–60 percent. Interestingly, bricks, which are made of clay, tend to have higher albedo values than the concrete pavers that have replaced them. Albedo values for bricks range from 23 to 48 percent, whereas concrete paver albedo values range from 17 to 27 percent (Table 3.1). By comparison, earth's overall albedo is 37 percent, according to NASA.

Thermal Emissivity

Energy also has **thermal** properties, which is felt as heat, and falls into the infrared range of the electromagnetic spectrum (Figure 3.1). **Thermal emittance** measures the ability of a material to emit, or shed, heat it has absorbed. **Emittance**, or **emissivity**, is also expressed as a value between zero and one, or expressed as a percentage. Materials which emit heat readily, such as wood, stay cooler and have higher emissivity, whereas those which hold their heat have relatively low emissivity.

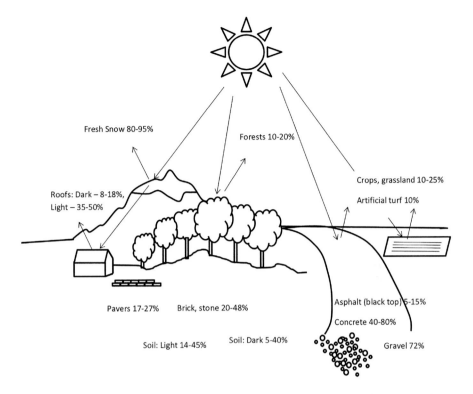

FIGURE 3.4 Albedo is the ability of surfaces to reflect sunlight. It may be expressed as a percentage from zero (a completely black surface that reflects no light) to 100 percent (a white surface that reflects all of it). (Illustration by the author.)

Windows made of "low-e" glass reflect thermal energy rather than allowing it to enter buildings, thereby reducing solar heat gain that might otherwise occur. One problem with low-e windows is that they reflect heat back onto other objects, such as the siding on neighboring buildings, shrubbery, and so on. Vinyl siding distorts at 160°F (71°C), and low-e windows have been known to reflect enough heat to distort vinyl siding. Forests, grasslands, water, bare soil, snow and ice, all have high emissivity.

Heat Capacity

Heat capacity is the ability of a material to store heat. Stone, asphalt, concrete, and steel have higher heat capacities than dry soil or sand. Impervious concrete and asphalt can reach peak summertime surface temperatures of 120°F–150°F (50°C–65°C) and can transfer heat downward to be stored in the pavement subsurface, where it is re-released as heat at night. The thicker the pavement is, the more heat it stores. Pavement represents a significant amount of cover, from nearly one-third to nearly one-half, in many urban areas.

TABLE 3.1
Albedo and Emissivity of Various Surfaces in the Landscape

Surface	Albedo (%)	Emissivity (%)
Asphalt (new – old)	5–10	90–98
Forests – coniferous	5–15	96–98
Soil – dark and wet	5–40	66 (black loam)
Artificial turf	10	NA[a]
Wood – pine	10	60
Forests – deciduous	15–20	98
Meadows	15–25	97
Soil – light and dry	15–45	38 (plowed field)
Concrete	17–27	85–95
Pavers	17–27	89 (light concrete)
Brick	23–48	93
Sand	24	72
Grass	25–30	98
Steel	55	12
Gravel	72	28
Fresh snow	80–95	98

[a] Not available.

Sources: Ahrens (2006), http://www.eoearth.org/article/albedo, Oke (1987) and Pon (1999).

SOLAR REFLECTANCE INDEX

The **Solar Reflectance Index (SRI)** was developed by the Lawrence Berkeley National Laboratory to express the effects of both reflectance and emittance of surfaces or materials in one number. The SRI value represents the temperature of a material when sunlight is hitting it, as compared to an all-black or all-white material. Lower SRI values indicate lower reflectivity, and therefore, more retention of the solar radiation.

According to the coolroofs.org website, The SRI:

> Is defined so that a standard black (reflectance 0.05, emittance 0.90) is 0 and a standard white (reflectance 0.80, emittance 0.90) is 100. Due to the way SRI is defined, particularly hot materials can even take slightly negative values, and particularly cool materials can even exceed 100. (Figure 3.5)

Since albedo has been shown to have a greater effect on maximum surface temperatures, the SRI is mostly a measure of that component. Additionally, most materials used in landscaping and construction have high emittance values, so there are limited options for modifying that aspect of heat gain as compared to changing albedo. The SRI values are used in both LEED and SSI accreditation. Both systems require an SRI of at least 0.29 for surfaces and structure coverings.

FIGURE 3.5 The Solar Reflectance Index (SRI) is a measure of both albedo and emittance. Both LEED and SITES™ accreditation require a minimum SRI value of 29.

HEAT LOSS AND HEAT TRANSFER

Besides albedo, emissivity, and heat capacity, two other properties influence how readily pavements absorb or lose heat. **Thermal conductivity** is the ability of a surface to transfer heat through it and to the underlying soil or other material. Surfaces with low conductivity will not transfer the heat to neighboring materials as quickly or as much compared to those with high conductivity. Metal surfaces tend to have high thermal conductivity, and thus transfer heat readily.

Convection is the transfer of heat by air movement. Wind can move heat away from a surface, and the rate of convection is related to the velocity of air movement, the temperature of the air passing over a surface, the total area of the surface exposed to air, and the surface roughness. Surface roughness contributes to air turbulence over the surface, which can increase convection and cooling. Roughness of a surface can also reduce solar reflectance. This subject is treated in more detail in the next chapter. In it, using plants to reduce convective heat loss caused by wind is discussed, as is the insulative property of plants placed close to a structure, and designing plant placement to funnel cooling breezes.

URBAN HEAT ISLAND

The term **urban heat island** describes a phenomenon that occurs in cities in which temperatures are higher than they are in surrounding rural areas (Figure 3.6). There are numerous components factored into this example of local climate change. It is local because its effects are limited to a local scale and range. The changes to climate are measured both directly and indirectly, and include changes in temperature, precipitation, and wind. In an article from NASA's Earth Observatory, a NASA team explains how Atlanta influences its own weather; temperatures in the Atlanta area reached 8°–10° higher than the outlying rural areas. Researchers attributed six thunderstorms one month to that temperature differential. They said that as temperatures rose during the day, air was drawn in from the surrounding areas. The incoming air converged in town and created an upward flow motion, pushing the hot air up and triggering thunderstorms. One climatologist, Dev Niyogi (2009), even attributes the severity of the 2008 tornado that struck Atlanta to the urban heat island effect.

Another effect from the localized climate change occurring during the Atlanta study was that ozone created due to the higher temperatures doubled. Ozone is a major contributor to smog. By reducing heat island effects, smog may also be reduced.

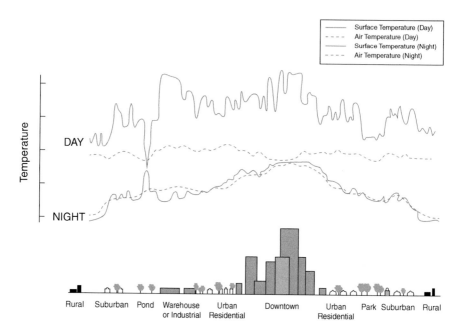

FIGURE 3.6 Urban heat island as evidenced by raised temperatures in a city as compared to surrounding areas. (Illustration courtesy EPA.)

A report by Pomerantz and colleagues at the Lawrence Berkeley National Laboratory (2000) analyzed the monetary benefits of reducing temperature from pavement albedo in Los Angeles. They concluded that more than $90 million per year could be saved in indirect costs of energy savings and smog reductions.

The urban heat island effect is measured and monitored using direct and indirect methods. Direct measurements include temperature measurements; indirect measurements include thermal imaging cameras. Temperature measurements can be transformed into isotherm maps and temperature graphs. Thermal imaging cameras give a vibrant visual image of temperature differentials using infrared photography (Figure 3.7). Infrared waves are mostly felt as heat and are not visible to the human eye.

Factors involved in the urban heat island effect include:

- Plant cover – **evapotranspiration** and shade provide cooling effects
- Plant architecture – can affect air flow in urban canyons
- Roof color – albedo/thermal emissivity can heat the air above them
- Paved areas – albedo/thermal emissivity can heat the air above them
- Weather – clear skies and calm winds can foster urban heat island formation
- Geographic location – bodies of water and topography can influence local wind patterns

The Sun and the Sustainable Landscape

FIGURE 3.7 Infrared photo of shade on house in the afternoon. Bar on right side of photograph indicates heat. White area in center of photograph corresponds to the hottest temperature in the picture, 146°F (63.3°C). Coolest area is in shade, at 100°F (37.8°C). (Photo by the author.)

Effects of the urban heat island effect include:

- Heat-related health effects and deaths
- Increased electrical demand in summer
- Increased carbon dioxide emissions from power plants
- Increased municipal water demand
- Unhealthy ozone levels
- Increased asthma and other respiratory problems

LANDSCAPING PRACTICES TO MITIGATE THE URBAN HEAT ISLAND EFFECT

Urban planning and design is an evolving field that is in the process of examining and incorporating new information as it becomes available. Many of the changes required to ameliorate the urban heat island effect are beyond the scope of the landscape industry. Yet plants and landscaping do play an important role in the effects of climate and weather in a range of settings, from urban to rural. In the urban environment, plants play a role individually and as a group such as in urban forests, parks, and greenspaces. Other examples of plants in the urban landscape include plants in containers, bioswales, street trees, and vegetated roofs and walls. There are many barren areas in the urban environment that provide an opportunity for adding vegetation. Studies conducted by the Lawrence Berkeley National Laboratory found that Chicago, Houston, Sacramento, and Salt Lake City had ~4 to 8 percent barren area overall. Furthermore, as urban areas expand, they often lose vegetated cover. For example, Houston lost 10 million trees each year between 1992 and 2000 due to development.

Plants can aid in mitigating the urban heat island effect by reducing temperatures. In addition to the cooling effect of plants, which helps reduce air-conditioning use, ground-level ozone, is also reduced. At ground level, ozone is an important air pollutant that contributes to respiratory ailments and can also be damaging to plants. Reduced air-conditioning use leads to reduced emissions and pollutants at the power plant. Since ozone forms more readily as temperatures rise, so a reduction in temperature results directly in a reduction of ozone.

Cooling Effect of Plants

The two main ways that plants reduce temperatures is through moisture loss via transpiration and by reducing solar heat gain by providing shade.

Transpiration

As plants take up moisture through their roots, they lose it through their leaves in a process called **transpiration**. Small pores in the leaves, called stomata, open and close depending on moisture availability. Stomata are usually located on the undersides of leaves. When the pores are open, plants are able to take in carbon dioxide and release oxygen as part of the process of photosynthesis. Thus, open stomata are a requirement and function of plant health and growth. Integral to this process, a humid microclimate around the leaf develops which extends outwards into the plant canopy. This humidity has a cooling effect on the surrounding air. Water is also lost from the soil surrounding trees and other plants through the process of evaporation. Together, the loss of moisture from soil and plants to the air is called **evapotranspiration**. The drier the air is, the greater the cooling effect is when moisture is added to it. Walking barefoot on grass in the summertime easily illustrates the cooling effect of evapotranspiration. Studies have shown that temperatures over grass sports fields are 2°F–4°F (1.1°C–2.2°C) cooler than air over bordering areas.

Estimates are that evapotranspiration can reduce summertime air temperatures by as much as 2°F–4°F (1.1°C–2.2°C) for a 5 percent increase in canopy cover. A study conducted by Saito and colleagues (1991) found that when individual building sites in an urban setting had greenspace, or a planted area, around them, temperatures were as much as 5°F (2.8°C) cooler than outside the greenspace. Turfgrass has been shown to be cooler than both asphalt and artificial turf in summer, an effect which is attributed to evapotranspiration. Tree groves have been shown to have peak air temperatures that are 9°F (5°C) cooler than surrounding areas over open terrain. Mature shade trees in suburban areas are 4°F–6°F (2.2°C–3.3°C) cooler than newer suburbs lacking tree cover.

Shade

In addition to the cooling effect of humidity from evapotranspiration, leaves also re-radiate some of the sun's radiation back into the atmosphere, trapping it in the plant canopy. In trees, for example, heat is held in the upper one-third of the crown, while the lower portion remains cooler. As a result of this effect, a shaded area under a tree can be as much as 15°F–25°F (8.3°C–13.9°C) cooler at mid-day as compared to unshaded areas. Grass under a shade tree is 10°F–14°F (5.5°C–7.8°C) cooler than unshaded turf, and air temperatures above blacktop can be 25°F (14°C) cooler when shaded. In summer, only about 10–30 percent of the sun's energy reaches the ground underneath the tree canopy. But in winter, as much as 80 percent of the sunlight is transmitted through many deciduous trees. Evergreens may still only allow 10 percent or so of the sun's energy through.

Plants can provide shading to structures, which can have either positive or negative effects on energy usage inside. In a simulated study, McPherson and Rowntree (1993) reported that shade from deciduous trees planted to provide shade to east walls caused the heating bill to increase to a greater extent than the reduction in cooling costs in Boston, Minneapolis, and Portland. The authors concluded that the potential energy costs of trees improperly located near buildings are greatest in cool climates, while their potential energy savings are greatest in warm climates. The simulation was then used to calculate costs and benefits of tree plantings, and they reported that a tree shading a west wall, regardless of the climate zone, provided about twice the energy savings of the same tree shading a similar east-facing wall. In actual measured, monitored studies, vegetation was consistently found to reduce wall surface temperatures by around 30°F (16.7°C) and savings in electricity from reduced use of air conditioning ranged from 10 to 80 percent, depending on geographic location and plant type (turf, trees and shrubs, forest).

In a separate study, Gordon Heisler (1986) determined that a sugar maple reduced the amount of sunlight hitting a wall, regardless of whether it had leaves or not. With leaves, the tree reduced irradiance on a south-facing wall by around 80 percent with leaves, and 40 percent when leafless. A London plane tree of similar size gave slightly lower irradiance reductions. When compared to a computer model placing trees on the west side or south side, greater benefits were seen with trees placed to the west of the house, particularly in sunny climates. One caveat when using shade trees is that winter shading is less desirable than summer shading in temperate climates. In Heisler's study, he discovered that trees with a taller trunk planted on the south side of the house will provide better benefits overall as compared to a tree with a shorter trunk, probably due to the shading that is provided in summer by the higher branches that allow more sunlight through in winter.

In numerous studies, Akbari and colleagues measured temperatures on roofs and walls on two buildings and found reductions of 20°F–45°F (11°C–25°C) due to shade from trees. Sandifer and Givoni (2002) looked at the effect of vines on a wall and found reductions of up to 36°F (20°C). In a study looking at the heat build-up in parked cars, Scott and colleagues (1999) discovered that shade trees could reduce the temperatures by about 45°F (25°C).

Some examples of plants providing shade and reducing heat build-up are:

- Plant a row of shrubs along a sidewalk or driveway or other landscaped area
- Plant vines on a trellis to shade a desired area where space is limited
- Select plants with high or low crown structure, depending on the sun's angle at a particular time of day when shading is desired

- Use taller trees with a broad crown to shade the roof of a house or other air-conditioned building
- Place plants in such a way to allow cooling breezes to pass through an area
- Use deciduous rather than evergreen trees on the south side to shade in summer but allow winter sun through
- Plant trees to provide shade over parking areas – in median strips and around the perimeter
- Plant trees to shade the west or southwest sides of buildings
- Strategically locate trees or shrubs to reduce or eliminate solar heat gain through windows
- Plant trees in playgrounds and play areas
- Plant trees on the perimeter of sports fields

Case Study: Landscaping and Cooling

Parker, J.H. 1983. Landscaping to reduce the energy used in cooling buildings. *J. of Forestry*. February: 82–84.

In developing best management practices for reducing energy needs for cooling buildings, John Parker proposed a set of practices he termed "precision landscaping". To optimize results through design and implementation, two factors must be considered: local climatic data and energy utilization of the structure in question. The cooling effects of plants are provided by blocking solar radiation from the building, adjacent ground, and the foundation of the building; creating cool microclimates near the structure, through evapotranspiration, and channeling or blocking air flow around the building.

Parker suggests establishing a plan that addresses peak load demand in a geographical location. He reasons that reducing energy demand during the peak load period can delay the need for new power plants. Since electric rates are usually higher during the peak load period, larger savings will be realized if reductions can be made then.

He uses south Florida as an example, giving August 6 as the center of the seasonal peak demand period. For other areas of the south and the United States, August 1 is a more appropriate date. On a daily basis, peak demand occurs in late afternoon when people return from work and increase home air-conditioner use. Therefore, designing a landscape that reduces solar radiation on the west-facing side of a building, particularly the windows, would result in the greatest savings.

A case study to test these ideas was performed on a mobile home that served as a day-care center on the Florida International University campus. Measurements were made on morning and afternoon ambient air temperatures and air-conditioning use both before and two years after installation of an energy conservation landscape. Ambient air temperature was not significantly different, but air-conditioning use was reduced by 58 percent after the landscaping was in place.

Shading Air Conditioners

Some people suggest planting shrubs or trees to shade an air-conditioning unit, thereby increasing its efficiency. Care must be taken, however, not to impede air flow around the unit. If adequate air flow cannot be achieved, it is best not to use plants in this manner. Ideally, the air conditioner should be protected from heating from direct sun. If plants are not possible, and the unit is receiving direct sun, a screening structure designed to allow air flow may be a good alternative. In any case, make sure there is 2 ft of clearance all around the equipment, and at least 5 ft over it. A shaded unit requires up to 10 percent less energy to operate than an unshaded unit.

Green Roofs

Green, or vegetated, roofs have been coming into increased usage for the past decade or so. They are particularly popular in larger cities, such as Chicago and New York City. A green roof is specifically designed to accommodate plants, a growing medium, filtration layers, root restriction, and an impermeable membrane on a roof. A study looking at green roofs in various climates around the world during peak hot temperatures showed that green roofs and walls reduced local temperatures by as much as 4°F–20°F (2.2°C–11.1°C). In warmer climates they have shown increased effectiveness at reducing local temperatures.

Some green roofs can tolerate foot traffic, while others cannot. They must be built to hold the additional weight of soil or soilless media and plants, as well as people, if that is desired. Some of the benefits of green roofs are:

- Rooftop cooling by absorbing ultraviolet radiation
- Absorb noise pollution
- Absorb air pollution
- Create habitat for birds, butterflies, and other insects
- Absorb carbon dioxide and emit oxygen
- Reduce heat island effect

For green roofs that are built to tolerate foot traffic, they can provide patio space or even a park-like environment for humans. Green roofs are discussed in more detail in Chapter 7. Their resurgence and increase in popularity began as a response to a need to reduce stormwater runoff, especially during major rainfall events.

Cooling Paved Surfaces

Researchers have estimated that every 10 percent increase in solar reflectance could decrease surface temperatures by 7°F (4°C). They also predicted that if cities increased the reflectance of their paved areas from 10 to 35 percent, the air temperature could potentially be reduced by 1°F (0.55°C). A 2007 paper estimated that increasing pavement albedo in cities worldwide, from an average of 35 to 39 percent, could achieve reductions in global carbon dioxide (CO_2) emissions worth about $400 billion.

Researchers at The Lawrence Berkeley National Laboratory and others are conducting ongoing research into cool pavement technology. In addition to surface reflectivity, it is possible that permeable paving materials, including permeable concrete and permeable asphalt, are cooler due to the moisture content and/or to reduced heat transfer below the surface. Open-grid paving can help to reduce solar reflectance in an area. Used on a large scale, such areas can help to reduce the urban heat island effect. Some municipalities are encouraging the use of this pavement material both to aid in reducing the heat island effect as well as to reduce stormwater through this permeable paving.

White concrete pavers are popular in green building and landscaping due to their high reflectance values. For example, Hanover Architectural Products offers a white paver (Glacier White Prest®) having a reflectance value of 0.655, emittance value of 0.97, and an SRI of 81. Several other materials are being researched, many of which are already developed, including:

- Asphalt modified with high albedo materials
- Modified concrete pavement using white cement
- Resin-based reflective pavements
- Porous asphalt
- Rubberized asphalt
- Pervious concrete
- Brick pavers with sand filler
- Plastic, metal, or concrete open-lattice pavers
- Chip seals with high-albedo aggregate

In addition to the above listed materials, high albedo treatment materials applied after installation or during routine maintenance can be used on some surface materials. **Whitetopping** can be ultra-thin (2 in. or less) or thick (4 in. or more) concrete that can be applied to new asphalt or when resurfacing. It may contain fibers for added strength. The solar reflectance of whitetopping is similar to that of concrete. **Microsurfacing** is a thin seal used on roads, airport runways, and parking lots to increase the albedo of asphalt. It has a solar reflectance of over 35 percent.

Researchers in Japan are investigating the use of permeable pavements for effectively reducing temperatures. In particular, they have tried sprinkling the pavement with reclaimed wastewater on roads in Tokyo. They found that this practice decreased the road surface temperature by 14.4°F (8°C) during the daytime and by 5.4°F (3°C) at night. An additional benefit of cool pavement is that their useful life may be extended over that of traditional pavements. Simulations of asphalt pavements that were 20°F (11.1°C) cooler showed that they lasted ten times longer, and that pavements that were 40°F (22°C) cooler lasted a hundred times longer than their conventional counterparts. Such temperature reductions are in the realm of possibility and would add to the cost benefit analysis and life cycle assessment of these products.

Another benefit of permeable paving that applies mainly to highways and may not apply in landscape situations is that their open porous structure reduces tire noise by 2–8 decibels. This benefit may decline over time, and the surfaces may not be as durable as conventional surfaces. As with many of these newer pavement technologies, much research remains to be done, and new products may come into the market as municipalities and others seek more alternatives. Furthermore, an increased interest in recycling materials may lead to new developments not yet on the horizon.

OTHER CONSIDERATIONS

In addition to using plants for cooling and cool pavements for reducing heat in the landscape, the landscape professional should keep in mind some basics about house orientation and sun exposure. An understanding of these concepts can aid in the landscape design and placement of plants for specific cooling purposes.

STRUCTURE ORIENTATION

The sun's angle changes throughout the year and throughout the day. As a result, some areas in the landscape receive more light and heat than others at various times. The east- and west-facing sides of a building receive morning and afternoon light respectively, while being shaded at the alternate times. The south side of a building is more vulnerable to low light angles in winter than spring or fall. On the north side of a building, summer sunlight will only strike on the northeast and northwest corners in the morning and afternoon, respectively, although this varies somewhat according to latitude.

SUN EXPOSURE

A building's exposure to the sun can be influenced by the direction it faces, as well as the amount of shade provided by trees or nearby structures. In the morning the air tends to be cooler, especially in the shade. However, the west side of the building will be exposed to the hot sun all afternoon. The west side of the building tends to be the hottest during the warmer months. During the cooler months, the sun is at a lower angle in the sky, and there is less direct sun on the structure as a result.

REDUCING HEAT GAIN IN SUMMER

Trees and shrubs can be strategically placed in order to reduce heat gain in summer by providing shade on paved areas, building surfaces, roofs, and windows. Turf grass and other plants can cool the air around them both by absorbing heat from the sun and releasing moisture through transpiration.

Trees can

- lower energy bills by 25 percent
- lower AC bills –15 to 50 percent
- lower heating bills –25 to 40 percent
- reduce air temperature by up to 25 percent under the tree

INCREASING HEAT GAIN IN WINTER

Designing landscape plantings to allow for heat gain in winter consists largely of avoiding shading the residence or other structures with trees. Large evergreen trees, such as pines, spruces, and firs should be placed at adequate distance from buildings to permit winter sunlight to come through. This is particularly important if active or passive solar energy features are located on-site. South-facing windows are an important component of passive solar heating. The brighter, less gloomy ambience inside the building provided by winter sunlight should not be underestimated, either. Deciduous trees should be selected and placed in such a way as to maximize summer protection and minimize winter obstruction of sunlight.

OPTIMIZING SOLAR INCIDENCE FOR WARMTH IN WINTER

December 21st is the official date of the winter solstice in the northern hemisphere. On this date, the sun is at its lowest in the sky, due to the inclination of the northern hemisphere of the earth away from the sun. At this low angle, more sunlight is able to enter a building, as long as there are no obstructions. In order to benefit from solar heat gain in winter, shading from fences, other buildings, evergreen trees, or other obstructions should be avoided. Other than this, the landscaper is somewhat restricted in what they can do to aid in heat gain in winter.

SUMMARY

The sun emits energy across a wide spectrum of wavelengths. The most important of those concerning sustainable landscaping fall into the visible light and infrared ranges.

Solar radiance and heat from the sun play an important role in both the landscape as well as the house. Heat gain occurs when heat enters a building. Windows and hardscaped areas of the landscape are subject to heat from the sun. Shade provided by trees or structures can modify the heat build-up on a driveway or other large paved area.

Light colors reflect more light and absorb less heat than dark colored surfaces. Albedo, surface reflectivity, and solar reflectance are used interchangeably, and are measures of reflected solar radiance. Thermal emittance measures the ability of a material to emit absorbed heat. Heat capacity is the ability of a material to store heat. Heat is emitted in the infrared range of the spectrum.

The urban heat island is a phenomenon in cities in which temperatures are higher than they are in surrounding rural areas. Thermal imaging cameras give a vibrant visual image of temperature differentials using infrared photography.

Plants cool the air through a process known as evapotranspiration, which can reduce summertime air temperatures, and through shading. Greenspace contributes to cooling in urban areas, as do areas covered by turfgrass. Leaves in trees can trap some of the sun's heat within a tree canopy. Shading a west wall provides greater energy savings than shading an east-facing wall. Air-conditioning use may be reduced due to the shading effect of plants on walls. Vines can be used for cooling where space is limited.

The Sun and the Sustainable Landscape

A shaded air-conditioning unit requires up to 10 percent less energy to operate than an unshaded unit. Green, or vegetated, roofs and walls also provide a cooling effect, with increasing effectiveness in warmer climates.

As the sun's angle changes throughout the day and throughout the year, some areas in the landscape receive more light and heat than others at various times. Building exposure will determine whether overheating in summer on the west-facing side is problematic, or whether too much shade on the south side in winter presents a problem. Trees and shrubs can be strategically placed in order to reduce heat gain in summer by providing shade on paved areas, building surfaces, roofs, and windows. Evergreen trees must be used cautiously so that they do not provide unwanted shade in the winter. South-facing windows can contribute to passive solar heating as long as they are left clear from obstructions.

REVIEW QUESTIONS

1. At what wavelengths on the electromagnetic spectrum is heat from the sun felt?
2. Differentiate between surface reflectance and emissivity.
3. Order the following list from greatest to least albedo: asphalt, deciduous forest, concrete, grass, wet bare soil.
4. The sun is at its lowest angle in the sky in the northern hemisphere on what date? And its highest angle on what date?
5. What are the three factors involved in the urban heat island effect?
6. What are the two main ways that plants reduce temperatures in the landscape?
7. Which side of a house tends to be hottest during summer in the northern hemisphere?
8. What is the energy savings for a shaded versus unshaded air-conditioning unit?
9. What is microsurfacing and what benefit does it provide?
10. Discuss how green roofs and walls counteract the urban heat island effect.

ENRICHMENT ACTIVITIES

1. Make a list of the features in a local landscape setting and research the albedo, emissivity, and SRI of each item.
2. Identify areas of heat build-up in a local landscape and develop a design plan that could reduce the heat build-up.
3. Record temperature differences in various landscape settings: turf in shade, turf in sun, a building in shade and full sun. Compare the temperature of different landscape materials: wood, concrete, gravel, plants.

FURTHER READING

Ahrens, C D and R. Henson. 2018. *Meteorology Today: An Introduction to Weather, Climate, and the Environment*, 12th ed. Cengage, Boston, MA, 656 pp.

Akbari, H., A.H. Rosenfeld, and H. Taha. 1990. Summer heat islands, urban trees, and white surfaces. *ASHRAE Trans.* 96(1): 1381–1388.

Akbari, H., D. Kurn, S. Bretz, and J. Hanford. 1997. Peak power and cooling energy savings of shade trees. *Energy Build.* 25: 139–148.

Akbari, H., M. Pomerantz, and H. Taha. 2001. Cool surfaces and shade trees to reduce energy use and improve air quality in urban areas. *Solar Energy* 70(3): 295–310.

Akbari, H. and S. Menon. 2007. Global cooling: effect of urban albedo on global temperature. *Paper for the Proceedings of the International Seminar on Planetary Emergencies*, Erice, Sicily.

Akbari, N., S. Davis, S. Dorsano, J. Huang, and S. Winnett (eds.) 1992. *Cooling Our Communities: A Guidebook on Tree Planting and Light-Colored Surfacing*. EPA, Washington, DC. https://escholarship.org/uc/item/98z8p10x. Retrieved May 27, 2019.

Alexandria, E. and P. Jones. 2008. Temperature decreases in an urban canyon due to green walls and green roofs in diverse climates. *Build Environ.* 43(4): 480–493.

Barlag, A.B. and W. Kuttler. 1990/91. The significance of country breezes for urban planning. *Energy Build.* 15–16: 291–297.

Brahich, C. 2007. Green roofs could cool warming cities. *New Scientist*. http://www.newscientist.com/article/dn12710-green-roofs-could-cool-warming-cities.html. Retrieved May 27, 2019.

Cathcart, T. 2002. *Regenerative Design Techniques*. Wiley, New York, 410 pp.

Coseo, P. and L. Larsen. 2015. Cooling the Heat Island in compact urban environments: the effectiveness of Chicago's green alley program. https://doi.org/10.1016/j.proeng.2015.08.504. Retrieved May 27, 2019.

Dwyer, J.F., E.G. McPherson, H.W. Schroeder, and R.A. Rowntree. 1992. Assessing the benefits and costs of the urban forest. *J. Arboric.* 18: 227–234.

Encyclopedia of Earth. http://www.eoearth.org/article/albedo. Retrieved May 27, 2019.

Gao, Y., D. Shi, R.M. Levinson, R. Guo, C. Lin, and J. Ge. 2017. Thermal performance and energy savings of white and sedum-tray garden roof: a case study in a Chongqing office building. *Energy Build.* 156. doi: 10.1016/j.enbuild.2017.09.091. Retrieved May 27, 2019.

Halusa, G. and L.A. Remer. 1999. NASA team finds Atlanta influences its own weather. *ScienceDaily*, 29 March 1999. www.sciencedaily.com/releases/1999/03/990325104705.htm. Retrieved May 27, 2019.

Hanover Architectural Products. Solar reflectivity and green roofs. https://www.hanoverpavers.com/images/PDFs/1202-ReflecFactSheet.pdf. Retrieved May 27, 2019.

Heisler, G.M. 1986. Energy savings with trees. *J. Arboric.* 12: 113–125.

Honjo, T. and T. Takakura. 1990/91. Simulation of thermal effects of urban green areas on their surrounding areas. *Energy Build.* 15–16: 443–446.

Huang, J., H. Akbari, and H. Taha. 1990. The wind-shielding and shading effects of trees on residential heating and cooling requirements. *ASHRAE Winter Meeting*, American Society of Heating, Refrigerating, and Air-Conditioning Engineers, Atlanta, GA.

Huang, J., H. Akbari, H. Taha, and A. Rosenfeld. 1987. The potential of vegetation in reducing summer cooling loads in residential buildings. *J. Clim. Appl. Meteorol.* 26: 1103–1106.

International Association for Urban Climate. http://www.urban-climate.org/ Retrieved May 27, 2019.

Kurn, D., S. Bretz, B. Huang, and H. Akbari. 1994. The potential for reducing urban air temperatures and energy consumption through vegetative cooling. *ACEEE Summer Study on Energy Efficiency in Buildings*, American Council for an Energy Efficient Economy, Pacific Grove, CA.

Lebwohl, B. 2009. Dev Niyogi on urban sprawl and storm intensity. http://earthsky.org/earth/dev-niyogi-on-urban-sprawl-and-storm-intensity. Retrieved May 27, 2019.

McPherson, E.G. Energy saving potential of trees in Chicago. https://webs.csu.edu/cerc/documents/EnergySavingPotentialofTreesInChicago.pdf. Retrieved May 27, 2019.

McPherson, E.G., J.R. Simpson, and M. Livingston. 1989. Effects of three landscape treatments on residential heating and cooling requirements. *ASHRAE Trans.* 96, Part I: 1403–1411.

McPherson, E.G. and R.A. Rowntree. 1993. Energy conservation potential of urban tree planting. *J. Arbor.* 19(6): 321–331.

Nelson Jr., W.R. 1980. Designing an energy-efficient home landscape. *Univ. Ill. Cooperative Ext. Serv. Circu.* 1178: 13pp.

NOAA. Solar position calculator. https://www.esrl.noaa.gov/gmd/grad/solcalc/. Retrieved May 27, 2019.

Oke, T.R. 1987. *Boundary Layer Climates*, 2nd ed. Methuen & Co. Ltd., London.

Parker, J.H. 1983. Landscaping to reduce the energy used in cooling buildings. *J. For.* 82–84: 105.

Pomerantz, M. 2017. Are cooler surfaces a cost-effect mitigation of urban heat islands? *Urban Clim.* 24: 393–397. doi: 10.1016/j.uclim.2017.04.009. Retrieved May 26, 2019.

Pomerantz, M., B. Pon, H. Akbari, and S.-C. Chang. 2000. The effect of pavements' temperatures on air temperatures in large cities. *Paper LBNL-43442, Lawrence Berkeley National Laboratory*, Berkeley, CA.

Ritter, M.E. 2018. The physical environment: an introduction to physical geography. https://www.earthonlinemedia.com/ebooks/tpe_3e/title_page.html. Retrieved May 27, 2019.

Rosenfeld, A.H., J.J. Romm, H. Akbari, and M. Pomerantz. 1998. Cool communities: Strategies for heat islands mitigation and smog reduction. *Energy Build.* 28: 51–62.

Saito, I., O. Ishihara, and T. Katayama. 1991. Study of the effect of green areas on the thermal environment in an urban area. *Energy Build.* 15(3–4): 493–498.

Sandifer, S. and B. Givoni. 2002. Thermal effects of vines on wall temperatures: comparing laboratory and field collected data. *SOLAR 2002, Proceedings of the Annual Conference of the American Solar Energy Society*, Reno, NV.

Scott, K., J.R. Simpson, and E.G. McPherson. 1999. Effects of tree cover on parking lot microclimate and vehicle emissions. *J. Arboric.* 25(3): 129–142.

United States EPA. Heat island effect. http://www.epa.gov/heatisld/resources/glossary.htm#h. Retrieved May 27, 2019.

Walker, L. and S. Newman. Landscaping for energy conservation. https://extension.colostate.edu/topic-areas/yard-garden/landscaping-for-energy-conservation-7-225/. Retrieved May 27, 2019.

Wong, E. Cool pavement. In: Reducing urban heat islands: compendium of strategies. http://www.epa.gov/heatisld/resources/pdf/CoolPavesCompendium.pdf. Retrieved May 27, 2019.

Wong, E. Heat island reduction activities. In: Reducing urban heat islands: compendium of strategies. http://www.epa.gov/heatisld/resources/pdf/ActivitiesCompendium.pdf. Retrieved May 27, 2019.

Wong, E. Trees and vegetation. In: Reducing urban heat islands: compendium of strategies. http://www.epa.gov/heatisld/resources/pdf/TreesandVegCompendium.pdf. Retrieved May 27, 2019.

Wong, E. Urban heat island basics. In: Reducing urban heat islands: compendium of strategies. http://www.epa.gov/heatisld/resources/pdf/BasicsCompendium.pdf. Retrieved December 22, 2010.

Yamagata, H., M. Nasu, M. Yoshizawa, A. Miyamoto, and M. Minamiyama. 2008. Heat island mitigation using water retentive pavement sprinkled with reclaimed wastewater. *Water Sci Technol.* 57(5): 763–771.

Yasunobu ASHIE. Study on the characteristics of solar radiation in the geometrically complex urban spaces by using a spectroradiometer. https://heatisland.lbl.gov/sites/default/files/cuhi/docs/210900-ashie-doc.pdf. Retrieved May 27, 2019.

Zajicek, J.M. and J.L. Heilman. 1991. Transpiration by crape myrtle cultivars surrounded by mulch, soil, and turfgrass surfaces. *HortScience* 26(9), 1207–1210.

4 The Wind and Energy Conservation

OBJECTIVES

Upon completion of this chapter, the reader should be able to:

- Explain how the wind affects heat loss in structures
- Demonstrate how to use plants to effectively control wind speed
- Design effective wind breaks
- Describe how breezes in summer can reduce cooling costs
- Design landscapes that provide wind funnels in summer
- Identify landscape plant usage that is beneficial to energy consumption in nearby structures
- Identify landscape components that adversely affect energy usage in nearby structures
- Demonstrate how to correctly place plants for energy efficiency

TERMS TO KNOW

Dead air space
Green space
Leaf scald
Leeward
Microclimates
Shelterbelt
Transpiration
Tree funnel
Urban boundary layer
Urban canopy layer
Urban canyon
Windbreak
Windward

INTRODUCTION

Wind affects house energy use most importantly in winter. In winter months the air is cold and tends to seep through cracks and openings in the home, allowing warm air to escape; this is not really a problem in summer. As a matter of fact, breezes in summer tend to have a cooling effect in the human environment. The presence of trees in the urban environment influences the movement of air

FIGURE 4.1 Urban canyons refer to highly built environments in large cities like this one in Chicago, Illinois. (Photo by the author.)

in **urban canyons** (Figure 4.1) by providing turbulent mixing from above due to surface roughness of the vegetation. Trees in the city also affect the transport of pollutants through the air.

Doorways and entryways permit major heat loss in winter and heat gain in summer. This can be addressed by designing the house to have a protected area at the points of entry. Plants, fencing, and walls are landscaping features than can create a buffer zone in entry areas, and also provide protection from the sun and wind.

In the landscape, wind affects plants by causing them to lose more water through **transpiration**. If plants lose water too quickly, their stomata will close, which prevents photosynthesis from taking place, as the leaves can no longer take up carbon dioxide. This can cause stunted growth. In winter, wind can cause evergreen plants to lose too much water, which leads to drying out and possible death. **Leaf scald** may occur when it is sunny out while the ground is frozen. Again, the plants lose water, but with the frozen ground, roots are unable to replace it as it is lost. As a result, leaves dry up and turn brown.

The Sustainable Sites Initiative™ addresses wind and energy consumption by providing credits for minimizing building heating and cooling requirements with vegetation by providing a one- or two-row windbreak.

This chapter examines ways to modify the effects of wind in the landscape using plants. Techniques include planting on a slope to trap cold air, using plants to insulate near a structure, creating tree funnels to channel cooling breezes at appropriate times during the year, and using windbreaks to change wind patterns in winter.

The Wind and Energy Conservation

FIGURE 4.2 Cold air gets trapped by plants or other obstructions as it moves down a slope. (Illustration by the author.)

TRAPPING COLD AIR ON A SLOPE

Cold air, which is denser and heavier than warmer air, migrates downward on sloped areas. Air moves up or down valleys and canyons accordingly. Cold air moves down a slope at night, and as the air warms during the day it moves back up the slope.

Sometimes a home or building is sited on sloped property. If the structure is downslope from the surrounding landscape, it is possible to reduce the movement of cold air toward the structure by dense vegetation such as evergreen trees or shrubs. This will serve as a cold air trap. If the structure is upslope, then it is desirable to allow the cold air to flow freely down the slope and away from the structure. If flowering or fruiting trees or shrubs are planted on a slope, it is important to place them in such a fashion as to allow cold air to drain away rather than getting trapped. Trapping the cold air in this case increases the risk for freeze damage to flower buds (Figure 4.2).

PLANTING FOR INSULATIVE PROPERTIES

Dead air space is created when two layers of material are placed together close enough to restrict air flow, yet do not touch. This is the situation that is created when dense shrubbery is planted near a structure (Figure 4.3). The recommended spacing is 12 in. If planted too closely, the humid microclimate that is created near the foliage may result in mold growth on the structure. If too much space is left, then there will be undesired air movement. The purpose of creating this dead air space is for the insulating effect it provides in both summer and winter.

THE COOLING EFFECTS OF WIND

Tree funnels can be designed to enhance breezes and channel winds in the desired direction. Such a funnel would consist of a row of trees or shrubs placed in a southwest to northeast orientation (Figure 4.4). This arrangement will capture southerly

FIGURE 4.3 Dead air space can provide insulation to a house if planted properly. (Illustration by author.)

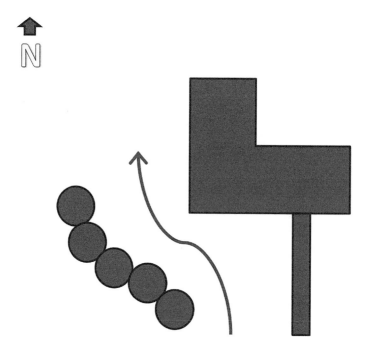

FIGURE 4.4 Tree funnels can direct cooling summer breezes to move stagnant air. (Illustration by the author.)

The Wind and Energy Conservation

FIGURE 4.5 Windbreak in western Iowa. (Photo courtesy Tim McCabe, USDA Natural Resources Conservation Service.)

breezes and direct them to the north side of the house, where the air is often more stagnant. This may provide the added benefit of reducing mold formation that often occurs on north-facing sides of houses.

Berms can also be designed for the wind funnel effect. If enclosure or privacy is desired, but space is limited, an open-slat fence will still allow air to flow through the area. This alleviates overheating in summer as well as stagnant air, which can contribute to disease problems, especially if humidity is present.

WINDBREAKS TO REDUCE HEAT LOSS

A **windbreak** or **shelterbelt** is a planting that is designed to intercept and redirect prevailing winds (Figure 4.5). They are used in climates where the winters are cold, to prevent heat loss from structures. They are oriented perpendicular to the prevailing wind, which tends to be west, northwest, or north. The best windbreaks provide protection from wind from both the west and the north. As wind speed increases, the rates at which cool air infiltrates a building also increases.

Case Study: Wind and Energy Efficiency

Dewalle, D. R., G. M. Heisler, R. E. Jacobs. 1983. Forest home sites influence heating and cooling energy. *J of Forestry.* February: 84–88.

The effect of deciduous and evergreen trees on heating and cooling needs in a building can be very different. Thus, the placement of tree planting needs to be taken under careful consideration to maximize energy savings. In a study conducted on small mobile homes in Pennsylvania, deciduous trees providing shade in summer resulted in 75 percent energy savings. However, the same shade in winter reduced heating needs by only 8 percent. Denser shade, provided by pine trees, actually increased heating energy needs in winter by 12 percent.

Mobile homes near pine forests did not benefit from the trees at low wind velocity. Under such conditions, the temperature differential between inside and outside air was the determining factor on heat loss from the homes. The pine trees actually had the effect of reducing outside air temperatures enough to increase heat loss from the homes. However, at higher wind velocity, the pine trees provided marked improvement in reducing heat loss from the homes.

The main factors affecting heat loss in winter were the wind speed, effectiveness of the windbreak, and home construction. The authors of this study concluded that heat loss from the mobile homes was similar for full-sized homes with similar construction. Concerning wind break effectiveness, they found that the wind velocity near the forest home sites was 40–80 percent lower than out in the open. They estimated that a properly designed windbreak could provide a 10 percent reduction in savings for heating.

Residential structures are sited on lots too small for effective windbreaks to be installed. However, studies show that wind speed is reduced in residential neighborhoods due to a combination of buildings and landscapes. Urban canyons are formed when there is nothing to break up the wind as it moves through the spaces formed by tall buildings. Trees, **green spaces**, and other plantings affect the urban heat island as well as pollution dispersal from vehicle emissions and particulate matter. Urban planning and design can incorporate the effects of vegetation on wind and air flow for maximum energy efficiency. Such planning is critical because of the difficulty in adding plants to the urban environment after development has been completed.

In contrast, rural sites often have plenty of space to install fully developed windbreaks. For these locations, it is important for the windbreak to be properly designed and installed; otherwise, the desired benefits will not be realized. Beside the costs in energy loss, an incorrectly executed windbreak can also result in unwanted snow drifts. Conversely, windbreaks can be designed specifically to prevent snow drifts on sidewalks, roads, parking lots, and other traffic areas.

Energy Usage in Winter

Energy consumption of buildings produces more carbon dioxide than cars in the United States, accounting for 38 percent of total emissions. American households spent $780 on natural gas for heating or $940 on electricity in the winter of 2010–2011. Researchers at Oak Ridge National Laboratory have found that retrofits and better designs could eliminate as much as 200 million tons of carbon dioxide emissions every year from buildings alone. Windbreaks in rural areas and trees and green spaces in urban areas play an integral part in energy reduction in homes and other structures.

While it is difficult to predict the energy savings in having a wind break on any particular property, properly designed windbreaks have been shown to reduce energy consumption in winter by 25 to 40 percent. In Canada, a study comparing two electrically heated trailers, one in a well-sheltered farmyard and the other in a completely unsheltered one, resulted in a 27 percent reduction in energy consumption.

The Wind and Energy Conservation

In a similar study in Pennsylvania, a single row of white pines reduced space heating energy needs in a small mobile home by 18 percent. Windbreaks are more effective at reducing energy needs in windy climates. It has been estimated that in the northeastern United States, energy savings can reach 10–15 percent, and in the north central United States, energy savings may reach 15–25 percent. Economic analysis indicates that over a 20-year life span of a windbreak, the value of reduced energy needs for heating would exceed the cost of windbreak establishment and maintenance.

In addition to reducing fuel consumption, properly designed wind breaks can reduce snowdrifts near roads, driveways, sidewalks, and other open areas. The potential cost savings is high. In Minnesota, a cost-benefit analysis of living snow fences found that for every $1 spent, there was a $17 return. The savings from reduced fuel consumption was combined with less economic disruption due to closed roads. Improved winter driving conditions are yet another important benefit.

DESIGNING THE WINDBREAK

A windbreak is planted perpendicular to the prevailing wind and causes the air to move up and over. There is a **windward** and **leeward** side of a wind break (see Figure 4.6). The leeward side is on the opposite side from the wind. Immediately next to the windbreak on the leeward side a vacuum is created. In this area, snow, dust, and debris can be deposited as they get sucked in by the vacuum that forms when air passes over the top of the windbreak. Due to the accumulation of snow in this space, it is important to place the wind break at least 50 ft (15 m) from driveways, buildings, and traffic areas. The triangular area on the leeward side of the windbreak represents the "quiet zone" where wind speed is reduced. If air on the windward side is turbulent, the windbreak is less effective, and wind speed may recover at a shorter distance on the leeward side than otherwise would occur.

Height

The height of the windbreak affects the distance over which the wind is carried. As wind moves up over a windbreak, the velocity of the wind is reduced. Wind should not be completely blocked, but some wind should be allowed to pass through.

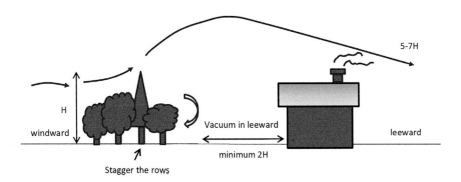

FIGURE 4.6 Windbreak showing windward and leeward sides, height, distance, wind speed. (Illustration by the author.)

Porosity of the windbreak is the most important feature in windbreaks that are long relative to their height. Windbreaks that are not very porous create more turbulence downwind, which can allow the wind to regain its original speed in a shorter distance out from the windbreak as compared to medium-dense windbreaks. The optimal porosity of a windbreak allows 40 percent of the wind to pass through, while deflecting 60 percent.

The horizontal extent of the windbreak effect is proportional to the height of the windbreak. The maximum protection of a windbreak is provided downwind at a distance of five to seven times the height of tallest tree in the windbreak. Thus, a windbreak with 30 ft (9 m) tall trees provides protection 150–210 ft (45–65 m) downwind. At further distances, protection is still afforded, but wind speed begins to increase again.

Shape and Size

Looking from the side, a rectangular profile is better than a pyramidal one for optimal cross-wind protection. Yet, the height of the windbreak should be varied, not one uniform height. This creates rough edges that break up the wind pattern. Trees should not be limbed up, but rather, in a windbreak, branches should reach all the way to the ground.

The number of rows in a windbreak varies, depending on available space. If deciduous plants are used, four to five rows should suffice, whereas, if evergreens are used, only two to three rows are needed. A windbreak works most effectively if its length is equal 11.5 times its mature height. For example, a windbreak that is 30 ft (9 m) tall should be 345 ft (115 m) long. The standard design is an L-shape, with one leg on the west side, and the other on the north side (Figure 4.7).

Plants for a Windbreak

A good windbreak should include at least two to three different species, and may combine deciduous and evergreen plants. Shrubs may be used in conjunction with trees. Plants should be selected for their ability to tolerate prolonged exposure to winter winds. One row should include a dense species, such as Norway spruce (*Picea abies*), white spruce (*Picea glauca*), Colorado spruce (*Picea pungens*), or Douglas fir (*Pseudotsuga menziesii*). Additional rows could include white pine (*Pinus strobus*),

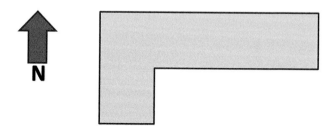

FIGURE 4.7 An L-shaped windbreak protects against both westerly and northerly winter winds. The length of a windscreen should equal 11.5 times its height ($L = 11.5\,H$). (Illustration by the author.)

TABLE 4.1
Trees for Windbreaks

Name	Species	HT (FT)	W (FT)	Spacing (W/IN-BETW)
		Evergreens		
Arborvitae	*Thuja occidentalis*	30–60	10–15	12/16
Colorado spruce	*Picea pungens*	80–150	20–30	16/16
Douglas fir	*Pseudotsuga menziesii*	150–300	12–18	16/16
Norway spruce	*Picea abies*	80–100(150)	25–35	12/16
Ponderosa pine	*Pinus ponderosa*	80–120	20–25	12/16
Red cedar	*Juniperus virginiana*	40–50	8–12	12/16
White pine	*Pinus strobus*	100–150	50–60	16/16
White spruce	*Picea glauca*	130–160	10–20	16/16
		Deciduous		
Arrowwood	*Viburnum dentatum*	10	10	6/6
Blackhaw viburnum	*Viburnum prunifolium*	10–15	8–12	6/16
Crabapple	*Malus* spp.	15–25	15–25	16/16
Hazelnut	*Corylus americana*	3–8(15)	5–10	6/16
Redbud	*Cercis canadensis*	30	30	6/16
Red twig dogwood	*Cornus sericea*	8–10	6–10	6/16
Winterberry	*Ilex verticillata*	15	15	6/16
Witch Hazel	*Hamamelis virginiana*	15–20	20–25	6/16

Austrian pine (*Pinus nigra*), and ponderosa pine (*Pinus ponderosa*). For small areas, arborvitae (*Thuja occidentalis*) and red cedar (*Juniperus virginiana*) work well. Plants that provide food and cover for wildlife may be incorporated into the planting. Berry and nut-bearing trees and shrubs may be used. Some examples include blackhaw and arrowwood viburnums (*Viburnum prunifolium* and *V. dentatum*, respectively), junipers (*Juniperus* spp.), hazelnut (*Corylus* spp.), crabapples (*Malus* spp.), redbuds (*Cercis* spp.), hollies (*Ilex* spp.), red twig dogwood (*Cornus sericea*), and common witchhazel (*Hamamelis virginiana*). Table 4.1 provides a list of windbreak plants for various hardiness zones in the United States.

WIND IN THE URBAN LANDSCAPE

A great deal of study has been conducted concerning the effect of built-up areas on the formation of **microclimates**. In addition to the urban heat island effect discussed in Chapter 3, wind is another climatic factor that interacts with the urban interface. Oke (1978, 1988) has described and defined **urban canopy layers** as those areas between the ground level and the roofs of buildings and **urban boundary layers** as those areas between the roofs and extending a few 100 m overhead.

Urban canyons are formed when there is nothing to break up the wind, as it moves through the spaces formed by tall buildings. Wind-tunnel studies have shown that tall buildings force air up and over, resulting in an increase in wind speed as it passes over the top of the building. Wind speed also increases around the corners of the building. Furthermore, some of the wind is deflected downwards when it hits the building, resulting in strong downward pressure on the windward side of the building. Oke discusses the following effects of tall buildings on the comfort and safety of pedestrians:

- In cold weather the wind chill factor makes wind around buildings feel even colder.
- Increased shade around tall buildings causes an increased loss of body heat.
- The whirling eddies around buildings causes leaves, dust, and debris to swirl around people.
- The force of wind increases with the square of its speed, so that a four-fold increase in wind speed results in a sixteen-fold increase in force – this has resulted in sometimes fatal injuries as a result of being blown over by the wind.

Oke thinks the best solution to these problems is not to build tall buildings in the first place. However, retrofitting the urban environment with plants has been used successfully in the past. The Harvard Graduate School of Design conducted wind-tunnel studies on a scale model of downtown Dayton, Ohio, and determined that addition of trees and other plants could help reduce wind speed in the city during winter months. The design group found that the branches of the trees slowed the wind significantly, as it was forced downward on the windward side of the building. Robinette (1977) addresses the practice of modifying air flow around buildings using plants, stating that depending on how they are used, plants can aid or reduce air infiltration in a building and can even change the direction of airflow within a building.

City planners, architects, and landscape designers need to think in advance about the effects of providing such amenities proactively, if they want to keep the costs down. It is much more effective to design buildings with wind effects in mind, or at the very least, to incorporate plants into the design from the beginning, so that installation and associated costs can be included in the construction of the overall project.

THE WIND AND URBAN POLLUTION

Trees, green spaces, and other plantings can help to ameliorate the effects of urban pollution. In the leeward side of a building, the downward pressure of wind causes dust, particulate matter, leaves, trash, debris, and pollutants from vehicle emissions and other sources to be dispersed. This situation is complicated by the presence of other tall buildings in the vicinity, as the low pressure area around one building can draw in air from the higher pressure zone of a nearby building. According to Oke, in narrow streets, air exchange is limited, whereas, in wider, more open areas, street-level flushing of pollutants can occur.

SUMMARY

Wind affects energy use in homes and buildings most importantly in winter. Cold air seeps through cracks and allows warm air to escape. Summer breezes have a cooling effect in the human environment. Trees in the urban environment provide turbulent mixing of the air due to surface roughness of the vegetation. Trees also affect the transport of pollutants through city streets.

Wind causes plants to lose more water through transpiration which can lead to stunted growth. Winter winds can cause evergreen plants to dry out. Leaf scald sometimes occurs on sunny days when the ground is frozen, as roots are unable to take up water, and the leaves dry up and die. The Sustainable Sites Initiative addresses wind and energy consumption by providing credits for minimizing building heating and cooling requirements with vegetation.

Cold air migrates downward on sloped areas. It is possible to reduce the movement of cold air by dense vegetation such as evergreen trees or shrubs. Flowering or fruiting trees or shrubs should be planted in such a fashion as to allow cold air to drain away rather than getting trapped. Dead air space can be created by planting dense shrubbery 12 in. from a structure, and this will provide insulation against heat or cold.

Tree funnels and berms enhance and direct cooling breezes in the desired direction. Open-slatted fences allow air to flow through smaller areas while still providing privacy.

Windbreaks are designed to intercept and redirect prevailing winds in winter to prevent heat loss from structures. Prevailing winds in the United States are from the north and west in winter. Windbreaks provide energy savings and can also provide other savings when used to prevent snow drifts on select areas. When designed for roads and highways, windbreaks can have a significant positive impact on winter driving conditions.

Windbreaks should be perpendicular to the prevailing wind. On the leeward side of the windbreak, snow, dust, and debris can be deposited. Due to this accumulation, windbreak should be planted at least 15 m from driveways, buildings, and traffic areas. As wind moves up over a windbreak, the velocity of the wind is reduced. Wind should not be completely blocked, but some wind should be allowed to pass through. The maximum protection of a windbreak is provided downwind at a distance of five to seven times the height of the tallest tree in the windbreak. At further distances, protection is still afforded, but wind speed begins to increase again.

The side profile of a windbreak determines wind pattern, with a varied profile providing the best results. Windbreaks should have four to five rows of deciduous trees or two to three rows of evergreens and include two to three different species. Deciduous and evergreen trees and shrubs may all be combined in a windbreak. Plants bearing nuts and berries will encourage and support wildlife. The windbreaks should be 11.5 times longer than their height at maturity. An L-shape design is standard.

Urban environments interact with wind in several ways. Built-up areas can obstruct the wind, cause it to increase speed, force pollutants to street-level, and serve as a hazard to human health and safety. Trees, parks, and other plantings can ameliorate these negative effects and should be considered in urban planning decisions early in the development process.

REVIEW QUESTIONS

1. Discuss how trees can be used to provide cooling from summer breezes.
2. Discuss the ways in which wind affects plants.
3. Name two ways the Sustainable Sites Initiative provides credits for minimizing building heating and cooling requirements with vegetation.
4. Name three ways you can modify the effects of wind in the landscape using plants.
5. Describe how plants are used to trap cold air on a slope.
6. What is dead air space? How can you create it with plants?
7. Under what conditions is it desirable to manipulate wind? Discuss how this is done.
8. What kind of energy savings can be realized using wind breaks?
9. Besides energy savings, what is another benefit of wind breaks?
10. Name three plants that work well in a windbreak, and list their mature height and spread.

ACTIVITIES

1. Using a resource for windbreaks in your region, design a windbreak for a rural property. Include at least three different species, at least one evergreen and one deciduous species. Include plants that provide food for wildlife.
2. Evaluate a local landscaped area with one or buildings on it for wind movement. Determine whether plants could be used for energy efficiency in the buildings. Design one or more plantings that could provide energy savings.

FURTHER READING

Balogun, A.A., J.O. Adegoke, S. Vezhapparambu, M. Mauder, J.P. McFadden, and K. Gallo. 2009. Surface energy balance, measurements above an exurban residential neighborhood of Kansas City, Missouri. *Boundary-Layer Meteorol.* 133: 299–321.

Barlag, A. and W. Kuttler. 1990/91. The significance of country breezes for urban planning. *Energy Build.* 15–16: 291–297.

Colorado State Extension. Windbreak planting and design. http://www.coopext.colostate.edu/Adams/sa/windbreak.htm. Retrieved July 27, 2009.

DeWall, D.R. and G.M. Heisler. 1983. Windbreak effects on air infiltration and space heating a mobile home. *Energy Build.* 5(4): 279–288.

Dewall, D.R. and G.M. Heisler. 1988. Use of windbreaks for home energy conservation. *Agric. Ecosyst. Environ.* 22–23: 243–260.

DeWalle, D.R., G.M. Heisler, and R.E. Jacobs. 1983. Forest home sites influence heating and cooling energy. *J. For.* 81: 84–88.

Energy Information Administration. 2005. Residential energy consumption survey: energy consumption and expenditures tables. http://eia.gov/emeu/recs/recs2005/c&e/space-heating/pdf/alltables1-13.pdf. Retrieved December 21, 2010.

Frizell, J.A. 1983. Landscape designers must put energy conservation in their plans. *Am. Nurseryman.* 157: 65–71.

Grossman-Clarke, S., J.A. Zehnder, T. Loridan, and C.S.B. Grimmond. 2010. Contribution of land use changes to near-surface air temperatures during recent summer heat events in the Phoenix metropolitan area. *Am. Meteorol. Soc.* 49: 1649–1664.

Heisler, G.M. 1990. Mean wind speed below building height in residential neighborhoods with different tree densities. *ASHRAE Trans.* 96 Part 1: 1389–1396.

Heisler, G.M. 1991. Computer simulation for optimizing windbreak placement to save energy for heating and cooling buildings. *In Trees and sustainable Development, The Third National Windbreaks and Agroforestry Symposium Proceedings*, Ridgetown College, Ridgetown, ON, pp. 100–104.

Heisler, G.M. and D.R. Dewalle. 1988. Effects of windbreak structure on wind flow. *Agric. Ecosyst. Environ.* 22–23: 41–69.

Heisler, G.M., S. Grimmond, R.H. Grant, and C. Souch. 1994. Investigation of the influence of Chicago's urban forests on wind and air temperature within residential neighborhoods. In: *Chicago's Urban Forest Ecosystem: Results of the Chicago Urban Forest Climate Project.* E.G. McPherson, D.J. Nowak, and R.A. Rowntree, Eds., USDA Forest Service, Northeastern Forest Experiment Station, 19–40. Can also be viewed at: http://www.nrs.fs.fed.us/pubs/qtr/qtr_ne186.pdf. Viewed January 11, 2012.

Honjo, T. and T. Takakura. 1990/91. Simulation of thermal effects of urban green areas on their surrounding areas. *Energy Build.* 15–16: 443–446.

Huang, J., H. Akbari, and H. Taha. 1990. The wind-shielding and shading effects of trees on residential heating and cooling requirements. *ASHRAE Trans.* 96 Part 1: 1403–1411.

Mason, S. 2000. Essential elements for windbreak design. http://web.extension.uiuc.edu/champaign/homeowners/0-1118.html. Retrieved July 27, 2009.

Meier, A. 1990/91. Strategic landscaping and air conditioning savings: a literature review. *Energy Build.* 15–16: 479–486.

Miller, P. 2009. Saving energy: it starts at home. *Nat. Geog.* 60–79.

Mizuno, M., M. Nakamura, H. Murakami, and S. Yamamoto. 1990/91. Effects of land use on urban horizontal atmospheric temperature distributions. *Energy Build.* 15–16: 165–176.

Nelson Jr., W.R. 1980. Designing an energy efficient home landscape. *Univ. Ill. Coop. Ext. Service. Circ.* 1178: 13 pp.

Oke, T.R. 1978. *Boundary Layer Climates.* Halsted Press, London, 372 pp.

Oke, T.R. 1988. Street design and urban canopy layer climate. *Energy Build.* 11: 103–113.

Pitt, L., D.G.J. Kissida, and W. Gould Jr. 1980. How to design a windbreak for residential landscaping. *Am. Nurseryman.* 152(10): 10–11.

Robinette, G.O. 1977. *Landscape Planning for Energy Conservation.* Environmental Design Press, Reston, VA, 224 pp.

Shashua-Bar, L. and M.E. Hoffman. 2000. Vegetation as a climatic component in the design of an urban street: an empirical model for predicting the cooling effect of urban green areas with trees. *Energy Build.* 31: 221–235.

Thurow, C. 1983. Improving street climate through urban design. American Planning Association. Planning advisory Service Report Number 376. 34 pp.

United States Department of Energy. Landscape windbreaks. http://www.energysavers.gov/your_home/landscaping/index.cfm/mytopic=11950. Retrieved December 21, 2010.

USDA Natural Resources Conservation Services. Conservation practices that save: windbreaks/shelterbelts. http://www.nrcs.usda.gov/technical/energy/windbreaks.html. Retrieved July 1, 2010.

U.S. Department of Energy, Energy efficiency and renewable energy. Energy savers. http://www.energysavers.gov/your_home/landscaping/index.cfm/mytopic=11950. Retrieved December 21, 2010.

Westerberg, U. and M. Glaumann. 1990/91. Design criteria for solar access and wind shelter in the outdoor environment. *Energy Build.* 15–16: 425–431.

Wilmers, F. 1990/91. Effects of vegetation on urban climate and buildings. *Energy Build.* 15–16: 507–514.

5 Water Issues

OBJECTIVES

Upon completion of this chapter, the reader should be able to:

- Describe the water treatment process
- Identify the sources of water supply
- Explain the components of water in the environment
- Discuss the difference between point- and non-point-source pollution
- Identify the two common nutrient contaminants in water
- Explain the effects of excess nutrients in the water supply
- Identify pesticides found in urban streams and groundwater
- Identify some of the risks associated with pesticides in water
- Describe the association between landscape use of fertilizers and pesticides and water pollution
- Identify techniques to prevent runoff from fertilizers and pesticides
- Explain the purpose of bioremediation and phytoremediation
- Identify the components of a constructed wetland

TERMS TO KNOW

Aquifer
Bioremediation
Constructed wetland
Endocrine disruptors
Effluent
Emergent species
Eutrophication
Ground water
Hydrology
Hypoxia
Non-point-source pollution
Organochlorine
Phytoremediation
Point-source pollution
Polycyclic aromatic hydrocarbons (PAHs)
Potable water
Runoff
Surface water
Watershed
Water table
Wetland

INTRODUCTION

In this chapter we will learn about the water cycle and briefly discuss water that is treated to ensure it is safe for human consumption. We will also look at ways in which our water supply becomes contaminated. We will place particular focus on human causes of contamination and relate that information to landscaping practices. There are both causes and solutions to be found in landscape practices. Thus, there are practices which can be altered to reduce or eliminate some sources of pollution, and other practices which can aid in remediating the problem. These will be explored and discussed in this chapter.

THE WATER CYCLE

The water cycle, also called the hydrologic cycle, is based on the understanding that the amount of water on the planet is a constant. The water cycle includes the differing forms water can take, from gaseous to liquid to solid, and describes how it cycles from one location to another over the course of time. The oceans represent the largest quantity of water containing around 97 percent of all of the world's water. Of the 3 percent that is fresh water, nearly 70 percent is tied up in glaciers and polar ice caps. Groundwater is another 30.1 percent, and surface water is only 0.3 percent of the fresh water. Most of the surface water is in lakes (87 percent), with 2 percent in rivers, and another 11 percent in swamps.

Rainfall provides valuable fresh water to gardens and landscapes. Rainwater catchment is a means of storing rain water that can later be used for irrigation. This is discussed in more detail in Chapter 6.

WATER SOURCES

Municipalities obtain their water from **surface water** or **groundwater**. Surface water drains from large areas of land through streams and rivers into lakes or reservoirs formed by dams. The drainage area is called a **watershed** and is largely fed by precipitation such as rain and snow. Water that is discharged after use may have sewage or liquid waste in it and is called **effluent**.

Most of America's water supply comes from surface water (80 percent). Its drinking water comes from both groundwater and surface water. The sources vary depending on location and whether it is drawn from a well or supplied by another entity, such as a municipality. A large proportion of water is used for irrigation (Figure 5.1). Water for this purpose is derived from both groundwater and surface water sources. Much of the water used for thermoelectric power, or generating electricity, is derived from surface water. As its name implies, groundwater is found in the ground where it occupies soil pore spaces and cracks and fissures in rocks. The soil may be thought of as a sponge, and the water it contains is the groundwater. Some municipalities and many rural residences use wells to draw up the groundwater for use.

Aquifers are layers of permeable rock, gravel, and soil that contain appreciable amounts of water. Deep aquifers are sometimes protected by impermeable materials, whereas shallow aquifers may be found in sand and gravel beds. Some examples

Water Issues   89

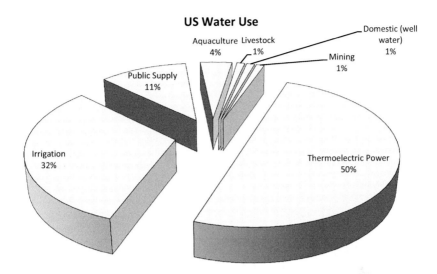

FIGURE 5.1 US water-use pie chart. (Illustration by the author.)

of aquifers in the United States include the Ogallala aquifer which underlies eight Plains states and supplies irrigation and drinking water to people in Colorado, Kansas, Nebraska, New Mexico, Oklahoma, South Dakota, Texas, and Wyoming. Four aquifers underlying the coastal plain of Alabama, Georgia, and South Carolina and partially into northern Florida combine to form the Southeastern Coastal Plain aquifer system. The Mahomet aquifer underlies a large portion of Illinois, providing millions of gallons of groundwater daily. The Spokane Valley-Rathdrum Prairie aquifer supplies water to eastern Washington and Idaho. Shallow aquifers in sand and gravel are more vulnerable to contamination by fertilizer nutrients and pesticides than deep aquifers.

The **water table** refers to the water level of the ground water. Wells are used to extract the water from the ground, and they must penetrate deeper than the water table in order to access the water. As precipitation from rain or snow-melt infiltrates through the ground, it recharges the groundwater. Nutrients and some pesticides can infiltrate through the ground, too, providing another source of contamination. However, this is more often a problem with shallow water tables than with deeper ones. Phosphorus tends to interact with other minerals in the soil and gets bound up and immobile, while nitrogen does not. The active ingredients in some pesticide residues are further broken down in the soil and will not migrate to the water table.

The level of the water table is an indication of the amount of water in an aquifer. If more water is withdrawn from the aquifer than is returned through recharge, the water table will be lowered and less water will be stored. Typically, water that is used for irrigation and for consumption is not sent back to recharge the aquifer. At least, that has been the practice for many places. However, municipalities and other water managers are beginning to rethink this practice. Natural recharge occurs through precipitation and snow-melt, and may occur through irrigation. In areas with growing populations other methods of recharge are being explored. Landscaping practices to aid in groundwater recharge are explored and discussed in Chapter 7, Stormwater Management.

POTABLE WATER

Potable water is clean water that is safe for consumption. Whereas some people obtain their water from a well (1 percent), most people rely on the public supply. Publicly supplied water is treated water so that it is safe for consumption. Before water is available for public consumption, it is filtered and possibly treated, depending on its source and local conditions. Some of the treatment methods used are ultraviolet light, filtration, reverse osmosis, deionization, and powdered activated carbon. Hard water is treated to remove calcium carbonate. There is a cost to water that is prepared in this way, and water treatment plants are limited in the quantity of treated water they can make available.

In addition to consumption of potable water, it is also used to irrigate plants inside the home and in the landscape, and for washing and bathing. Public drinking water quality is regulated by the US Environmental Protection Agency, which sets maximum concentration levels for many contaminants. To make the water safe, it often must go through a treatment system that tests for microorganisms, chemicals, radioactive substances, salts, or metals such as lead and removes them.

POLLUTED WATER

Streams, rivers, and lakes support a variety of aquatic life forms; thus, the pesticides and excessive nutrients in the water have notable effects on aquatic life. The two primary sources of water pollution from landscaping practices are fertilizers and pesticides. Applications of these substances in urban settings can result in water pollution through **runoff** into surface waters or infiltration into groundwater. This type of pollution cannot be traced to a single source, and is therefore called **non-point-source** pollution. **Point-source** pollution can be attributed to a source, such as a wastewater treatment plant or industrial waste from a known source being dumped into the water supply, such as a river or lake. Point sources have been regulated since 1972 by the Clean Water Act, and this has had a positive effect on preventing or at least limiting contamination of water supplies.

EXCESSIVE NUTRIENTS IN WATER

Excess nutrients in water cause a number of environmental problems. The two nutrients of concern are nitrogen and phosphorus. Excessive nitrate in drinking water can lead to the potentially fatal "blue baby syndrome". Hemoglobin, which normally carries oxygen in the bloodstream, is changed by the presence of nitrates, reducing its ability to carry oxygen. In addition to possible death, long-term respiratory and digestive problems may occur.

Elevated levels of nitrogen and phosphorus in surface water cause **eutrophication**, a condition that includes low oxygen levels in water (**hypoxia**), algae blooms, and fish kills. Occurrences of excess plants in water interfere with activities such as swimming, boating, and fishing; clog water pipes and filters; and cause foul odors and bad tasting water. In the Gulf of Mexico, hypoxic conditions adversely affect the fishing industry and the environment through the harm caused to fish and shellfish. In waters along the Atlantic coast, the toxic *Pfiesteria* organism thrives in areas where high nutrient concentrations occur.

According to the United States Geological Survey (USGS) report, "The Quality of Our Nation's Waters": "About twelve million tons of nitrogen and two million tons of phosphorus are applied each year as commercial fertilizer. Another seven million tons of nitrogen and two million tons of phosphorus are applied as manure". Whereas, the distribution of fertilizer use varies across the country, the highest application rates occur in the Upper Midwest. Other areas of high application rates occur on the east coast, throughout the southeast, and in agricultural areas of the west. Another important source of nutrients is the private septic systems in rural and residential areas. Figure 5.2 shows US fertilizer consumption over a recent 10-year period.

Approximately 90 percent of nitrogen and 75 percent of phosphorus found in the water supply originate from non-point sources, the major contributors of which are agriculture and landscaping. Excess nutrients in urban streams come from fertilizers applied on golf courses and lawns. The low water absorption rate of lawns— less than 10 percent of that of natural woodlands—is part of the problem due to the runoff that occurs during storms. The large areas of impermeable paving and rooftops help in washing away fertilizers that have been applied in the landscape. Waste from humans and animals is another source of nutrient contamination of waterways. When heavy rainfall occurs, wastewater treatment plants discharge large amounts of water directly into streams and rivers. In some municipalities, when excessive

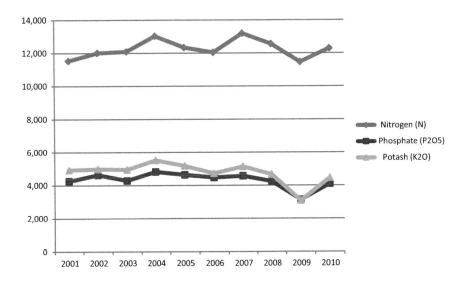

FIGURE 5.2 US fertilizer consumption. (Illustration by the author.)

amounts of stormwater flow into the **effluent** system, it is simply discharged into the nearby surface water – rivers, streams, and lakes. Release into oceans was once common, but has been prohibited since 1972 by the Clean Water Act.

Excess phosphorus has found its way into water in the past, due in large part to phosphates in laundry detergent. States began banning the use of laundry detergents with phosphates in them beginning in the 1970s, leading to a decrease of phosphorus contributions to almost negligible amounts. In Washington, automatic dishwasher detergent contributes about 10–12 percent of the phosphorus in municipal wastewater. Whereas treatment plants can remove much of the phosphorus in wastewater, they cannot remove all of it. The best way to reduce phosphorus in wastewater is to prevent it from entering there in the first place. In many states, the solution to the problem of phosphorus in water is to ban its use in various forms, including the use of fertilizer in agricultural and urban horticulture situations. States and municipalities are now limiting or banning the use of certain fertilizers for residential use, including Illinois, Maine, Maryland, Michigan, Minnesota, New Jersey, New York, Vermont, Virginia, Washington, and Wisconsin. Many others are considering such bans currently.

Pesticides in Water

Pesticides have been found in fish, bed sediment, and the water. Fish accumulate some pesticides by feeding on smaller invertebrate organisms or smaller fish that have fed on contaminated plants. Guidelines and standards have been set for aquatic life by the United States, Canada, international organizations, and other nations and tribes. Some states have developed their own standards, which may take precedence in some situations.

About two billion pounds of pesticides are applied each year in the United States. This rate has been steady for a number of years, following a rise in the 1970s due to increased herbicide use. About 70–80 percent of pesticide use is for agricultural purposes, and about 60 percent of those pesticides are herbicides. Insecticide use has declined in response to environmental concerns, resulting in reduced use of **organochlorine** insecticides (DDT), which are persistent in the environment and in the food chain. In their place, usage of less persistent insecticides has increased.

According to the US Geological Survey, a number of herbicides and insecticides are commonly found in our nation's water. In a study that distinguished between agricultural and urban streams, rivers, and groundwater, there were some differences and some similarities. Most urban streams had among the highest concentrations of insecticides, and they quite often exceeded guidelines set for aquatic life. In streams located in urban areas, insecticides were found more frequently than herbicides. About 30 percent of the sites studied had insecticide concentrations that exceeded human health guidelines for edible fish tissue. "Almost all urban streams had high or medium concentrations of the organochlorine insecticides compared with other sites". Furthermore, the study determined that, at 21 percent of urban sites, "Concentrations of pesticides in whole fish exceeded guidelines for the protection of fish-eating wildlife".

TABLE 5.1
Insecticides Commonly Used in Urban Areas through Water Quality Studies

Insecticide	Chemical Type	Additional Information
Carbaryl (Sevin, Tornado, Bugmaster)	Wide-spectrum carbamate	Moderately to very toxic
Diazinon (Dazzel, Gardentox, Kayazol, Knox Out)	Non-systemic organophosphate	EPA phase-out 2001–2004; Used on horticultural plants, pest strips, etc.
Malathion (Mercaptothion, Cythion, Dielathion, Fyfanon, Hilthion, Karbofos)	Organophosphate	Sucking and chewing insects on fruits and vegetables; mosquitoes, flies, household insects, animal parasites (ectoparasites), and head and body lice
Chlorpyrifos (Dursban, Lorsban, Paqeant, Scout)	Broad-spectrum organophosphate	Moderately toxic to humans. A skin and eye irritant

Other results in the USGS study determined that there was at least one pesticide found in nearly every stream and in more than half the wells sampled. Furthermore, many samples contained two or more pesticides. Of the 83 pesticides and breakdown products found, guidelines and standards have only been established for 43 of them. Overall, the most heavily used pesticides are also those most frequently found in water. Table 5.1 shows insecticides commonly used in urban areas that are found in water quality studies.

Human and Environmental Effects of Pesticides in Water

Pesticides pose numerous health risks to humans and the environment. Some of the human health risks associated with pesticides are increased cancer rates, low birth weight, and reproductive disorders, including low sperm counts and nervous system disorders. Pesticides known as **endocrine disruptors** interfere with natural hormones and adversely affect reproduction in humans and other mammals.

Some of the potential effects of these and other pesticides have not been thoroughly studied. Furthermore, interactions and possible synergistic effects of two or more pesticide ingredients have rarely been studied. Yet, when testing is conducted, there is often more than a single pesticide found at one time. Also, breakdown products may have toxicities that have not been studied either. Drinking water guidelines have been established for many, but not all, of the pesticides currently in use (Table 5.2).

Contributions from Urban Areas

Whereas agricultural practices are typically linked to runoff and subsequent water pollution, increasingly, residential areas are proving to be major contributors to the pollution experienced in many urban watersheds. Impermeable surfaces such as roofs and paved areas permit vast amounts of runoff to flow directly into streams and rivers.

Seasonal patterns of pesticides in streams correspond to the timing, frequency, and amount used, combined with the magnitude of runoff following major storms. A study conducted in King County, Washington, found a correlation between the

TABLE 5.2
Chemical Pollutants Found in Urban Water and Their Human Health Effects

Pesticide	Use	Known or Possible Risk	Notes
Alachlor	H	Reduced sperm count[a]	Is a chloroacetanilide
Atrazine	H	Breast cancer, low birth rates, reduced sperm counts, cancer, birth defects, menstrual problems[b]	
Cyanizine	H		
Carbaryl	I	Likely carcinogen, acetylcholinesterase inhibitor	Sold under the trade name Sevin. Is a carbamate
Chlordane	I	Prostate and testicular cancer	Used for termites. Banned in 1988
Chlorpyrifos	I	Acetylcholinesterase inhibitor, neurological effects, pervasive developmental disorders in children exposed in the womb, low birth weight, linked to autoimmune disorders and ADHD	Sold as Dursban and Lorsban. Banned for residential use, still used in agriculture
DDT	I	Endocrine disruptor, linked to diabetes	Persists in food chain
DEA	H		Breakdown product of atrazine
Diazinon	I	Acetylcholinesterase inhibitor, reduced sperm count, weakness, headaches, chest tightness, blurred vision	Banned on golf courses and sod farms in 1988 due to large bird kills, banned for residential use in 2004, used in agriculture
Dieldrin	I	Breast cancer; Parkinson's; damage to nervous, immune, and reproductive systems	Banned
Diuron	H	Diuron has a low acute toxicity to mammals even though it carries the signal word WARNING on the label. The signal word is applied because the compound can cause eye and throat irritation[a]	
Malathion	I	Skin and eye irritation, cramps, nausea, diarrhea, excessive sweating, seizures and even death[a]	
Metalochlor	H	Abdominal cramps, anemia, shortness of breath, dark urine, convulsions, diarrhea, jaundice, weakness, nausea, sweating, and dizziness[a]	
Prometon	H		
Simazine	H	Congestion of heart, lungs, and kidneys; low blood pressure; muscle spasms; weight loss; and damage to adrenal glands[a]	
Tebuthiuron	H	Skin, eye, or clothing contact with the herbicide should be avoided[a]	

(Continued)

TABLE 5.2 (*Continued*)
Chemical Pollutants Found in Urban Water and Their Human Health Effects

Pesticide	Use	Known or Possible Risk	Notes
2, 4 –D	H	Coughing, burning, dizziness, and temporary loss of muscle coordination. Symptoms of poisoning can be fatigue and weakness with perhaps nausea. On rare occasions there can be inflammation of the nerve endings with muscular effects following high levels of exposure[a]	Manufacturing process can result in contamination by dioxins

I, insecticide, H, herbicide

[a] Etoxnet, operated by Cornell University http://pmep.cce.cornell.edu/profiles/extoxnet/24d-captan/24d-ext.html

[b] Centers for Disease Control and Prevention https://ephtracking.cdc.gov/showAtrazineHealth.action

pesticides found in urban streams after rainstorms, and those recently purchased at local home and garden stores. Five of the pesticides found exceeded levels for protection of aquatic life set by the Environmental Protection agency (EPA). They were the insecticides diazinon, carbaryl, chlorpyrifos, lindane, and malathion. The active ingredient (2-(4-chloro-2-methyl phenoxy) propionic acid) in the herbicide Mecoprop (MCPP) and 2,4-D, commonly known by a number of different names, including Weedone and Weed-B-Gone (2,4-D Dichlorophenoxyacetic acid), which is widely used on lawns and other turf areas to control broadleaved weeds, were found in water samples from all twelve of the study sites, as was the insecticide diazinon. These three pesticides had the largest sales figures in the study. Not only are they used in residential areas, but are also used along rights of way and in recreational and industrial areas.

The US Geological Survey studies further show that more than 90 percent of water and fish samples from all streams sampled contain at least one pesticide. The pesticides that were found in stream and fish samples included those that are most commonly used in residential gardening and landscaping. The pesticides are found not only in surface water, but also in aquifers that supply many areas with drinking water and water for everyday use. What is further problematic is that pesticides are found in air, rain, snow, and fog. Table 5.3 shows a list of pesticides detected in urban streams in King County, Washington.

INSECTICIDES

Diazinon and carbaryl were the most common insecticides found in urban streams; and diazinon, chlorpyrifos, and malathion most commonly exceeded aquatic life guidelines. Insecticides were much less common in groundwater, but dieldrin and diazinon were present in high levels in some instances. Dieldrin had been used for termite control until 1987 and was banned for agricultural use in 1974. It accumulates

TABLE 5.3
Pesticides Detected in Urban Streams in King County, Washington, 1998–2003

Active Ingredient	Trade Name Example
Herbicide	
2,4-D	Weedone
Acetochlor	Guardian
Atrazine	AAtrex
Dicamba	Banvel
Dichlobenil	Casoron
Dichlorprop	2,4-DP
EPTC	Eptam
MCPA	Kilsem
MCPP	Mecoprop
Metolachlor	Dual
Napropamide	Devrinol
Oxadiazon	Ronstar
Prometon	Pramitol
Simazine	Princep
Tebuthiuron	Spike
Triclopyr	Treflan
Insecticide	
Carbaryl	Sevin
Chlorpyrifos	Dursban
Diazinon	Diazinon
Gamma-HCH	Lindane
Malathion	Malathion
Fungicide	
Pentachlorophenol	Penta

Source: USGS, By L.M. Frans, 1999. https://pubs.usgs.gov/sir/2004/5194/.

in the body, and is stored in fat. Exposure to moderate levels of dieldrin for a long time causes headaches, dizziness, irritability, vomiting, or uncontrollable muscle movements. Diazinon was banned for use in 2004.

HERBICIDES

Herbicides were found in many of the urban streams tested, and they "were moderately common in shallow groundwater beneath urban areas." The most common herbicides found there were simazine and prometon. Furthermore, herbicides were also found in shallow groundwater beneath urban areas. In one instance, in a shallow aquifer that is used for drinking water, the concentration of the herbicide atrazine exceeded the drinking water standard. Most deep aquifers had lower frequencies of herbicides and none exceeded drinking water standards.

OTHER POLLUTANTS

Polycyclic aromatic hydrocarbons (PAHs) found in pavement sealant that is coal-tar-based are the single largest source of PAH contamination to urban lakes. This group of compounds contains some known carcinogens, teratogens, mutagens, and toxins. The sealant is widely used in urban areas. The particles are able to run off in rain water and into storm drains after friction from vehicle tires abrades it into small particles. Sealcoat breaks down after only a few months after application, with about 5 percent of washing off the driving areas of the parking lots each year. Sealcoat manufacturers recommend reapplication every 2–4 years. In addition to the water contamination, dust in homes adjacent to parking lots with coal-tar-based sealcoat has been analyzed, and it has been found to have elevated concentrations of PAHs. Coal-tar-based sealcoat is most commonly used in the central, southern, and eastern United States.

PREVENTING AND TREATING CONTAMINATED WATER

There appear to be two ways to eliminate all pesticide runoff into urban surface and ground water: stop using toxic pesticides and prevent runoff. Dieldrin, formerly used for termite control, has been banned from use, as has DDT, made famous by the author of the book *Silent Spring*, Rachel Carson. One problem with both of these and a few other pesticides is their longevity, or persistence in the environment. When a new pesticide comes onto the market, it must be tested for persistence in the environment, and this information must be taken into account during the regulation process. Breakdown products must be examined for their persistence as well as their toxicity. Sometimes the breakdown products are as toxic as the parent chemical; sometimes, they are less toxic.

Preventing runoff from pesticide and fertilizer applications requires planning and knowledge of coming rainfall so that pesticide application prior to rain is avoided. Furthermore, practices that enhance infiltration of rainfall, such as use of permeable paving, rain gardens, green roofs, and bioswales, can all help. These practices and techniques are discussed in more detail in Chapter 7. In this chapter, constructed wetlands for treatment of water will be explained in detail.

REDUCING USE OF PESTICIDES IN THE LANDSCAPE

The use of organochlorine insecticides has decreased in the last several decades, due in large part to government bans. Yet, these chemicals are still detectable in fish many years later. The lesson is that less toxic pesticides and those that do not persist in the environment should be used; overall, usage should be better targeted to pests; usage prior to major storm events should be avoided; pesticides should be handled appropriately; and increasingly better methods for dealing with landscape pests should be devised. In Chapter 12, Integrated Pest Management is discussed as a better method for pest management in the landscape. A variety of techniques and practices are provided to complement, and in some cases replace, the use of chemical pest control.

Some of the reasons given for the high levels of pesticides found in our water are that they are used excessively and are not applied according to label directions. While horticultural professionals must be licensed in order to apply pesticides for other people, individuals applying pesticides at their own residences do not require a license, permit, or other documentation of training. As part of the licensing procedure, horticultural professionals must demonstrate that they understand how to read a pesticide label and agree to follow the instructions provided. As a matter of fact, a pesticide label is a legal document that must be strictly obeyed, regardless of who is using the pesticide. Some of the information provided on a pesticide label include the type of protective clothing that must be worn during mixing or application, proper disposal of remaining pesticide, how to clean pesticide containers, and so on. These details and others can help to minimize contamination of the environment. Since runoff appears to be a major culprit of water contamination, possible solutions to the problem should take that into account. Levels of pesticides in water tend to be highest during periods of greatest rainfall. In most areas this is May through July. Avoidance of pesticide application prior to major storm events should be practiced by professionals and homeowners alike.

BIOREMEDIATION AND PHYTOREMEDIATION

Bioremediation and **phytoremediation** are terms that describe the use of biological (bio-) organisms or plants (phyto-) to remove toxic substances from the environment. Some examples of bioremediation organisms are: the bacteria *Zooglea ramigera* – it can remove cadmium from wastewater, and the fungus *Penicillium digitatum* – it can absorb uranium from solutions of uranyl chloride. Some plants that have been shown to remove toxic chemicals from the environment include *Zea mays*, or corn: it removes lead from the soil. Two members of the mustard family, *Brassica juncea* and *B. carinata*, were found to be able to remove large quantities of chromium, lead, copper, and nickel in laboratory tests on artificial soil. Table 5.4 lists some recommended plants for phytoremediation.

WETLANDS AND CONSTRUCTED WETLANDS

A **wetland** is a place where water-saturated land dominates the landscape. It can be large or small, and usually has specially adapted plants and animals as part of its ecosystem. Deepwater swamps, estuaries, freshwater marshes, bogs, riparian wetlands, peatlands, tidal salt marshes, tidal freshwater marshes, and mangrove wetlands are various types of wetlands.

Some of the largest wetlands in the United States are the Hudson Bay lowland, the Prairie potholes located in the upper Plains, and the Mississippi and Mackenzie River Basins located in the central United States and Alaska, respectively. Throughout the

TABLE 5.4
Plants and Microorganisms That Are Used in Constructed Wetlands and the Contaminant or Nutrient They Remove

Plant or Microorganism	Species	Contaminant or Nutrient Removed
Cattail	*Typha* spp.	Nitrate, phosphorus, lead, cadmium, copper, nickel, cobalt, zinc
Bulrush	*Scirpus* spp.	N, P, bacteria, oil, organics
Reed	*Phragmites* spp.	N, P, bacteria (*E. enterococci, E. coli, Salmonella*)
Common three-square bulrush	*Scirpus pungens*	Several metals
Softstem bulrush	*Scirpus validus*	Pollutants, bacteria (*E enterococci, E. coli, Salmonella*)
Blue flag iris	*Iris versicolor*	Several nutrients
Duckweed	*Lemna, Spirodela, Wolffia,* etc.	Organic compounds from pesticides and pharmaceuticals
Water weed	*Elodea, Hydrilla, Egeria*	Several nutrients
Yellow flag iris	*Iris pseudocorus*	Bacteria (*E enterococci, E. coli, Salmonella*)
Mint	*Mentha aquatica*	Bacteria (*E enterococci, E. coli, Salmonella*)
Reed canary grass	*Phalaris arundinacea*	Nitrogen, mercury, can be weedy
Rushes	*Juncus* spp.	Cobalt, copper, manganese, nickel, zinc, nitrogen, phosphorus

Source: U.S. EPA http://www.epa.gov/owow/wetlands/pdf/hand.pdf

United States there are countless wetlands, from the very large to the very small. There is a wetland in nearly every county in the United States; however, many wetlands have been destroyed due to their negative image as a place harboring mosquitoes or as simply a low-lying area that could be better utilized if filled in and leveled out for roads, construction, farming, or other activities. The Swamp Lands Act of 1849 helped in the effort by states to drain wetlands.

Interestingly, scientists now believe that clearing wetlands in Florida in order to site citrus orchards further south to avoid winter frosts has actually triggered more freeze events than in the past. This is due to the fact that water cools more slowly than land at night. The land does tend to warm during the day, but on calm, clear nights, much of that warmth is lost to the atmosphere. During three freezes, in 1983, 1989, and 1997, researchers found

> a strong connection between areas that were changed from wetlands to agriculture during the 20th century, and those that experienced colder minimum and subfreezing temperatures over a longer time, in the current land-use scenario. Water typically doesn't cool as quickly as the land at night, which may explain why when wetlands are converted to croplands the area freezes more quickly and more severely

THE NATURE, FUNCTION, AND VALUE OF WETLANDS

A wetland ecosystem is a natural detoxifying environment and a biological water filter. Wetlands aid in groundwater recharge and discharge, reduce floodwater, help to retain and stabilize sediment, retain toxic substances, remove or transform nutrients, and support an abundance and diversity of plants, fish, and wildlife.

Wetland biological processes are complex interactions between **hydrology** (the movement of water), soil chemistry, nutrient cycling, and so on. The essential elements of life are recycled there: oxygen, nitrogen, phosphorus, carbon, and hydrogen. Plants in a wetland grow prodigiously and take up vast quantities of nutrients. Organic solids and sediments that enter into a wetland rarely leave it. Some wetland plants can also remove metals and pathogens.

As an additional benefit, wetlands provide habitat for a diverse array of wildlife, fish, and amphibians, which makes them a perfect environment in which to create nature preserves and recreational areas, and work well for sport hunting and fishing. Wetlands may also provide educational opportunities, demonstrating an ecosystem with which many are not familiar.

Wetlands need not be hundreds or even tens of acres in size. Small wetlands of only several square yards to one or two acres comprise the majority of wetlands in the United States. Even these very small spaces support a wide diversity of species.

COMPONENTS OF A WETLAND

The major components of a wetland system are: water, plants, animals, microorganisms, and a substrate. These are usually situated atop a base of low permeability. Both aquatic and terrestrial plants and animals may be found in a wetland because while the water is present during much or all of the year, it may not be present throughout the year. This variability is caused by several factors, such as rainfall, soil type, topography, vegetation, and others.

CONSTRUCTED WETLANDS

A **constructed wetland** is a built wetland system that mimics naturally occurring systems and is used to accomplish a specific goal, typically bio-filtering of wastes such as sewage waste or effluent. A constructed wetland is built in conjunction with a leaching field when it is to be used for sewage waste. Some municipal wastewater treatment facilities are now incorporating constructed wetlands into their operations as a means to reduce operating costs, as well as to avoid the high costs of constructing new, state-of-the-art facilities when the older facilities require upgrading.

Constructed wetlands have been used to treat sewage waste and effluent at the residential, industrial, community, and municipal levels. Other uses for constructed wetlands are treatment of:

- Agricultural wastewater
- Coal mine drainage
- Stormwater runoff
- Petroleum refinery runoff

- Compost and landfill leachates
- Fish pond discharges
- Industrial waste waters

Golf courses may be ideal locations for constructed wetlands as part of their water management system. They would provide a place for nutrients and pesticides to be filtered out of runoff while providing a means of storing excess stormwater from heavy rainfall. In residential developments that incorporate a golf course, there is the added benefit of filtering runoff from lawns and landscapes in the neighborhood. Some golf courses are incorporating constructed wetlands in conjunction with the Audubon Cooperative Sanctuary Program. More on this program is discussed in Chapter 2.

Constructed Wetlands Design

The design of a constructed wetland incorporates the ideas and knowledge of engineers, microbiologists, and wetlands ecologists. Landscape architects, parks and recreation planners, city planners, and wildlife biologists may all have a role in a given project. Constructed wetlands range in size from a 100-gallon tank servicing a small greenhouse to a size large enough to serve a municipal wastewater treatment facility. Individual homes may have a constructed wetland as may industrial sites or parks. Some examples are Gateway National Recreation Area in New York, King County, Washington, Florida, and Maryland. The size of a constructed wetland to service a residential property is usually based on the number of bedrooms and the percolation rate of the soil. Temperature plays a role, too. Figure 5.3 shows a constructed wetland at the edge of a landscape pond at Aquascape company headquarters in St. Charles, Illinois.

FIGURE 5.3 Aquascape constructed wetland in St. Charles, Illinois. (Photo by the author.)

Case Study: Tres Rios, Phoenix Arizona.

Reference: Gelt, Joe. Constructed Wetlands: Using Human Ingenuity, Natural Processes to Treat Water, Build Habitat. http://ag.arizona.edu/azwater/arroyo/094wet.html. Viewed December 28, 2011.

In Arizona, constructed wetlands are being used increasingly to treat wastewater. Since 1990, the number of constructed wetlands has increased from 4 to 26, with 24 additional projects planned or under construction. They serve a variety of needs, from individual residences to large, multi-use facilities. The Tres Rios project outside of Phoenix is a demonstration project that treats effluent from the nearby wastewater treatment plant. In addition to Phoenix, multiple nearby cities participate in the demonstration project, which is testing the effectiveness of the wetlands in treating the effluent on this fairly large-scale. The cities of Phoenix, Mesa, Tempe, Scottsdale, and Glendale are hoping to meet tougher federal clean water standards without incurring the millions of dollars in cost that would be required to build a new water treatment facility.

The water that flows to the Tres Rios project has already gone through preliminary cleansing at the nearby water treatment facility. However, constructed wetlands in general may include the primary treatment in their design. Such primary treatment wetlands would be off-limits to the public or to wildlife, due to the nature of the solids that must be filtered out. Tres Rios is a three-step filtering system. Water first enters a marsh, then flows to an open pool, then back again to a marsh. Vegetation includes bulrushes, cattails, and reeds. Nesting places and habitat support wildlife and water fowl. Fish-eating birds such as the blue heron, night heron, and egrets can be seen stalking along the edges of the water.

Other Arizona constructed wetlands include the Pintail Lake on National Forest Service land in Show Low; Tucson's Sweetwater Wetlands project; the Jacob Lake Inn constructed wetland that treats wastewater from campgrounds, cabins, and a laundry; and Kingman's constructed wetland that handles three million gallons per day.

Case Study: Humboldt Bay, Arcata, California

Reference: Suutari, Amanda. Arcata, California, USA – Constructed Wetland: A Cost-Effective Alternative for Wastewater Treatment. http://www.ecotippingpoints.org/our-stories/. Viewed December 27, 2011.

The community of Arcata, California got a rude awakening when they learned that their aged wastewater treatment plant required replacing. The state of California and the federal government proposed a multi-million dollar regional sewage plant. The people of Arcata would have been required to foot about $10 million of the total bill. In addition, upkeep of the plant would easily run into the hundreds of thousands of dollars.

At the time, in the 1970s, understanding of the natural processes that work together to treat raw sewage was not widespread. However, several Arcatans, notably fisheries science professor Dr. George Allen, and ecological engineer Robert Gearheart, conceived of a plan to use a wetland to treat the municipal wastewater. They were joined by City Counselor Dan Hauser and Public Works Director Frank Klopp in proposing the following design: the primary sedimentation facilities that were already in place, an oxidation pond facility, and three newly constructed marshes. Effluent from the marshes would flow into a lake prior to being released into the Bay.

One of the marshes would be located at a site of a former lumber mill decades earlier. Lumber, concrete, other residue from the saw mill had to be removed. A second marsh was former pastureland adjacent to a closed landfill. Marsh plants were allowed to grow in naturally in this marsh. Earth-moving was required to create places that would be shallower or deeper, to allow a variety of plants to thrive. Vegetation was introduced to one-third of this marsh in order to ensure deeper vegetation at the end of the process. Plants were obtained from a variety of sources, including a neighboring lake, as well as mail-ordered plants. Many challenges were faced with establishing the plants, but eventually, a method was worked out whereby the plant tops were cut off and the root system was tucked in securely to the soil.

The treatment area is now a restored waterfront area that provides habitat for birds and other wildlife. It consists of 154 acres of freshwater and saltwater marshes, tidal mudflats, and grasslands. It has become a treasured recreation area, while treating and disposing of the city's sewage.

Manuals and guides for constructed wetlands are available from several sources. The references cited at the end of this chapter list some of them. A lot of research and study has been conducted to determine best design practices for constructed wetlands. State and local resources should be used in order to ensure that the best design is used, as well as in proper compliance with local regulations.

SITING A CONSTRUCTED WETLAND

Wetlands should be constructed in low, gently sloped areas where water can flow by gravity. It should not be adjacent to the property line. Ideally, a wetland will be located close to the source of wastewater and large enough to accommodate current needs, with room for expansion if future needs are expected to increase.

NATURAL COMPONENTS OF A CONSTRUCTED WETLAND

Components of constructed wetlands vary somewhat, depending on local conditions and the intended purpose.

Water in a Wetland

Hydrology links all of the functions in a wetland, and as such plays a very important role. Water supports the living organisms that are responsible for the biological and chemical processes. Water levels can fluctuate due to rainfall, and water loss through evaporation and transpiration.

Substrates in a Wetland

Above the relatively impermeable base, rock, gravel, sand, soil, and organic matter are used to provide a substrate for the water and living organisms. Chemical processes depend on the conditions in the substrate, in which oxygen is displaced by water, creating anaerobic conditions. Some pollutants, such as nitrogen and metals, are more readily removed under such conditions. The substrate provides ample surface area for the absorption of contaminants.

Plants in a Wetland

A constructed wetland is normally supplied with plants at initial establishment. Aquatic invertebrates and microorganisms develop naturally over time.

Because of their highly defined goal, fewer species are used in a constructed wetland than would be used in restoration of a naturally occurring wetland. Plants that are used in constructed wetlands must meet the following criteria:

- Ability to withstand 6–24 in. of water; this type of plant is known as an **emergent species** in the wetland environment.
- Vigorous lateral and vertical rooting. Deeper roots provide more surface area for microbial bacteria and for oxygenation of the root zone.
- Rapid reproduction and infilling.
- Local availability.

The three most widely used emergent aquatic species are invasive in some locales, and should be avoided or used with extreme caution. They are cattails (*Typhus* spp.), bulrushes (*Scirpus* spp.), and common reeds (*Phragmites* spp.). Cattails produce rhizomes, making them easy to propagate, and various species are found throughout the United States. Bulrushes remove nitrogen very efficiently, and they tolerate a wide pH range. Their roots are capable of penetrating 2.5–3 ft deep, thereby helping to oxygenate deep into a gravel subsurface. Common reeds are annual grasses with rhizomatous roots that grow to around 18 in. deep. They help to oxygenate the water, but introduced species and subspecies of *Phragmites* are invasive and should be avoided. Canna species are used in areas where cold hardiness is not required.

Built Components of a Constructed Wetland

Water, plants, animals, and microorganisms are components of both naturally occurring and constructed wetlands. The latter, however, require other components to meet the needs of wastewater treatment. A pump may also be required if the wastewater cannot be gravity-fed through the system. The primary components required to build a wetland are:

- A drainfield, lagoon, or wildlife habitat pond
- Septic tank
- Liners and berms
- Distribution and collection piping
- Wetland cell

The first steps in building a constructed wetland are to obtain a design and to secure permits from the appropriate regulating authority. The latter is usually the county health department.

In Massachusetts, John Todd has developed a system of contained aquatic cells that can be used to treat waste water. They may be housed within a greenhouse. He refers to his system as Eco-Machines™. The cells can also serve as a water garden that provides high quality water without the use of hazardous chemicals.

SUMMARY

People obtain water from either surface water, such as rivers, streams, and reservoirs, or from groundwater. The drainage area for water is known as a water shed, which is fed by precipitation. In addition to water for drinking, a great deal of water is used for irrigation of agricultural crops, while lesser amounts are used on golf courses and in greenhouses and nurseries.

Clean, potable water is safe for consumption and is used to water plants inside the home and in the landscape, as well as for washing and bathing. To make the water safe, it often must go through a somewhat costly treatment system to remove microorganisms, chemicals, salts, or other undesirable components.

Municipal water supplies come from surface water or groundwater. When water is in short supply, for whatever reason, landscape water use is one of the first targets for reduced water consumption. Water that is used on landscapes currently accounts for one- to two-thirds of residential water use.

The water cycle is based on the understanding that the amount of water on the planet is a constant. The water changes from gaseous to liquid to solid as it cycles from one location to another. Ninety-seven percent of the world's water is in the oceans. Of the remaining 3 percent, about a third of it is groundwater. Surface water is only 0.3 percent of the fresh water.

Groundwater may be stored in layers of permeable rock, gravel, or soil known as aquifers. The water table refers to the level of groundwater, and it may be deep, requiring deep wells to extract it, or shallow. Sometimes the water table is at the surface of the soil, and it may also seep out through steep or vertical slopes.

The two primary sources of water pollution caused by landscaping practices are fertilizer nutrients and pesticides. Runoff from stormwater can carry these components into urban streams. Due to the difficulty of tracing the exact source or location of the contamination, such pollution is referred to as non-point-source pollution.

Excess nutrients in the water, primarily nitrogen and phosphorus, can lead to problems with excessive algal growth known as eutrophication or increased nitrates in drinking water, a cause of blue baby syndrome.

Pesticides are found in urban streams more commonly during periods of high usage (spring) and following major storm events. Urban streams typically contained insecticides more frequently than herbicides in water quality studies. High or medium concentrations of organochlorine insecticides were almost always detected in urban streams compared to other sites.

Bioremediation and phytoremediation refer to the removal of toxic substances from the environment by biological organisms, such as bacteria, fungi, and plants. Wetlands, sometimes referred to as the earth's kidneys, provide a natural detoxifying environment. Constructed wetlands mimic naturally occurring ones, while purposely incorporating detoxifying plants and other organisms. Some municipal wastewater treatment plants are incorporating constructed wetlands into their design as a means of removing excess nutrients and allowing other toxic substances to filter out or become deactivated before moving into surface or groundwaters so that wildlife will not be harmed. Constructed wetlands are versatile in that they can be, and have been, used at the residential, industrial, community, and municipal levels.

REVIEW QUESTIONS

1. What is an aquifer?
2. What are the two sources of water for municipalities?
3. What are the two primary nutrients that are considered pollutants because they are often present in excess amounts?
4. List one adverse effect of each of the nutrients in question 2.
5. What is the difference between point-source and non-point-source pollution?
6. In urban streams, which are more prevalent, herbicides or insecticides?
7. What is the relationship between major storm events and pesticides in streams in urban areas?
8. Name three pesticides commonly found in urban streams or drinking water.
9. What function do natural wetlands serve?
10. List three major features of wetland plants.

ACTIVITIES

1. Using the National Atlas website, http://nationalmap.gov, compare the US population between 1990 and the present with Public Supply water use. Discuss the results.
2. Using the map of aquifers at the National Map website, find the aquifer nearest you. Determine whether it has changed in level over the past 10 years, and if so, what is the direction and rate?
3. Search the internet for a manual or guidelines for building a constructed wetland in your region. Identify the major components required and develop a design for a wetland on a residential site in your area.
4. Using the NAWQA (National Water Quality Assessment) study website (https://water.usgs.gov/nawqa/), find the study units located nearest you and research the findings for water in your area. Research changes that may have occurred over the years of the study.

FURTHER READING

Campbell, C.S. and M.H. Ogden. 1999. *Constructed Wetlands in the Sustainable Landscape.* Wiley & Sons, New York, NY, 270 pp.

Davis, L. Handbook of Constructed Wetlands: a guide to creating wetlands for: agricultural wastewater, domestic wastewater, coal mine drainage, stormwater in the Mid-Atlantic region. Vol. 1: General Considerations. http://www.epa.gov/owow/wetlands/pdf/hand.pdf. Retrieved May 30, 2019.

East Texas Plant Materials Center. 1998. Constructed wetlands for on-site septic treatment: a guide to select aquatic plants for low-maintenance micro-wetlands. National Environmental Services Center. (NESC) Item # WWBLOM37. (1-800-624-8301).

EPA. Constructed Wetland Handbook. http://www.epa.gov/owow/wetlands/pdf/hand.pdf Retrieved May 30, 2019.

Garcia-Sanchez, M., Zdeněk, K., Mercla, F., Aranda E., Tlustoš, P. 2018. A comparative study to evaluate natural attenuation, mycoaugmentation, phytoremediation, and microbial-assisted phytoremediation strategies for the bioremediation of an aged PAH-polluted soil. *Ecotoxicol. Environ. Saf.* 147: 165–174.

Glaser, A. 2006. Threatened waters: turning the tide on pesticide contamination. *Pesticides You* 25(4): 9. http://www.beyondpesticides.org/documents/water.pdf

Henderson, C., M. Greenway, and I. Phillips. 2007. Removal of dissolved nitrogen, phosphorus, and carbon from stormwater by biofiltration mesocosms. *Water Sci. Tech.* 55(4): 183–191.

Latimer, J. G., R.B. Beverly, C.D. Robackerr and O. M. Lindstrom. 1996. Reducing the pollution potential of pesticides and fertilizers in the environmental horticulture industry: I. greenhouse, nursery, and sod production. *HortTech.* 6: 96–140.

Latimer, J. G., R.B. Beverly, C.D. Robacker and O. M. Lindstrom. 1996. Reducing the pollution potential of pesticides and fertilizers in the environmental horticulture industry: II. Lawn care and landscape management. *HortTech* 6: 150–288.

Litke, D. W. 1999. A review of phosphorus control measures in the United States and their effects on water quality: U.S. Geological Survey Water-Resources Investigations Report 99–4007, 38 p.

Lopes, T. J. and S. G. Dionne. 1998. A review of semivolatile and volatile organic compounds in highway runoff and urban stormwater. US Geological Survey report 98-409. 67 pp.

McKenzie, C. 2004. Homeowner cost cutter: build your own constructed wetland. *Small Flows Q. Fall.* 5(4): 26–29.

NASA. http://www.nasa.gov/vision/earth/environment/wetland_freeze.html. Retrieved July 20, 2011. Retrieved May 30, 2019, but is being kept online for historical purposes only, and is no longer being updated.

National Groundwater Association. Groundwater Facts. http://NGWA.org. Viewed May 30, 2019.

National Atlas. http://nationalmap.gov/. Retrieved May 30, 2019.

National Atlas. Water use in the United States. https://waterdata.usgs.gov/nwis. Retrieved May 30, 2019.

Nowell, L. H. Moran, P. W., Schmidt, T. S., Norman, J. E., Nakagaki N., Shoda, M. E., Mahler, B. J., Van Metre, P. C., Stone, W. W. Sandstrom, M. W., Hladik, M. L. 2018. Complex mixtures of dissolved pesticides show potential aquatic toxicity in a synoptic study of Midwestern U.S. streams. *Sci. Total Environ.* 613–614, 1469–1488. doi:10.1016/j.scitotenv.2017.06.156.

Salt, D. E., N. Benhamou, M. Leszczyniecka, I. Raskin, and I. Chet. 1999. A possible role for rhizobacteria in water treatment by plant roots. *Int. J. Phytoremediation* 1: 67–79.

Schulman, R., D. E. Salt, and I. Raskin. 1999. Isolation and partial characterization of lead accumulating mutants of *B. juncea*. *Theor. Appl. Genet.* 99: 398–404.

Sprague, L. A., D. A. Harned, D. W. Hall, L. H. Nowell, N. J. Bauch, K. D. Richards. 2007. Response of stream chemistry during base flow to gradients of urbanization in selected locations across the coterminous United States, 2002–2004: US Geological Survey Scientific Investigations Report 2007-5083, 132 p. [downloadable copy available at http://pubs.usgs.gov/sir/2007/5083/]

Steiner, G.R. and J.T. Watson. General design, construction, and operation guidelines: constructed wetlands wastewater treatment systems for small users including individual residences. 2nd edition. Tennessee Valley Authority, Water Management resources Group. National Environmental Services Center. (NESC) Item # WWBLDM65. (1–800-624–8301).

Taylor, C., D. Jones, J. Yahner, M. Ogden, and A. Dunn. 1998. *Individual residence wastewater wetland construction in Indiana.* Purdue University, Lafayette, IN.

Taylor, G. D., T. D. Fletcher, T. H. F. Wong, P. F. Breen, and H. P. Duncan. 2005. Nitrogen composition in urban runoff-implications for stormwater management. *Water Res.* 39: 1982–1989.

Torno, H. C., J. Marsalek, and M. Desbordes. 1985. *Urban Runoff Pollution.* Springer-Verlag, New York, NY, 893 pp.

United States Geological Survey. Surface water use in the United States. http://ga.water.usgs.gov/edu/wusw.html. Viewed July 31, 2008.

US Geological Survey. Pesticides in the nation's streams and ground water, 1992–2001-a summary. 2006. https://pubs.usgs.gov/circ/2005/1291/. Retrieved May 30, 2019.

US Geological Survey. 1999. The quality of our nation's waters: nutrients and pesticides. US Geological Survey Circular, 1225. 82p.

US Geological Survey. 1999. Pesticides detected in urban streams during rainstorms and relations to retail sales of pesticides in King County, Washington. USGS Fact Sheet 097-99. April.

vanMetre, P.C., D. Alvarez, B.J. Mahlera, L. Nowell, M. Sandstrom, P. Moran. 2017. Complex mixtures of Pesticides in Midwest U.S. streams indicated by POCIS time-integrating samplers. *Environ. Pollut.* 220(Part A): 431–440.

Zaurov, D. E., P. Perdomo, and I. Raskin. 1999. Optimizing soil fertility and pH to maximize cadmium removed by Indian mustard from contaminated soils. *J. Plant Nutr.* 22: 977–986.

6 Water Conservation

OBJECTIVES

Upon completion of this chapter, the reader should be able to:

- Identify reasons for water shortages in the United States
- Explain the role of landscaping when there are water shortages
- Identify landscape practices that can reduce the use of potable water
- Implement water-wise gardening techniques
- Discuss rainwater harvesting techniques
- Recognize the correct technique for applying mulch
- Create a checklist to ensure properly functioning irrigation systems
- Explain the benefits and dangers of using gray water

TERMS TO KNOW

Black water
Dripline
Effluent irrigation
Evapotranspiration (ET)
Gray water
Guard cells
Stomata
Transpiration
Turgor
Water-wise gardening
Xeric
Xeriscaping

INTRODUCTION

We usually think of deserts or droughts when we think of lack of water. But due to population growth and increased urbanization in many areas, that is no longer the case. Municipalities supply potable water by first treating and testing it to ensure it is safe for human consumption. A safe water supply that is delivered to individual homes and businesses is an important component of the infrastructure that many people take for granted. However, when water is in short supply, due to seasonal drought, multi-year drought, or simply because demand exceeds the capabilities of a system, landscape water use is one of the first targets for reducing water consumption. Perhaps it is because water that is used on landscapes currently accounts for a third of residential water use, totaling nearly seven billion gallons of water each day. This chapter examines ways to reduce water usage in the landscape while still achieving a high aesthetic quality.

The Sustainable Sites Initiative addresses water conservation in several prerequisites and credits. The practices they encourage are addressed by the practices discussed herein. For example, The Sustainable Sites Initiative rewards use of non-potable water for irrigation and water features when possible; managing and cleansing water on-site; and water conservation practices; and they encourage the use of drought-tolerant plants.

There are several issues involved in water use in the landscape of which practitioners should be aware. They include the source and quality of the water, plant water requirements, irrigation system factors, and water features in the landscape, such as ponds and fountains. The practices discussed in this chapter are rainwater harvesting, irrigation efficiency, drought-tolerant plants, water-wise gardening – also known as xeriscaping, mulching to conserve moisture, and use of gray water in the landscape.

PRECIPITATION

Rainfall provides valuable fresh water to gardens and landscapes. Rainwater catchment is a means of storing rain water that can later be used for irrigation. How much rainfall is enough? That depends on landscape needs.

In arid, or desert regions, annual rainfall is 10 in. (24 cm) or less each year. In the eastern United States, 40–60 in. (96–152 cm) of rainfall per year is common. Snow and fog can also serve as valuable sources of water for plants. Various areas of the country are affected by precipitation in different ways. For example, coastal areas tend to receive more fog than many other places. Mountains typically have a rainy side and a dry side, depending on the direction of movement of moist air over them. The desert southwest receives around 10 in. (24 cm) of rain per year, whereas the mid-west may receive 40 in. (96 cm) or more rainfall in a typical year. Accordingly, landscape plants in these regions may require more or less water as a supplement to rain water. If landscape plants have evolved in the region where they are used, they may be well-adapted to surviving without supplemental irrigation.

DROUGHT AND WATER SHORTAGE

The USGS (United States Geological Survey) defines drought as "A condition of moisture deficit sufficient to have an adverse effect on vegetation, animals, and man over a sizeable area". The US Drought monitor (http://droughtmonitor.unl.edu/) tracks drought conditions for the entire country (Figure 6.1).

In the eastern and southeastern states, rainfall may be ample enough to satisfy agricultural and horticultural needs. However, many municipalities in these states are implementing water-use guidelines for sustainable use of water. Why are these areas suffering from a water shortage? It is in part due to increasing populations putting demand on water treatment systems at greater rates than cities can afford to keep up with. To some extent, industries are also making greater demands than can be met by municipal supplies.

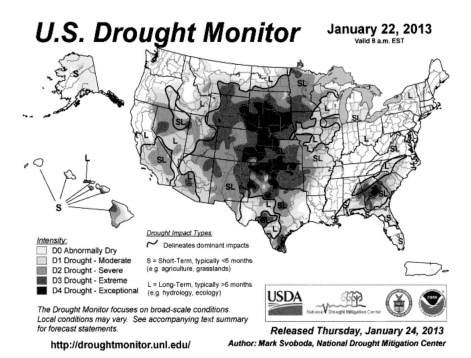

FIGURE 6.1 US Drought monitor map.

In the some parts of the western states, lack of rainfall and arid conditions are the norm. However, recent evidence indicates that human-caused climate changes related to greenhouse gases are contributing to changes in the water cycle in these areas. Specifically, reduced snow melt in the mountains, more precipitation in the form of rainfall rather than snow, and drier summer conditions are occurring, and these factors are expected to lead to water shortage problems in the near future. In addition to these changes, which have occurred during the last half of the twentieth century, the western United States already supports a large population which continues to grow in spite of a history of water-supply problems in some areas.

PLANT WATER REQUIREMENTS

Plants need water for growth, and are approximately 80–90 percent water. They take up water mainly through their roots and lose it through pores on the leaves called **stomata** in the process of **transpiration**. Stomata must be open in order for water loss to occur, and thus the process of photosynthesis in which carbon dioxide is taken up and oxygen is given off is also able to occur while stomata are open. Stomata open and close in response to **turgor** pressure in the two **guard cells** that form the stomata. Guard cells that are filled with water have high turgor pressure; when the water level in the plant falls, the cells become flaccid and stomata close.

Plant water use varies by species and size of the plants, but water loss is also affected by the surrounding environment. Dry air draws moisture out of leaves. Wind carries the moisture away, leading to continued moisture loss. If the soil is also dry and the stomata close, this contributes to a stressful situation for plants because in addition to inadequate moisture for healthy functioning and growth, the plants cannot photosynthesize or respire.

Evapotranspiration (ET) is the combined loss of soil water from loss through evaporation and plant loss through leaves in the process known as transpiration; wind, heat, dry air, and direct sunlight on leaves increases water loss through transpiration. Humid, moist conditions reduce water loss through transpiration. The rate of ET can be used to schedule irrigations both for timing and for length of time or quantity of water to apply. Evapotranspiration calculations can be complex for landscape situations because of the variety of plant species, various plant ages, differing plant growth habits, and many other plant-related variables. Other factors which can contribute to a complex ET calculation in landscapes are:

- Soil conditions
- Rainfall (predicted and actual)

In any case, irrigation is a helpful aid to ensure plants receive adequate water and do not go through periods of drought stress.

RAINWATER COLLECTION SYSTEMS

In areas that experience seasonal drought, water collection systems can provide a supply of irrigation water as a supplement to rainfall and to reduce or eliminate the need to use potable water. The types of systems used for such purposes will usually have smaller storage tanks compared to systems that are designed to address long-term water needs.

Rain barrels are commonly available in 50–100 gallon (189–378 L) sizes. These are placed above ground and typically are tied into the gutters of a house. Many people find that once they have installed a rain barrel, it fills too quickly in common rainfall events. This leads to considerations for larger storage tanks. Larger storage tanks may be placed above or below ground level. Water storage tanks are sometimes called cisterns (Figure 6.2). There are considerations that apply to each storage type: above-ground or below-ground. In any case, a pump may be necessary to transfer the water to desirable locations in the landscape.

ABOVE-GROUND WATER STORAGE

Above-ground water storage tanks do not require excavation to install, nor do they need to be structurally as strong as those that will be installed underground. As a result, they tend to be less expensive overall.

One disadvantage of above-ground tanks is freezing temperatures which make them inoperable in winter. Another problem is getting water into the tank. The water usually will come from roofs and paved areas. The larger tanks may be taller than

FIGURE 6.2   Above-ground corrugated metal cistern. (Photo by the author.)

a building, but even those that are shorter than the roofline or laid on their sides will still be higher than the pavement, requiring a pump to move the water into the storage tank. Larger tanks take up a lot of space and may be unsightly, or there may simply not be enough room or a desirable location for placing such tanks.

Above-ground storage tanks are usually made from either plastic or corrugated steel. Shipping costs can be expensive, so local sourcing is important. Whereas plastic tanks range in size from 55 to 12,000 gallons, corrugated steel tanks come in 500 to 500,000 gallon sizes. Although plastic and steel tanks share the benefits and disadvantages listed above, most corrugated tanks must be assembled on-site, adding labor costs to the overall price.

BELOW-GROUND WATER STORAGE

Below-ground tanks do not have the visibility issues of above-ground ones, but they do require that underground space be available. It is easier to get water into them by simple gravitational force, and the water level can be kept below the frost line to avoid freezing problems. Underground storage tanks are usually made of some type of plastic, such as polyethylene, polypropylene, or fiberglass. Whereas larger sizes are available, 1,700 gallon size tanks are common, and they may be modular, allowing multiple tanks to be used together. For unique situations, fiberglass cisterns can be made-to-order, although they can be costlier than other plastic types. In general, below-ground tanks are costlier than above ground types, and are costlier to install.

The American Rainwater Catchment Systems Association (ARCSA) maintains a website (http://www.arcsa.org/) with many more resources. They provide educational seminars and professional certification programs.

Aquascape is a water gardening products and parts distributor who has moved into water collection systems, in part due to the prolonged drought experienced in the southeastern United States in the early- to mid-2000s. The AquaBasin™ water reservoir from Aquascape sits underground, beneath a decorative fountain. Water collects in the reservoir and is pumped directly through the center of the fountain via connective plumbing hidden within the system. By placing decorative rocks around the fountain, the top of the basin is effectively camouflaged. A variety of basin sizes are available. On a larger scale, collapsible water storage blocks are set under permeable pavers in decorative patios with a water pump installed to access the stored water. Rainfall conveniently passes through the pavers, and any number of storage blocks can be installed to accommodate local needs.

CALCULATING RAINFALL AMOUNTS

Calculations can be done to determine the amount of storage required for a given amount of rainfall over a given area. One inch (2.54 cm) of rainfall on a $1\,\text{ft}^2$ ($0.09\,\text{m}^2$) area is equal to 0.62 gallons (2.34 L) of water. Thus, a catchment area of $1,000\,\text{ft}^2$ ($93\,\text{m}^2$) would provide 623 gallons (2,358 L) of water from a 1 in. (2.54 cm) rainfall.

For sloped roofs, the square foot (square meter) area is based on the footprint. Thus, measuring around the perimeter of a building should give an accurate value for the square foot (square meter) catchment area available for rainfall harvesting.

IRRIGATION AND WATER-USE EFFICIENCY

The use of irrigation in landscaping varies by locale, but the water used for landscaping comes under increasing scrutiny during times of water shortages. This is true regardless of location. Malfunctioning sprinkler heads and other sources of wasted water in irrigation systems are coming under tighter regulation in many areas (Figure 6.3). Irrigation systems should be regularly and routinely checked for malfunctions that lead either to unnecessary water loss, failure to supply adequate water to target plants, changes in plant placement, or other potential problems. Some municipalities are implementing stricter regulations with regards to irrigation efficiency. It is only prudent for the landscape irrigator to be mindful of the issues and respond accordingly.

Some actions that can be taken toward more efficient irrigation are:

- Use rain sensors to prevent irrigation during rainfall
- Repair leaks
- Clean emitters
- Check fittings and components
- Create irrigation zones and place plants with similar requirements together in a zone
- Use soil moisture sensors to schedule irrigation

FIGURE 6.3 Automatic sprinklers may malfuntion, causing wasted water.

The Irrigation Association maintains a website with a number of useful tools, technical papers, and other resources for professional irrigators (see Further Reading section at the end of this chapter). One particularly useful tool is the ET calculator, which would permit the user to calculate water use based on local climate conditions as well as specific plant requirements. They formulated a calculation with the help of the American Society of Civil Engineers to provide a standardized approach to calculating reference ET. The IA-endorsed standardized equation:

- Is applicable to both agricultural and landscape irrigation.
- Facilitates the use and comparison of crop and landscape coefficients for a wide range of climates and locations.
- Provides guidelines for calculating ET in regions with limited climatic data.

The calculation is complex, and is designed only for a large area covered by a single crop of plants. It includes factors that account for:

- Standardized reference crop for short or tall surfaces
- Net radiation at the crop surface
- Soil heat flux density at the soil surface
- Mean daily or hourly air temperature
- Mean daily or hourly wind speed at a given height
- Vapor pressure

DROUGHT-TOLERANT PLANTS

Drought-tolerant plants use water more efficiently or require less water than those that are not drought tolerant. Some plants, such as the palo verde tree (*Cercidium* spp.), California buckeye (*Aesculus californica*), and ocotillo (*Fouquieria splendens*) evade drought by dropping their leaves or otherwise becoming dormant during dry periods. Other plants have adapted to **xeric**, or arid, environments by having a specialized process of photosynthesis known as Crassulacean Acid Metabolism (CAM) or by having structural changes. Small leaves, leathery leaves, leaves covered with trichomes, and gray-colored or whitish leaves provide various advantages. In some cases, the smaller surface area results in reduced moisture loss. In other cases, the whitish color of the leaf results in higher albedo, and thus, reduced leaf temperature. Thick, leathery leaves lead to reduced water loss, as do more numerous trichomes.

A good landscaping practice would be to use plants whose water requirements could be met by local rainfall amounts and avoid those which require irrigation. Many plants will need supplemental irrigation during establishment, but this should be minimized by planting during the rainy season (fall through spring) providing only one season of supplemental watering.

WATER-WISE GARDENING

Water-wise gardening incorporates several principles of design to minimize water use. Components of this practice include using plants whose water needs match the typical rainfall of the geographic location, minimizing the use of irrigation, and maximizing irrigation efficiency (Table 6.1).

A method of water-wise gardening known as **xeriscaping** was developed by the Denver, Colorado Water Department. Denver is located on the dry eastern side of the Rocky Mountains, in a region known as a rain shadow, meaning that most of the precipitation along the Rocky Mountain range is lost as air rises on the western side of the mountains and cold air condenses. Most of the population of Colorado lives on the dry side of the Rocky Mountain range. Thus, water conservation is an important component of sustainable landscaping there.

TABLE 6.1
A Select List of Landscape Plants Suitable for Xeriscaping

Common Name	Botanical Name	Hydrozone	USDA Hardiness Zone
	Trees		
Sweet acacia	*Acacia farnesiana*	M	9–10
Florida maple	*Acer barbatum*	M	7–9
Deodar cedar	*Cedrus deodara*	L/M	7–9

(*Continued*)

TABLE 6.1 (*Continued*)
A Select List of Landscape Plants Suitable for Xeriscaping

Common Name	Botanical Name	Hydrozone	USDA Hardiness Zone
Cornelian cherry	*Cornus mas*	M/H	4–8
Loquat	*Eriobotrya japonica*	M	8–10
Arizona ash	*Fraxinus velutina*	L/M	5
Golden-rain tree	*Koelreuteria paniculata*	L/M	6–9
Poplar	*Populus deltoides*	L	2–9
Oak, live	*Quercus virginiana*	M	7–10
Locust, black	*Robinia pseudoacacia*	M	4–8
Chinese elm	*Ulmus parvifolia*	L/M	5–9
Shrubs			
Manzanita	*Arcstaphylos manzanita*	L	8–10
Artemesia, wormwood	*Artemesia*	L/M/H	Varies
Bird-of-paradise bush	*Caesalpinia gilliesii* (*Poinciana gilliesii*)	M	9–11
Redbud, western	*Cercis occidentalis*	M/H	7–9
Desert almond	*Prunus fasciculata*	L/M	6–9
Bluebeard	*Caryopteris incana*	M	7–9
Bougainvillea	*Bougainvillea* species	L/M	8–10
Butterfly bush	*Buddleia davidii*	L	5–9
Grapeholly, Chinese	*Mahonia fortune*	M	8–9
Rosemary	*Rosmarinus officinalis*	L/M	8–11
Juniper	*Juniperus* species	L/M	3–8
Turf Grasses			
Buffalograss	*Buchloe dactyloides*	L	Warm season
Blue grama	*Boutelova gracilis*	L	Warm season
Perennials			
Achillea, Yarrow	*Achillea fillipendula, millefolium*		3–9
Agave	*Agave* spp.	L	Min. 50°F
Aloe	*Aloe arborescens*	L	Min. 50°F
Jupiter's beard, red valerian	*Centranthus ruber*	L	5–8
Siberian Iris	*Iris reticulata*	L	4–9
Red-hot poker	*Kniphofia*	M	5, 6–9
Kansas gayfeather	*Liatris pycnostachya*	L/M	4–9
Russian sage	*Perovskia atriplicifolia*	L/M	6–9
Portulaca	*Portulaca grandiflora*	L	Annual
Sedum	*Sedum* species	L	Varies
Mexican sunflower	*Tithonia rotundifolia*	L	Annual
Yucca	*Yucca* species	L	Varies

Case Study: Stapleton's Central Park in Stapleton, Colorado; Aurora Municipal Xeriscape Garden, Kuiper High Plains Garden.

References: Designscapes Colorado. http://www.designscapescolorado.com/projects/stapleton-xeriscape-garden. Viewed Dec. 27, 2011; Aurora Municipal Xeriscape Garden. https://www.auroragov.org/AuroraGov/Departments/AuroraWater/WaterConservation/OutdoorWater. Viewed December 27, 2011.

The Stapleton Xeriscape garden is one of the demonstration gardens provided by Denver Water. Pictures of the garden can be viewed on the website of Designscapes Colorado (see reference in Further Reading section at the end of the chapter). The Xeriscape Garden at Stapleton uses a precisely controlled irrigation regime in combination with native and low-water-use plants.

The Aurora Municipal Xeriscape Garden comprises six acres of rolling hills. Signs have been installed to guide the visitor through the seven principles of xeriscaping. The small amount of irrigation that is done uses reclaimed water (gray water) from the Sand Creek Wastewater Reclamation Plant. A variety of low-water-use plants have been installed and are individually labeled.

The Kuiper High Plains Garden has subsurface irrigation trials on turf plots and a no-water garden. Various trials are ongoing and change from time to time. This garden can be viewed at its location next to the Griswold Water Treatment Plant in Aurora, Colorado.

Xeriscaping involves the use of planting zones so that low-, medium-, and high-water requiring plants are grown together in the landscape. This allows their water needs to be met in a more efficient manner than if they were spread throughout the landscaped area. Irrigation zones can then be designed accordingly. There are seven principles that make up the practice of xeriscaping. They are:

1. Plan and design for water conservation, beauty, and utility
2. Improve the soil
3. Limit turf to practical sizes or select alternatives to turf
4. Water efficiently with proper irrigation methods
5. Select appropriate plants and group according to similar water needs
6. Mulch to reduce evaporation
7. Proper maintenance

As can be seen from the list above, the principles of xeriscaping are the same principles that should be followed in any sustainable landscape design.

Water-wise gardening is part of a broader landscape practice known as "right plant, right place". Wetter and drier areas of a landscape would be noted in a landscape audit (see Chapter 16) so that appropriate plants can be placed in them. In a general sense, plants that are not adapted to the rainfall of a region should be avoided. In the same manner, plants having specific water needs should only be grown where those needs can be met naturally. Dry, sandy soil cannot retain enough moisture for plants that are adapted to wet conditions, nor can drought-loving plants tolerate wet conditions. A thorough knowledge of plant water-use requirements is necessary for proper plant selection and placement. Irrigation needs vary by species, plant size and architecture, and environmental factors such as humidity (Montague et al. 2004). Nevertheless, trees could be grouped into general water-use categories.

Case Study: The Water Conservation Garden at Cuyamaca College in El Cajon, California

The Water Conservation Garden (WCG) began as an idea during the multi-year drought in southern California in the early 1990s. Cuyamaca College provided the necessary land supplemented with expertise from its horticulture department. They partnered with two local water districts, Helix and Otay, to form a tax-exempt authority. They were later joined by the San Diego Water District, the Sweetwater Authority, and the city of San Diego.

Their stated mission is to "Educate and inspire through excellent exhibits and programs that promote water conservation and the sustainable use of related natural resources". To that end, they have demonstration gardens that include a meadow garden, a cactus and succulent garden, a container garden, a native plant garden, and others. There is also a compost exhibit and an exhibit comparing seven different turf species and their water requirements.

An important component of the WCG is education. For example, they provide a WaterSmart landscape guide (available through their website). The guide provides a six-step guide for creating a water-smart landscape. Another example is the link from their website to a YouTube video that features Ms. Smarty Plants, who teaches school children about water conservation practices in gardening.

Reference: The Water Conservation Garden. http://www.thewatergarden.org. Viewed 4/6/20.

MULCH

Mulch is a covering that helps retain soil moisture in planted areas. It may be organic or inorganic. Some examples of commonly used mulch include wood ships, straw, rubber chips, and various sizes of rock. Landscape fabric is sometimes used under mulch to improve weed-control effectiveness. Plastic may be used as mulch, but is usually restricted to use in vegetable gardens. It is impermeable, and thus plant roots do not have access to water unless drip irrigation is installed beneath it.

How to Apply Mulch

Two to four inches of mulch are recommended for use in borders and beds. With less than that, good weed control and moisture retention are not achieved. With more than that, tree and plant roots cannot get adequate oxygen to their roots. Eventually the roots will grow upwards in thicker layers of mulch, which causes them to be more vulnerable during dry periods. Mulch should not be applied in such a way as to allow it to touch the trunk of a tree. It should be spread to the radius of a circle defined by the outer branches, or the **dripline**, of the tree.

EFFECTIVENESS OF DIFFERENT MULCHES

A lot of research has been done on mulches. There is an increasing interest in using them for a variety of reasons. One is the fact that there is a lot of urban wood waste that can be recycled into landscape mulch. Another reason is the increasing availability of recycled rubber used as mulch. The variety of mulching materials available raises the question of how effective the different materials are in achieving the usual goals of water retention and weed suppression. An additional concern is carbon sequestration and the effect organic versus inorganic mulches have on tree photosynthetic rates. And yet another concern is undesirable soil heating due to mulch layers.

Chalker-Scott (2007) conducted a survey of mulch research and reported that permeable materials that allow water to infiltrate are preferred. Plastic mulches or other impermeable materials should be avoided. Organic (plant-based) mulches were found to conserve water better than inorganic (stone, gravel) ones, but both types were better than synthetic (fabric, plastic) mulches. All mulches were better than bare soil at conserving soil moisture. Not surprisingly, studies have shown that cover crops, or living mulches, compete with landscape plants for water. Coarse organic mulches hold water like a sponge and could retain water before plant roots are able to obtain it, releasing it through evaporation eventually.

Mulches have been shown to moderate soil temperature, a factor which is seen as a benefit in hot summer months where mulch keeps the soil cooler than bare soil. However, mulch color may affect the air temperature above it, resulting in an increase of water loss from landscape plants through transpiration. Zajicek and Heilman (1990) demonstrated that crepe myrtle trees surrounded by pine bark mulch lost more water through transpiration than trees grown on bare soil or those surrounded by turfgrass.

Stinson and colleagues (1990) reported that weed control by fifteen types of mulch was highly effective when compared to bare soil. Light reduction by the mulches is an important factor providing weed suppression. This is an important consideration when trying new materials for use as landscape mulch.

Some wood mulches may also leach allelopathic chemicals that prevent weed growth. In a study of "pure mulches" containing material from a single tree species, Percival and colleagues (2009) reported that some pure mulches provided benefits not seen with other mulch types. Transplant survival rates of beech trees were significantly improved, and post-transplant growth of hawthorn, apple, and pear trees were all greatly improved. Mulches were derived from European beech (*Fagus sylvatica*), common hawthorn (*Crataegus monogyna*), silver birch (*Betula pendula*), common cherry (*Prunus avium*), evergreen oak (*Quercus ilex*), and English oak (*Q. robur*).

Landscape fabrics should not be used as permanent mulch around woody plants. Their fine, feeder roots tend to grow upwards into the fabric. If the fabric is ever removed, these roots are extensively damaged.

Water Conservation

GRAY-WATER USE

Gray water is non-toilet wastewater that has already been used for washing activities. This water comes from sinks, tubs, showers, washing machines, and dishwashers. It may be contaminated with microorganisms that may be harmful to humans, but may be safely used to irrigate landscape plants. The preferred gray-water sources are from bathroom sinks, showers, and tubs. Water from kitchen sinks and dishwashers often has grease in it. Wastewater from diaper laundry must be avoided due to the possibility of contamination by fecal matter. This is particularly true when using the gray water on a vegetable or fruit garden.

Case Study: Gray-Water Use in Santa Barbara, Danville, and Castro Valley, California.

Reference: Whitney, Alison, Richard Bennett, Carlos Arturo Carvajal, and Marsha Prillwitz. 1/17/99. Monitoring Graywater Use: Three Case Studies in California. http://www.water.ca.gov/wateruseefficinecy/docs/. Viewed December 28, 2011.

The California Department of Water Resources (DWR) conducted a gray-water-quality study in conjunction with the City of Santa Barbara and East Bay Municipal Utilities District. Three cities participated in the study: Santa Barbara, Danville, and Castro Valley. Soil samples were tested in the areas where the water had been applied from a subsurface irrigation line. They were tested for chemical properties that could affect plant growth either directly or indirectly: boron, pH, sodium, chloride, calcium, magnesium, electrical conductivity, and calculated sodium.

Results indicated that the pH at three sites except one sample were 7.1 or higher. This supports the recommendation that acid-loving plants be avoided in areas where gray water is used. Additionally, it was determined that there was a slight to moderate danger of emitters clogging due to a build-up of precipitated lime in the irrigation line. All of the soil samples tested in the soft or moderately hard range for calcium carbonate. This result indicated that the water would probably not form a lather or foam up when used.

Salinity, as tested by the electrical conductivity test, was not a problem in any of the samples. Sodium levels were problematic in samples from Castro Valley, but not the other two cities. However, the sodium adsorption ratio readings were well within safe parameters for all samples. All other chemical properties were well within safe levels for healthy plant growth. In the case of the slightly high sodium levels at the Castro Valley site, leaching with potable water or sufficient rainfall would take care of this problem.

Water that contains fecal matter and urine is **black water** and must be avoided for landscape or other purposes without adequate treatment. In Chapter 5, a system known as constructed wetlands is discussed, whereby black water can be treated in a far less expensive way than sewage treatment plants provide. Constructed wetlands use specific plants in a well-defined design, and may be incorporated into the landscape if designed and installed properly.

The use of gray water for outdoor needs not only increases water efficiency, but also reduces energy usage. Gray-water systems are available for purchase and come in a variety of sizes. The smallest are relatively simple systems for treating less than 400 gallons of water per day, and may be installed by homeowners or professionals.

Municipalities and states regulate the use of gray water. They are covered in detail at this website: http://www.oasisdesign.net/greywater/law/index.htm. Whereas numerous states have similar laws regulating gray-water systems (fifteen are included on their site), the website author recommends following the approach Arizona has taken. In Arizona, a three-tier system has been implemented. Each tier addresses different sizes of gray-water systems.

Gray-water systems must meet a list of reasonable requirements and are covered under a general permit without the builder having to apply for any special permit. The second tier is for systems that generate more than 400 gallons per day, or for commercial, multi-family, and institutional systems. They require a standard permit. Third tier systems treat over 3,000 gallons a day. Regulators consider each of them on an individual basis.

The Arizona law avoids proscribing design specifics, but some of the requirements are

- Avoid human contact with gray water and soil irrigated by it.
- Gray water from the residence is used only on the property for gardening, lawns, composting, or landscape irrigation.
- Gray water is not used on edible plants.
- The gray water was not used to clean car parts or otherwise contaminated with hazardous chemicals.
- Storage tanks for gray water have covers or lids to restrict access and prevent mosquito breeding.
- The system is sited outside of a floodway.
- Piping is clearly labeled as non-potable water.
- Surface irrigation is only allowable by flood or drip irrigation; flood irrigation should be contained in basins or swales.
- Kitchen sink water should be applied subsoil or contained with a rat-proof outlet shield.

The guidelines listed above are self-explanatory and seek to avoid human and animal health problems, or environmental issues. Other limitations may be imposed by municipalities or counties.

In San Luis Obispo County reusing gray water for landscape irrigation or water features has been legal since the early 1990s, during a drought there. The water had to meet state standards and be inspected. At the time a permit and fee were required, but currently a permit is no longer required. Nevertheless, all gray-water systems still have to meet the latest plumbing code.

TIMING OF GRAY-WATER USAGE

Gray water may be used at times of drought or year-round. Under conditions of normal rainfall, when landscape plants are receiving adequate water, supplemental irrigation should not be required. Many landscape plants require an inch of water per week, on average. More is needed for establishing new plants or for water-loving plants. Less is needed for drought-tolerant plants and established plants having extensive root systems. When supplemental water is needed, a rule of thumb is to apply a half gallon of water for every square foot of landscaped area. Thus, for a 1,000 ft^2 area, 500 gallons of water may be used weekly.

PROBLEMS TO AVOID

Soaps, bleach, and other additives should be carefully selected when gray water will be used for irrigation purposes. Soaps that are high in sodium, a water "softener", should be avoided or minimized. There are many brands of biodegradable and environmentally friendly cleaning products available. The homeowner should be aware of the products they are using in order to avoid or minimize problems with sodium build-up in the soil. A pH of 7.5 or higher is an indication of high sodium levels. This can be counteracted by the application of calcium sulfate (gypsum) at a rate of 2 pounds per 100 ft^2 monthly. Leaching by rainfall or alternating gray water with fresh water can also help alleviate the problem. Ammonia-based cleaners are acceptable in gray water, as they contain a nitrogen source that is beneficial to plant growth.

GRAY-WATER DELIVERY

Gray water can be hand-carried to the desired areas, or plumbing may be installed to deliver it directly from the tub, shower, washing machine, and so on. This work should only be performed by a plumber, and must be inspected to ensure it doesn't break local ordinances or sanitary codes.

PLANT SAFETY CONCERNS

Effluent irrigation uses gray water from sewage treatment plants. This is common practice in the southwestern United States on golf courses. Effluent irrigation changes soil properties, requiring different management practices in order to maintain good-quality turf. Some of the problems are high salt content and the presence of plant nutrients in varying amounts. Whereas salts in the water can be detrimental to plant growth, plant nutrients are beneficial, as long as they are not present in excess amounts. Researchers at the University of Arizona found that municipal effluent used on turfgrass, bermudagrass, and perennial ryegrass produced a good-quality turf and did not require as high levels of fertilizer as compared to potable well water. Some of the advantages of using effluent stated by the authors are:

- The climate in Arizona is favorable for growing turfgrass year-round, permitting year-round use of waste water.
- Nitrates and phosphates in effluent have low potential for pollution of groundwater even when applied in large quantities.
- As groundwater declined in some areas, the use of effluent would reduce dependence on potable water for irrigation purposes and could provide some groundwater recharge.

- Wastewater is available at a fairly constant rate and quality throughout the year in many metropolitan areas.
- Turfgrass requires nutrients at fairly high levels, thus nutrients in wastewater can reduce reliance on commercial fertilizer.

The researchers did find that bermudagrass seed did not germinate as well on unfertilized plots receiving the effluent as unfertilized plots received potable water. However, the plots receiving effluent showed improved seedling establishment the following month. When fertilizer was added, the seedlings in plots receiving potable water surpassed those in the effluent irrigated plots. This result was seen again in plots of turfgrass receiving the two types of water and varying levels of nitrogen. Overall, the effluent irrigated plots produced good-quality turf, although supplemental nitrogen was required to maintain quality in the winter for the perennial ryegrass.

Researchers in Canada compared the nutrient source derived from municipal solid waste on growth of two nursery and two turfgrass species with other, more traditional nutrient sources, and discovered that by correcting for imbalances in essential plant nutrients, the municipal solids-derived nutrient source shows good potential. An added benefit of using nutrients from wastewater production is that such nutrients are otherwise lost resources.

Case Study: Santa Fe and Albuquerque, New Mexico

Reference: Pushard, Doug. The tale of two cities – billions conserved. Harvesth2o.com, December 2011. http://www.harvesth2o.com/tale_of_two_cities.shtml. Viewed December 28, 2011. (Based on an article published in the *Green Fire Times*)

For more than a decade the New Mexico cities of Albuquerque and Santa Fe have had water conservation programs in place. Albuquerque is about 60 miles southwest of Santa Fe, with an average annual temperature of 56.2°F. The average annual temperature in Santa Fe is 6° cooler. Rainfall in Albuquerque is just under 10 in., on average, 5 in. less than Santa Fe's rainfall. Water conservation programs were initiated in 1994 (Albuquerque) and 1995 (Santa Fe). Albuquerque uses more water than Santa Fe, in part due to its warmer temperatures and lower rainfall, but also because it has a longer growing season (by 2 months). Albuquerque also had more commercial and industrial users during the time of this study.

Both cities implemented an array of practices covering both indoor and outdoor uses. Albuquerque had more rebates and incentives for outdoor water reduction than Santa Fe. These included partial rebates for multi-setting sprinkler controllers, a discount on rentals of sod-removing equipment, money toward the purchase of a rain sensor for irrigation systems, rebates on rainwater storage tanks, rebates for costs incurred in implementing a xeriscape, and a discount on the cost of an irrigation efficiency class. Albuquerque mandates use of non-potable (gray) water for irrigation. Non-potable water is piped in through dedicated city lines and the rate is 20% less than that for potable water.

Santa Fe gave rebates for water-conserving toilets and high-efficiency washing machine for laundry. They also provided rebates on rain barrels and cisterns. In both cases, water rates were adjusted upwards to serve as a disincentive.

Between 1995 and 2009 Albuquerque has reduced water consumption by 92 gallons per day per person. In 2009 they were using 159 gallons per day per person. Santa Fe has saved 70 gallons per day per person, for a total of 98 gallons per day per person in 2009.

Another facet of gray-water usage that is being investigated and tried include "sewer Mining" in Sydney, Australia. This practice involves tapping into wastewater lines, drawing off the gray water for use on the landscape, golf course, or other acceptable use, and allowing the rest of the waste to proceed to the treatment plant as before.

SUMMARY

Population growth and increased urbanization in many areas are putting increasing demands and strains on local water supplies. Clean, potable water is safe for consumption and is used to water plants inside the home and in the landscape, as well as for washing and bathing. To make the water safe, it often must go through a somewhat costly treatment system to remove microorganisms, chemicals, and salts, or other undesirable components.

Municipal water supplies come from surface water or groundwater. When water is in short supply, for whatever reason, landscape water use is one of the first targets for reduced water consumption. Water that is used on landscapes currently accounts for one- to two-thirds of residential water use.

The Sustainable Sites Initiative addresses water shortages in several prerequisites and credits. In the eastern and southeastern states, ample rainfall does not guarantee an adequate potable water supply. This is in part due to increasing populations putting demand on water treatment systems at greater rates than cities can afford to keep up with. In the western United States, climate changes are contributing to reduced snow melt in the mountains, more precipitation in the form of rainfall rather than snow, and drier summer conditions. All of these are expected to lead to water shortage problems in the near future. In addition, the western United States is an arid region that supports a large population, which continues to grow in spite of a history of water-supply problems in some areas.

Our water comes from the groundwater, aquifers, surface water, dams, and reservoirs. Before it is available for public consumption, water is often filtered and treated to remove biological contaminants or other pollutants.

The water cycle is based on the understanding that the amount of water on the planet is a constant. The water changes from gaseous to liquid to solid as it cycles from one location to another. Ninety-seven percent of the world's water is in the oceans. Of the remaining 3 percent, about a third of it is groundwater. Surface water is only 0.3 percent of the fresh water.

Plants are approximately 80–90 percent water. They take up water through their roots and lose it through stomata in the process of transpiration. Evapotranspiration

(ET) is the combined loss of soil water from loss through evaporation and plant loss through leaves in the process known as transpiration.

Drought-tolerant plants have a number of mechanisms that allow them to use water more efficiently or require less water than those that are not drought tolerant. Dropping leaves, becoming dormant during dry periods, CAM metabolism, and structural changes such as small leaves, leathery leaves, leaves covered with trichomes, and light-colored leaves are several adaptations of drought-tolerant plants.

Irrigation usage in landscaping should be monitored for efficiency to prevent waste. Irrigation systems should be regularly and routinely checked for malfunctions that lead to unnecessary water loss, failure to supply adequate water to target plants, changes in plant placement, or other potential problems.

Water-wise gardening incorporates several principles of design to minimize water use. Components of this practice include using plants whose water needs match the typical rainfall of the geographic location, minimizing the use of irrigation, and maximizing irrigation efficiency. Xeriscaping, developed by the Denver, Colorado Water Department involves the use of planting zones so that low-, medium-, and high-water requiring plants are grown together in the landscape. The principles of xeriscaping concur with general rules for sustainable landscaping, including the use of drought-tolerant plants and efficient irrigation.

Mulch helps retain soil moisture in planted areas, regardless of whether it is organic or inorganic. Proper mulch depth is 2–4 in. to provide good weed control and moisture retention, while allowing tree and plant roots to get adequate oxygen. When used around a tree, it is spread outwards to the dripline.

Gray water from sinks, tubs, showers, washing machines, and dishwashers can be used for secondary purposes as long as certain requirements are met. The preferred gray-water sources are from bathroom sinks, showers, and tubs, and gray water does not contain grease, fecal matter, or urine. The use of gray water not only increases water efficiency, but also reduces energy usage.

Municipalities and states regulate the use of gray water. Arizona has a three-tier system which provides a template for other states. The guidelines provided by the state seek to avoid human and animal health problems or environmental issues. Soaps that are high in sodium should be avoided or minimized. Homeowners need to be aware of the products they are using in order to avoid or minimize problems with sodium build-up in the soil.

Nursery and turfgrass plants can be grown safely and successfully using gray water, sewage effluent, and derived municipal solid waste, as long as nutrient and salt levels are monitored for safe and adequate levels. Reduction of fertilizer may provide an additional benefit.

REVIEW QUESTIONS

1. Describe the process by which plants lose water to the environment.
2. Why do areas with adequate rainfall sometimes have a shortage of clean drinking water?
3. Name three steps that can be taken to ensure irrigation efficiency.
4. Discuss the concept of xeriscaping and name the principles involved.

5. What are some adaptations plants have made to aid in reducing water loss?
6. How does mulch play a role in water efficiency in the landscape?
7. What is the difference between gray water and black water?
8. Under what conditions can gray water be used to irrigate landscape plants?
9. Discuss the importance of soap selection when gray water is going to be used.
10. Discuss concerns and safety factors of gray irrigation water on landscape plants.

ENRICHMENT ACTIVITIES

1. Using the US Drought monitor map, name the states or regions that experienced a drought this year (it can be done for individual months).
2. Over a period of time, record the precipitation for your area. Calculate monthly totals and compare this to norms for your location using local sources or a website.
3. Calculate the amount of rainfall that can be collected from a local residence per inch of rainfall, and compare this to the amount that could be collected from a local school building. Using rainfall figures for your location and water usage figures from each building, determine what percentage of the water needed could be harvested.
4. Design a xeriscape for a local residence or school building. Use only plants that do not require additional watering once they are established.
5. Conduct a literature review for water use of plants in your region. Compile a list of low-, medium-, and high-water consumers. If possible, relate each category of use with irrigation requirements to meet plant demands.

REFERENCES

Abdollahi, K. K., Z. H. Ning, T. Legiandenyi, and A. Negatu. 2009. Impact of different bio-based mulches on the urban soil CO2 flux in southeastern Louisiana. In: *The Landscape below Ground III: Tree Root Development in Urban Soils*. G. Watson, L. Costello, B. Scharenbroch, and F. Gilman, Eds. International Society of Arboriculture. Champaign, IL.

ActewAGL education website. The urban water cycle. http://www.actewagl.com.au/Education/water/UrbanWaterCycle/default.aspx. Retrieved July 14, 2010.

Barnett, T. P., D. W. Pierce, H. G. Hidalgo, C. Bonfils, B. D. Santer, T. Das, G. Bala, A. W. Wood, T. Nozawa, A. A. Mirin, D. R. Cayan, M. D. Dettinger. 2008. Human-induced changes in the hydrology of the western United States. *Science*. 319: 1080–1083.

Calhoun, S. 2009. Designing water-thrifty gardens. *Am. Gardener*. 88: 20–24.

Campisano, A., Butler, D., Ward, S. Burns, M.J., Friedler, E., DeBusk, K., Fisher-Jeffes, L.N., Ghisi, E., Rahman, A., Furumai, H., and Han, M. 2017. Urban rainwater harvesting systems: Research, implementation and future perspectives. *Water Res.* 115: 195–209.

Carver, S. 2008. Water-wise landscaping can improve conservation efforts. *Landsc Manag*. May/June Supplement Livescapes. 4: 8.

Cathcart, T. 2002. *Regenerative Design Techniques*. John Wiley & Sons, New York, NY, 410 pp.

Chalker-Scott, L. 2007. Impact of mulches on landscape plants and the environment – a review. *J. Environ. Hort.* 25(4):239–249.

Colorado Water Wise. http://coloradowaterwise.org/ Viewed January 1, 2012.

East Bay Municipal District. 1995. Water wise gardening. http://www.ebmud.com/watercon/garden.html. Retrieved April 30, 1998.

Eberle, W. M. and J. G. Thomas. 1981. *Some Water-Saving Ways*. Kansas State University, Manhattan, KS, 4 pp.

Green Water Systems. http://greenwatersystems.us/graywatersystem.htm. Retrieved Aug. 3, 2009.

Harris, R. W. 1992. Water management. In *Arboriculture*. Prentice Hall, Englewood Cliffs, NJ, 674 pp.

Hayes, A. R., C. F. Mancino, and I. L. Pepper. 1990. Irrigation of turfgrass with secondary sewage effluent: I. Soil and leachate water quality. *Agron. J.* 82: 939–942.

Hayes, A. R., C. F. Mancino, W. Y. Forden, D. M. Kopec, and I. L. Pepper. 1990. Irrigation of turfgrass with secondary sewage effluent: II. Turf quality. *Agron. J.* 82: 943–946.

Ingels, J. 2009. *Landscaping Principles and Practices*, 7th ed. Cengage-Delmar, New York, NY. 573 pp.

Irrigation Association. http://www.irrigation.org/. Viewed June 12, 2019.

Jackson, R. 2012. Hello, again: The ancient practice of water harvesting is being brought back to life in the United States. *Lawn Landsc.* http://www.lawnandlandscape.com/ll1212-alternate-sustainable-water-sources.aspx viewed 12/29/2012.

Krizner, K. 2008. Smart water solutions. *Landsc Manag.* May/June Suppl Livescapes.

Martin, E., S. G. Buchberger, D. Chakraborty. 2015. Reliability of Harvested rainfall as an auxiliary source of non-potable water. *Proc. Eng.* 119: 1119–1128.

Matos, C., C. Santos, S. Pereira, I. Bentes, M. Imteaz. 2013. Rainwater storage tank sizing: case study of a commercial building. *Int. J. Sustain. Built Environ.* 2(2): 109–118.

Meche, M. S., H. J. Lang, and A. L. Kenimer. 1998. Bioremediation of greenhouse effluent using constructed wetlands. *Hortic. Sci.* 33: 492.

Michitsch, R. C, C. Chong, B. E. Holbein, R. P. Voroney, and H.-W. Liu. 2007. Use of wastewater and compost extracts as nutrient sources for growing nursery and turfgrass species. *J. Environ. Q.* 36: 1031–1041.

Montague, T., R. Kjelgren, R. Allen, and D. Wester. 2004. Water loss estimates for five recently transplanted landscape tree species in a semi-arid climate. *J. Environ. Hort.* 22(4): 189–196.

National Water Harvesters Network. Rainwater harvesting. Rainwaterharvesting.org. Retrieved June 13, 2019.

Percival, G. C., E. Gklavakis, and K. Noviss. 2009. Influence of pure mulches on survival, growth and vitality of containerized and field planted trees. *J. Environ. Hort.* 27(4): 200–206.

Robinette, G. O. and K. W. Sloan. 1984. *Water Conservation in Landscape Design and Management*. Van Nostrand Reinhold Co., New York, NY, 258 pp.

Shaw, D. A., D. R. Pittenger, M. McMaster. 2005. Water retention and evaporative properties of landscape mulch. *Proceedings of 26th Annual Irrigation Show*, Phoenix, AZ, Nov. Irrigation Assoc., Falls Church, VA. http://groups.ucanr.org/CLUH/files/67347.pdf. Retrieved January 6, 2011.

Smith, A. K. 2008. The case for efficient irrigation. *The American Nurseryman.* June 15, pp. 64–5.

Southeastern Colorado Water Conservation District. http://www.secwcdxeriscape.org/ Retrieved January 3, 2011.

Stinson, J. M., G. H. Brinen, D. B. McConnell, and R. J. Black. 1990. Evaluation of landscape mulches. *Proc. Florida State Hortic. Soc.* 103: 372–377.

University of Massachusetts. Recycling gray water for home gardens, in plant culture and maintenance. http://www.umassgreeninfo.org/fact_sheets/plant_culture/gray_water_for_gardens.html Retrieved 7/18/08.

US Drought Monitor. http://droughtmonitor.unl.edu/. Viewed July 29, 2011.

USGS. Water science in schools. http://ga.water.usgs.gov/edu/waterdistribution.html. Retrieved January 4, 2011.

Walter, I. A., R. G. Allen, R. Elliott, D. Itenfisu, P. Brown, M. E. Jensen, B. Mecham, T. A. Howell, R. Snyder, S. Eching, T. Spofford, M. Hattendorf, D. Martin, R. H. Cuenca, and J. L. Wright. 2002. The ASCE standardized reference evapotranspiration equation. DRAFT. http://www.kimberly.uidaho.edu/water/asceewri/main.pdf. Retrieved January 9, 2011.

White, J.D. 2008. When the well runs dry: managing water before it becomes a crisis. *GrowerTalks*. August, pp. 42–43.

Yu, Z. L. T., J. R. Deshazo, M. K. Stenstrom, Y. Cohen. 2015. Cost–benefit analysis of onsite residential graywater recycling: a case study on the city of Los Angeles. *J. Am. Water Works Assoc.* 107(9), E436-E444 doi: 10.5942/jawwa.2015.107.0124.

Zajicek, J. M., and J. L. Heilman. 1991. Transpiration by crape myrtle cultivars surrounded by mulch, soil, and turfgrass species. *Hortic. Sci.* 26(9): 1207–1210.

7 Managing Excess Water in the Landscape

OBJECTIVES

Upon completion of this chapter, the reader should be able to

- Explain how excess water poses a problem in landscapes
- Discuss policies concerning stormwater runoff
- Identify landscape practices for managing excess water
- Evaluate a variety of methods of installing vegetated roofs
- Explain the similarities and differences between swales, French drains, and rain gardens
- Describe the urban water cycle
- Discuss medium-scale rainwater collection systems
- State the benefits of permeable paving
- Identify various permeable paving options

TERMS TO KNOW

Bioretention
Combined sewer overflow (CSO)
Extensive green roof
French drain
Green roof
Green wall
Illicit discharge
Intensive green roof
Municipal separate storm sewer system (MS4)
Permeable pavement
Rain garden
Swale

INTRODUCTION

Too much water in the landscape is a problem during or after rainfall. When too much water is retained in the landscape, it may be attributable to a poorly designed drainage system or poorly draining soil. A poorly designed drainage system is one that does not allow water to drain away from a house or structure, or that does not allow water to drain off of a property in an appropriate manner. Poorly draining soil may be compacted, has high clay content, or lacks adequate soil structure.

A number of factors contribute to poorly draining soil in the urban landscape, e.g., construction activities. An additional source of too much water in the landscape can be attributed to paved surfaces and roofs (Figure 7.1). Roofs and driveways account for about 70 percent of impervious surfaces in developed neighborhoods. Parking lots and other paved areas also contribute to the problem. In many places, paved retention ponds and drainage basins are installed to collect and remove water during storm events. This water often flows directly into neighboring streams, rivers, and lakes. According to the US Environmental Protection Agency (EPA), a typical city block generates nine times more runoff than a woodland area of the same size (Figure 7.2).

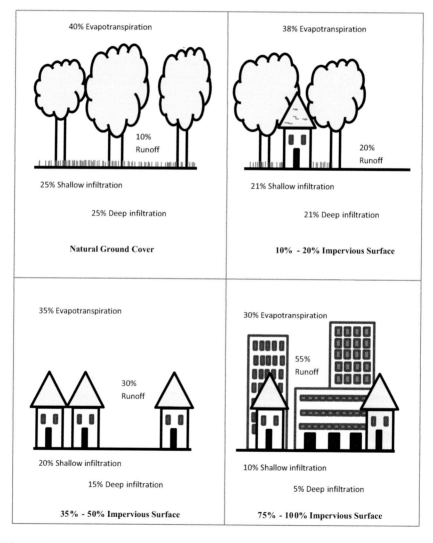

FIGURE 7.1 Infiltration in natural and paved areas. (Illustration by the author.)

Managing Excess Water in the Landscape

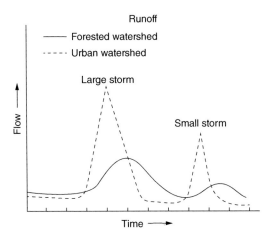

FIGURE 7.2 Runoff from forests and urban areas. (Illustration by the author.)

STORMWATER RUNOFF

When stormwater runs off a property, it usually ends up in a storm sewer or in surface water such as streams and rivers. Too much water running off of a landscape and into the storm sewer is a sustainability issue because some municipalities have a combined storm sewer and wastewater sewer system. Releases from these systems are called **combined sewer overflow** or **CSO**. During incidences of high rain, excess stormwater goes into the sewer system and causes an excess that the treatment tanks cannot hold. At such times, the wastewater and stormwater together are released back into the environment in order to reduce the amount of water flowing into the treatment tanks. For example, in New York City municipal sewage and stormwater use the same plumbing system. When the volume entering the pipes exceeds the capacity of the wastewater treatment plants, the excess is released into the East and Hudson Rivers. Too much water released into rivers, streams, and other surface waters leads to bank erosion, the deepening of a streambed over time, and excessive sediment deposits in streams.

Another reason stormwater runoff is a sustainability issue is that it often contains toxic materials from pesticides, herbicides, and driveway spills. It also contains toxic substances from pavements, such as polycyclic aromatic hydrocarbons or PAHs, which are toxic, mutagenic, teratogenic, and probably carcinogenic. Coal tar sealant used to seal asphalt pavements, including driveways, is the single largest source of PAHs in urban lakes in the United States. Pet wastes, improperly disposed-of household chemicals, automobile fluids, de-icing materials, and other substances also contribute to polluted water in the environment. Finally, when rainwater passes over hot pavements, it can be heated to a temperature that is unhealthy for aquatic life. In *The National Water Quality Inventory: 2000 Report to Congress*, urban runoff was listed as a leading contributor to impaired water quality in surface waters. It is also a contributing factor to contaminated groundwater. For updates on the nation's water quality, online mapping tools, research, and related information can be viewed at the United States Geological Survey (USGS) website.

Urban runoff is treated in two separate categories by the EPA: point sources and nonpoint sources. Point sources include municipal wastewater treatment, manufacturing plants, and other large entities. Nonpoint sources can be residences, small business, and other small entities that release contaminants, either purposely or inadvertently, as in the case of driveway spills, for example. Nonpoint sources are small contributors to the problem, yet in aggregate, they are responsible for a large amount of the pollution in surface and groundwaters.

The US EPA, which is responsible for enforcing provisions of the Clean Water Act, maintains extensive information about water quality on its website: http://www.epa.gov/. They provide guidelines and resource materials in an effort to reduce nonpoint-source pollutants from surface and groundwaters. Keeping contaminated water out of surface and groundwater is a national enforcement initiative of the EPA. The Green Infrastructure Research Program has been designed by the EPA to assist municipalities and other entities to meet water quality mandates.

Case Study: Stormwater Management – Several Examples Cited on InterNACHI and EPA Websites.

References: InterNACHI (International Association of Certified Home Inspectors), Constructed wetlands: The economic benefits of runoff controls. http://www.nachi.org/constructedwetlands.html. Viewed Dec. 27, 2011.

U.S. EPA. 1995. Economic Benefits of Runoff Controls. http://www.epa.gov/owow/NPS/runoff.html. Viewed Dec. 28, 2011.

According to the International Association of Certified Home Inspectors (InterNACHI), housing developments with bodies of water can increase property values by as much as 28 percent. In several studies cited on the topic, ponds or other features required to manage runoff from neighboring properties could be designed to be aesthetically pleasing and improve profitability within the development. In some cases cited in a similar EPA article, existing wetlands were enhanced prior to the development of the subdivision. In all cases, waterfront property was more valuable than property further away from the water.

Some examples of water features cited in these articles were enhancement of existing wetlands, wet ponds designed to manage runoff, and constructed wetlands. For example, in Highland Park, Illinois, in a community known as Hybernia, 122 single-family residences were constructed around a constructed pond/stream system and 27 acres approved as a state nature preserve. In Atlanta, Georgia, the Hyatt Regency Ravina hotel complex has a designed stream and waterfall system with linked wet ponds. Recirculating pumps keep the water flowing throughout the system, purpose of which is to manage runoff from the property. In Alexandria, Virginia, a condominium development, Virginia Chancery, on the Lake has a 14-acre urban runoff detention area. Instead of the usual detention pond one might see on commercial properties, there is a wet pond surrounded by a walking trail, a gazebo, and a fishing pier.

Many other examples are provided in the websites listed in the section "Further Reading" at the end of this chapter.

TABLE 7.1
Illicit Discharge of Water in Urban Settings

Generating Site

Apartments	Untreated effluent, failing septic systems
Homes	Untreated effluent, failing septic systems
Car washes	Vehicle wash water
Restaurants	Waste water
Airports	Motor oil, fuel spills, antifreeze
Landfills	Anything that can escape if non permeable liner fails or is absent
Gas stations	Motor oil, fuel spills, antifreeze

There are multiple ways for contaminated water to enter the surface and groundwater. Any discharge into a storm drain that is not entirely composed of stormwater is considered as an **illicit discharge**. Some sources of illicit discharges are shown in Table 7.1.

The EPA administers the National Pollutant Discharge Elimination System (NPDES) to address these problems. This permit program controls water pollution by regulating point sources that discharge pollutants into waters of the United States.

The Sustainable Sites Initiative provides credits for several issues related to stormwater runoff. They include protecting (and restoring) riparian and wetland buffers, managing water onsite, cleansing water onsite, and designing stormwater management features to be a landscape amenity.

THE URBAN WATER CYCLE

Typically, we think of the water cycle in terms of precipitation, infiltration, runoff, storage, and evaporation or transpiration. This is the natural cycling of water between earth and its atmosphere that has occurred for millions of years. However, there is an urban water cycle, which describes the route that water takes from the time it enters the urban system (falls by precipitation into collection ponds, lakes, rivers, or reservoirs) until it leaves through a variety of routes. In one route, the water is purified for use and supplied through a network of pipes. After use, it is collected in a sewer system, and treated and released back into the environment.

In another route, water from precipitation runs off of roofs and other impervious surfaces into drainage ditches, and flows directly into surface waters such as rivers or streams. This type of system is known as a **Municipal Separate Storm Sewer System** or **MS4**.

SOLUTIONS TO EXCESS WATER IN THE LANDSCAPE

Solutions for managing excess water in the landscape range from the simple to the complex. For example, some recent responses to stormwater management include the use of rain barrels, redirecting downspouts to garden beds, cisterns, dry wells, installing **French drains** and rain gardens, design of swales, planting a canopy of trees and shrubs, and constructing rainwater harvesting features. The French drain (named for Henry French) permits water infiltration through the use of a gravel-filled pit that collects excess rainwater. Vegetated, or green, roofs are yet another response. Blue roofs are non-vegetated and are explicitly designed to detain water. Although they are not a landscaping option, per se, they may be a part of the range of responses available for stormwater management. Other small non-landscape-related responses include re-using water for non-potable uses such as toilet flushing.

The EPA has many suggestions for managing excess water in the landscape and in the home. The overall effort is directed at all levels, from states to municipalities to communities. One aspect of the EPA approach is to aid in development of cost-effective, sustainable, environmentally friendly infrastructure. In their words, "Green Infrastructure management approaches and technologies infiltrate, evapotranspire, capture and reuse stormwater to maintain or restore natural hydrologies".

In the larger context, it would be best if natural areas could be maintained as such. Natural areas, such as woodland and forests, floodplains and other low lying areas, riparian areas, and wetlands, can all contribute positively to stormwater retention and infiltration. Municipalities and other development entities should strive to incorporate such natural entities into the urban environment. In addition to the ecosystem services they provide, they are also a magnet for human recreation and relaxation.

DRAINAGE

Drainage on an urban site may be either planned for or ignored by the developers of the site. Typically, gutters direct rainwater from the roofs of structures to storm sewers. In many situations, gutters simply allow rainwater to accumulate near the foundation of the building, which is not a good situation for the building. However, the primary alternative is to route the rainwater using perforated drainage tubing to a nearby storm sewer. On a larger scale, retention or detention ponds collect rainwater off of parking lots or other large-scale building sites.

LANDSCAPE SWALES

One easy remedy for improper or inadequate drainage is to design and install a **swale**. A swale is formed by creating a depression in the land which will carry the water away from the property. Swales can be planted with vegetation ranging from turf grass to prairie grasses and forbs (herbaceous flowering plants).

RAIN GARDENS

Rain gardens are planted areas that have become a widely used method to allow rainwater to permeate into the ground rather than run off into surface waters. Another term that refers to the usage of plants for water infiltration is **bioretention.** The EPA has suggested building rain gardens as part of their Green Infrastructure

Managing Excess Water in the Landscape

Development program. Many states have informational materials and programs in place to support the development of rain gardens in their communities.

The benefits of rain gardens include

- Increased water infiltration into the soil
- Reduced mosquito breeding
- Decreased flooding
- Reduced stream bank erosion
- Reduced stream bed incision
- Less surface water contamination
- Groundwater replenishment
- Enhanced wildlife habitat
- Increased property values
- Reduced carbon footprint

Siting a Rain Garden

Rain gardens should be sited no closer than 10 ft from a structure. They also should not be placed within 25 ft of a septic system. Do not place a rain garden in areas of compacted soil. Ideally, the rain garden will aid in draining water away from such spots. If water poured into a hole in a poorly draining spot in the yard does not infiltrate within 24 hours, it is not a good location for a rain garden.

Rain gardens should be in full to part sun if at all possible. It is also best to avoid placing rain gardens on a slope. If there is no alternative, then the rain garden must be deeper overall to account for the slope, and a berm must be placed at the low end of the garden in order to prevent water from running over and past the rain garden. The design must allow water to infiltrate in the area of the rain garden.

Calculating the Area Required for a Rain Garden

In order to calculate the area required for a rain garden, one must account for these factors: amount of rainfall to be accommodated, total catchment area, and depth of rain garden. The catchment area may be a driveway or parking lot, in which case, the area is easily obtained. If the catchment area is the roof of a house, then just the footprint of the roof is used. Thus, for a house that is 30 ft across the front and 30 ft from front to back, with a single-ridged roof, the catchment area for one side of the roof will be 15 × 30 ft = 450 ft. The roof on one side of the house plus the yard on that side are added together to calculate the total catchment area (see Figure 7.3).

The depth of a rain garden is dependent on the soil type. In general, sandy soil will drain faster than clay or silty soil. Thus, a rain garden in sandy soil does need to be as large. To see how fast the soil drains water, dig a hole 12 in. deep and 12 in. wide, and fill it with water. Mark the water line with a dowel or stick. Measure the time required for the water level to drop. Quick-draining soils will require shorter measurement intervals than slowly draining soils. Using these measurements, calculate the per-hour drainage rate. This is the required depth for the rain garden. The maximum depth for rain gardens is 12 in.

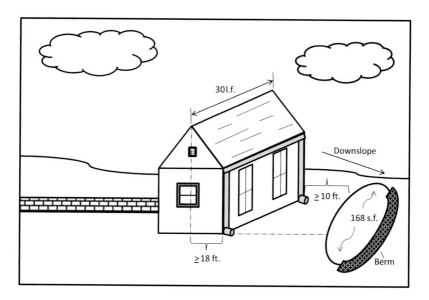

FIGURE 7.3 Calculating size of rain garden. (Illustration by the author.)

Many rain gardens are designed for one-inch per hour of rain. However, this may be adjusted up or down depending on local conditions. Using the numbers provided here, the following calculation shows the required size of the rain garden:

$$450 \text{ Catchment area} \times 1 \text{ in. rain} = 90 \text{ sq. ft. of garden}$$

5 in. drained per hour.

How to Build a Rain Garden

Once the location and the size for the rain garden are known, it is important to have any buried utilities marked before digging begins. Spring is the best time to install a rain garden, although fall plantings may work. If fall planting is used, follow the same guidelines for your area for fall planting of herbaceous perennials. The rain garden should be lower than the surrounding area. If it is located downslope from the surrounding area, that is ideal. However, if it is going to be sited in a level area, then it should be dug out to create a depression that sits 6 in. below the surrounding area.

The rain garden must be well-draining in order to allow infiltration to occur. Thus, amending the soil may be necessary if it has greater than 10 percent clay, which applies to all but the sandiest of soils. A good blend consists of 20 percent organic matter such as compost, partially composted leaves, grass, or other plant materials, 50 percent sandy soil, and 30 percent topsoil.

Plants for Rain Gardens

The ideal plants for rain gardens tolerate a range of moisture extremes, from very wet to very dry. Even though the name "rain garden" implies wetness, there will be periods of drought when the rain garden will be dry. Such periods may be lengthy and may occur during the hot season. In addition, the rain garden soil has been amended to be very well-draining, so it will not hold moisture well during these droughty periods.

Geographic region will limit the species that can be used successfully in a rain garden. Lists have been developed by many states, municipalities, and various agencies, and these are readily available on the Internet. Table 7.2 lists some examples of plants that work well in rain gardens.

RAINWATER COLLECTION

Rainwater collection can be achieved by a number of methods, including barrels, above- or below-ground storage tanks, and retention or detention ponds. The smallest, simplest method is installation of a rain barrel, usually in conjunction with a

TABLE 7.2
Plants for a Rain Garden

Woody Plants	Species
Alder	*Alnus incana* ssp. *Rugosa*
American elderberry	*Sambucus canadensis*
Arrowwood viburnum	*Viburnum dentatum*
Buttonbush	*Cephalanthus occidentalis*
Dwarf fothergilla	*Fothergilla gardenia*
Pawpaw	*Asimina triloba*
Pond apple	*Annona glabra*
Red maple	*Acer rubrum*
River birch	*Betula nigra*
Serviceberry	*Amelanchier laevis*
Summersweet	*Clethra alnifolia*
Sweet gum	*Liquidambar styraciflua*
Sweetbay	*Magnolia virginiana*
Wax myrtle	*Myrica cerifera*
Winterberry	*Ilex verticillata*
Herbaceous Plants	
Blue lobelia	*Lobelia syphilitica*
Cardinal flower	*Lobelia cardinalis*
Cinnamon fern	*Osmunda cinnamomea*
Drooping sedge	*Carex pendula*
Lady fern	*Athyrium filix-femina*
Obedient plant	*Physostegia virginiana*
Oswego tea, beebalm	*Monarda didyma*
Red Louisiana iris	*Iris fulva*
Swamp milkweed	*Asclepias incarnata*
Swamp sunflower	*Helianthus angustifolius*
Tussock sedge	*Carex stipata*
White turtlehead	*Chelone glabra*

downspout. This permits the collection of runoff from a roof. Sizes of rain barrels range from ten to over a hundred gallons. For irrigation purposes, the recommended rate is one-half gallon of water for each square foot to be watered.

More details on rainwater collection, storage, and irrigating with rainwater are provided in Chapter 6. Irrigating with rainwater should not pose any problems assuming that the water doesn't contain any contaminants.

PERMEABLE PAVEMENT MATERIALS

Permeable pavement materials allow water to infiltrate into the soil, recharging groundwater (Figure 7.4). As discussed in Chapter 6, permeable pavement can also be used as a top layer over a rainwater harvesting system. It comes in numerous forms, made from a variety of materials. More traditional materials include porous

FIGURE 7.4   Permeable paving reduces surface runoff. (Photo by the author.)

asphalt, pervious concrete, and permeable interlocking concrete pavement. Newer products are made from recycled rubber and plastic, held together with an adhesive. Two examples of such materials are the poured product Flexi-Pave and pre-molded AZEK Pavers that fit into a grid system for ease of laying. Grass pavers can be made from interlocking concrete or from high-strength plastic having voids where topsoil is placed and turf is planted on top. Some of the plastic products are made with post-consumer plastic.

Case Study: Permeable Paver: Dayton, Ohio.

Reference: Thurow, Charles. 1983. Improving street climate through design. Am. Planning Assoc., Chicago chapter.

Dayton, Ohio, set up the Dayton Climate Project to test a broad range of techniques for improving the climate quality of the city. It was a cooperative effort among several universities and private groups. The original climate project has been underway for over four decades. The landscape-related strategies included planting more vegetation, using permeable paving with turf interplanted in paver spaces, and shading sidewalks with trees (and awnings). A 30,000 ft^2, 80-car green parking lot was built in the downtown area utilizing open-space pavers, with grass planted in the spaces provided.

The results of the parking lot were that runoff from the turf-planted pavement was reduced by 65 percent, and ambient temperature above the parking lot was significantly cooler in hot weather. Unfortunately, the cost of the pavers offset this savings, since they cost $1.73 per ft^2, as compared to $0.67 for asphalt. Overall, the lot provided several benefits. Thus, the reduction in runoff meant that smaller storm sewers could be used, which translated into a 30–40 percent savings in construction costs of new lots. Furthermore, maintenance costs for the vegetated pavers were lower than those for the asphalt.

GREEN ROOFS

Green roofs are sometimes called vegetated roofs. Their primary purpose is to reduce stormwater runoff. A well-designed green roof having only 3–6 in. of media and shallow rooted can retain rainwater for up to an hour before runoff from the roof occurs. As an added benefit, green roofs can help reduce cooling needs in the summer and heating in winter. At the Peggy Notebaert Center in Chicago, there has been a 20 percent savings in cooling summer costs and smaller amount of heat savings in winter. The rubber membranes used on green roofs are expected to last twice as long as, or longer than they would on a traditional roof. This is due in large part to the elimination of ultraviolet radiation exposure on the membrane, and day–night extreme temperature fluctuations which cause fatigue and cracking of the membrane.

Modern green roof technology began in Germany as a response to stormwater runoff. In the United States, green roof research was begun at several universities around the same time. In 2000, Dr. Brad Rowe at Michigan State University collaborated with the Ford Motor Company to install a 10.6-acre green roof at the company's truck assembly plant in Dearborn, Michigan. Numerous universities now have green roof research programs in order to determine the best designs and plant materials for their region. The US EPA and DOE (Department of Energy) provide informational materials about green roofs, as referenced in the section "Further Reading" at the end of the chapter. Green Roofs for Healthy Cities maintains a website as well.

Case Study: Ballard Public Library Seattle Washington

Reference: Paladino & Company. February 2006. King County Green Roof Case Study Report. King County Department of Natural Resources & Parks. http://www.seattle.gov/dpd/cms/groups/pan/@pan/@sustainableblding/documents/web_informational/dpdp_020117.pdf. Viewed Dec. 28, 2011.

The Ballard Public Library in Seattle, Washington, is one of several locations where green roofs were installed and monitored for feasibility as a sustainable design strategy. Other sustainable design strategies in the library include daylighting, photovoltaics, operable windows, and recycled materials.

The green roof was installed with the express purpose of reducing stormwater runoff. Drought tolerant plants and a drip irrigation system were installed on this 18,000 ft^2 extensive (4-in. thick media topped with a coconut fiber mat) roof. The structural load is 50 pounds per square foot. The roofing material is 2.3 mm rubberized asphalt membrane. It is covered with a root barrier and Styrofoam deck insulation. Molded panels of recycled materials with drainage channels comprise the drainage layer, and an air layer made of filter fabric provides aeration. Grasses, mosses, herbs, and sedums were planted, having been selected for their low-maintenance requirements.

Originally, birds pulled out the small plants, but they eventually grew in to resist removal by birds. Creeping-type weeds have been problematic, due to their root systems growing into the coconut fiber mat. The mat itself has to be cut and removed in order to remove the weeds. Removing the weeds from the roof has been a challenge, especially during the beginning stages.

Other green roofs installed in King County are on the Seattle City Hall, the Dexter Regulator Station, the Henderson Regulator Station, the Seattle Justice Center, King County International Airport, and Evergreen State College. Design recommendations for green roofs that resulted from the abovementioned projects include the following:

- Perform a structural feasibility study to ensure the weight load can be accommodated for a minimum of 2 in. of media.
- Coordinate the various personnel involved – architect, roofing contractor, landscape designer, horticulturist, and/or product manufacturer.
- Include at least one team member with prior green roof experience.
- Plan for plant dormancy in cases where aesthetics is a concern.
- Ensure appropriate plant selection for the climatic conditions and soil type.

- Provide easy access for maintenance and weed disposal.
- Specify drought tolerant plants to eliminate irrigation requirements.
- Use transplants rather than seeded plants to reduce problems with birds eating the seedlings.
- Provide adequate roof slope for drainage, to avoid ponding of water and to promote healthy plant growth.
- Develop a maintenance manual prior to roof installation.
- Train all personnel in proper maintenance of the plants.
- Maintain aggressive weed control practices.
- Utilize a variety of plant species.

Green Roofs and Media Depth

Green roofs can be categorized into two groups based on the depth of the media: intensive and extensive. **Extensive green roofs** are typically 2–4 in. deep. Drought tolerant plants such as sedums are often used on these shallow-media roofs. **Intensive green roofs** are deeper, up to 2 ft or more. They can accommodate plants that require media deeper than 4 in.

Green Roof Design

Green roofs must be built much sturdier and stronger than non-vegetated roofs. Proper engineering is necessary to prevent structural problems. Roofs built to hold one hundred pounds per square foot is common if they will have plants and people on them. For holding only the plants, which takes into account 2–4 in. of media and layers of required materials, an extensive green roof may be built for sixty pounds per square foot. For deeper media to hold larger plants, such as trees, the strength requirements are even greater. In the city of Chicago, a permit is required to install a green roof. Traditional green roof construction includes several layers of materials, including a rubber membrane, root barrier, soilless media, and plants (Figure 7.5).

As green roofs have increased in number, innovation has led to a variety of installation practices. Original designs included multiple layers of materials to provide roof covering, a root barrier, a drainage layer, media for plant roots, and plant material.

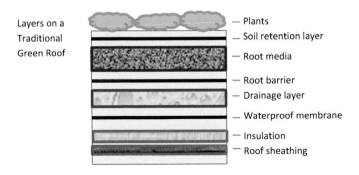

FIGURE 7.5 Traditional layers on a green roof. (Illustration by the author.)

FIGURE 7.6 Modules for vegetated roof. (Photo by the author.)

Newer designs incorporate media-filled planted modules that can be placed directly on top of a roof covering. These can be handled singly after the roofing membrane is installed. This practice allows for easier installation that does not require the level of expertise many traditional designs required (Figure 7.6).

Plants for Green Roofs

The plants that will work well on a green roof vary widely based on the location. To survive on a rooftop, plants must be able to persist over a long period of time, which requires self-regeneration, by seed or vegetatively. Extensive roofs have only a few inches of root media to grow on. This media is porous and lightweight, and therefore does not hold water well or for long periods. Thus, plants must be able to tolerate short bursts of available water and long periods of drought. Temperatures on roofs can fluctuate widely, from very cold in winter to extremely hot in summer. Plants may be exposed to a lot of wind on rooftops. Table 7.3 lists some of the plants that have done well in various green roof research programs.

Weed seeds are introduced by wind and birds. In addition to selecting appropriate plants, 100 percent coverage is desired in order to prevent or minimize weed infestation.

GREEN WALLS

Green walls are sometimes known as vertical gardens. They are often discussed in relation to green roofs. However, they do not provide stormwater amenities that green roofs do. They can actually be less sustainable when they require irrigation to provide adequate water for the plants, which they may not otherwise have access to under natural conditions. Green walls that are made of retaining wall blocks with spaces designed in the block to allow for plants do work well for aesthetic purposes. They may also provide gardening space that is otherwise not available.

TABLE 7.3
Plants for Green Roofs

Botanical Name	Common Name
Sedum acre	Goldmoss stonecrop
Sedum album	White stonecrop
Sedum kamtschaticum	Orange sedum
Sedum ellacombeanum	Orange stonecrop
Sedum spurium "Royal Pink"	Caucasian stonecrop, two row stonecrop
Sedum middendorfianum	Chinese Mountain stonecrop
Sedum pulchellum	Widowscross
Sedum reflexum	Jenny's stonecrop
Sedum spurium "Coccineum"	Two-row stonecrop

Plants were at Michigan State University in extensive media 10 cm (4 in.) deep, of 60 percent heat-expanded slate, 25 percent USGA grade sand, 5 percent aged compost, and 10 percent Michigan peat moss.

SUMMARY

Excess water in the landscape is a concern, particularly in urban environments, and requires special efforts at stormwater management to avoid several problems. Sources of excess water in the landscape include poorly draining soil, impervious surfaces, and lack of vegetation.

Excess stormwater is often directed into surface drainage and ends up in streams and rivers, where it leads to unwanted channel incision, bank erosion, and excessive deposits in streams. Toxic materials are washed into the runoff, including herbicides, insecticides, fertilizers, PAHs, pet fecal matter, and improperly disposed-of household chemicals. In addition, the water may be heated to a temperature that is unhealthy for aquatic life. Urban runoff comes from both point and nonpoint sources. The US EPA is responsible for enforcing provision of the Clean Water Act, including illicit discharge that contributes to contaminated water.

The urban water cycle describes the route that water takes from the time it enters the urban environment through precipitation until it leaves through a variety of routes. A number of solutions have been implemented in efforts to manage excess water in the landscape. Maintaining natural areas, such as woodlands, forests, floodplains, riparian areas, and wetlands is the best solution due to the ecosystem services they provide, as well as opportunities for human recreation.

Drainage is an important component of a well-developed urban site. Landscape swales and rain gardens provide a means by which stormwater can infiltrate into the soil and eventually replenish groundwater below. Rainwater harvesting systems can provide irrigation water storage and may be integrated with a permeable paving system. Permeable paving can also be installed without a rainwater harvesting system underneath it, in which case it serves to allow water infiltration instead of surface runoff. Green roofs were originally designed in response to stormwater and have been shown to absorb up to an hour's worth of rainfall before runoff begins to occur. Research is ongoing into specific designs, plant selection, and media specifications.

REVIEW QUESTIONS

1. What problems are created when too much water from sewer systems is released into surface waters?
2. Discuss the runoff characteristics of woodland compared to an urban environment.
3. Name three sources and three examples of illicit discharge into the urban water system.
4. What is the single largest source of polycyclic aromatic hydrocarbons in urban lakes?
5. Contrast and compare point and nonpoint sources of contaminants of surface and groundwater.
6. Name three management solutions to excess water in the urban landscape.
7. Compare a landscape swale and a rain garden.
8. Discuss the important features of choosing a site for a rain garden.
9. Name the benefits of permeable paving.
10. Discuss media depth of green roofs with respect to roof strength and plant selection.

ENRICHMENT ACTIVITIES

1. Visit your local wastewater treatment plant, and interview the manager. Find out whether your city has a combined or separate sewer stormwater system. Based on your discussion, write a short paper about the history of wastewater treatment in your city.
2. Identify options for onsite stormwater management in your town that would reduce the amount going into the storm sewers. Use the EPA Green Infrastructure website to research recommendations for your region.
3. Calculate the amount of rainfall that can be intercepted by a flat roof on a nearby building. Do the same for a parking lot.
4. Design a rain garden for a local landscape.

FURTHER READING

Azek Building Products. https://www.azek.com. Retrieved July 3, 2019.
Brelot, E., B. Chocat, and M. Desbordes. Innovative technologies in urban drainage: selected Preceedings (sic) of NOVATECH 2001 the 4th International Conference on Innovative Technologies in Urban Drainage held in Lyon-Villeurbanne, France, 25–27 June 2001. IWA Publishing, London.
Broughton, J. 2001. Rain gardens: healthy for nature and people. *Chicago Wilderness*, Spring. 15–16 pp.
Dunnett, N. and A. Clayden. 2007. *Rain Gardens: Managing Water Sustainably in the Garden and Designed Landscape.* Timber Press, Inc., Portland, OR, 188 pp.

Durham, A. K., D. B. Rowe, and C. L. Rugh. 2007. Effect of substrate depth on initial growth, coverage, and survival of 25 succulent green roof plant taxa. *Hortic. Sci.* 42(3): 588–595.

Eco-roofs.www.eco-roofs.com. Viewed July 3, 2019.

Eksi, M., D.B. Rowe, I. S. Wichman, J. A. Andresen. 2017. Effect of substrate depth, vegetation type, and season on green roof thermal properties. *Energy Build.* 145: 174–187.

Flexi-pave. http://www.kbius.com/. Viewed July 3, 2019.

Getter, K. L. and D. B. Rowe. 2006. The role of extensive green roofs in sustainable development. *Hortic. Sci.* 41(5): 1276–1285.

Getter, K. L., D. B. Rowe, and J. A. Anderson. 2007. Quantifying the effect of slope on extensive green roof stormwater retention. *Ecol. Eng.* 31: 225–231.

Getter, K. L. and D. B. Rowe. 2009. Substrate depth influences sedum plant community on a green roof. *Hortic Sci.* 44(2): 401–407.

Getter, K. L., D. B. Rowe, G. P. Robertson B. M. Cregg, and J. A. Anderson. 2009. Carbon sequestration potential of extensive green roofs. *Environ. Sci. Technol.* 43(19): 7564–7570.

Green roofs for healthy cities. http://www.greenroofs.org/. Retrieved July 3, 2019.

Greengrid roofs. http://www.greengridroofs.com/. Retrieved July 3, 2019.

Henderson, C., M. Greenway, and I. Phillips. 2007. Removal of dissolved nitrogen, phosphorus, and carbon from stormwater by biofiltration mesocosms. *Water Sci. Technol.* 55 (4): 183–191.

Interlocking Concrete Pavement Institute. https://www.icpi.org/whychooseconcretepavers#-ADVANTAGES. Viewed July 3, 2019.

Kerkhoff, K. L. 2006. How to capitalize and reduce stormwater runoff in your landscapes. *Grounds Maint.* p. 70. http://grounds-mag.com/mag/grounds_maintenance_capitalize_reduce_stormwater/. Viewed June 15, 2020.

Krizner, K. 2008. Smart water solutions. *Landscape Management.* May/June. pp. 31–2

Lopes, T. J. and S. G. Dionne. 1998. A Review of Semivolatile and Volatile Organic Compounds in Highway Runoff and Urban Stormwater. U.S. Geological Survey, South Dakota, 67 pp.

Metropolitan Council of the Twin Cities Area. 2002. *Urban Small Sites Best Management Practices Manual.* Metropolitan Council., St. Paul, MN.

Monterusso, M. A., D. B. Rowe, D. K. Russell, and C. L. Rugh. 2004. Runoff water quantity and quality from green roof systems. *Acta Hortic.* 639: 369–376.

Ogle, D. G. J. C. Hoag. 2000. Stormwater plant materials a resource guide: detailed information on appropriate plant materials for the best practices. USDA Natural Resources Conservation Service, Boise, ID.

Presto Products Co. 2004. Geoblock® system general design and construction package. Presto Products company. 16 pp.

Robinette, G. O. and K. W. Sloan. 1984. *Water Conservation in Landscape Design and Management.* Van Nostrand Reinhold Co., New York, NY. 258 pp.

Rowe, D. B., J. Andersen, J. Lloyd, T. Mrozowski, and K. Getter. The green roof research at Michigan State University. http://hrt.msu.edu/greenroof/. Viewed July 3, 2019.

Rowe, D. B., M. A. Monterusso, and C. L. Rugh. Assessment of heat-expanded slate and fertility requirements in green roof substrates. *HortTech.* 16(3): 471–477.

Rowe, D. B. 2011. Green roofs as a means of pollution abatement. *Environmental Pollution* (in press). http://www.ncbi.nlm.nih.gov/pubmed/21074914. Viewed July 3, 2019.

Schmidt, R., D. Shaw, and D. Dods. 2007. The blue thumb guide to rain gardens. *Waterdrop Innovations LLC.* 81 pp.

Tapia Silva, F. O., A. Wehrmann, H. J. Henze, and N. Model. 2006. Ability of plant-based surface technology to improve urban water cycle and mesoclimate. *Urban For Urban Green.* 4: 145–158.

Taylor, G. D., T. D. Fletcher, T. H. F. Wong, P. F. Breen, and H. P. Duncan. 2005. Nitrogen composition in urban runoff—implications for stormwater management. *Water Res.* 39: 1982–1989.

Torno, H. C., J. Marsalek, and M. Desbordes. 1986. *Urban Runoff Pollution*. Springer-Varlag, New York, NY, 893 pp.

United States Department of Energy. Green Roofs. Federal Technology Alert. DOE/EE-0298. www.eere.energy.gov/femp. Retrieved July 3, 2019.

United States Geological Service. PAHs and coal-tar-based pavement sealcoat. http://tx.usgs.gov/coring/allthingssealcoat.html. Retrieved July 3, 2019.

USGS. National water quality assessment. https://www.usgs.gov/mission-areas/water-resources/science/national-water-quality-assessment-nawqa?qt-science_center_objects=0#qt-science_center_objects. Retrieved July 3, 2019.

Wetherbee, K. 2010. Rainscaping. *The American Gardener*, July/Aug.

8 Soil Health

OBJECTIVES

Upon completion of this chapter, the reader should be able to

- Explain the importance of soil health in landscaping
- Describe the effects of construction activities on soil health
- Explain how soil analysis fits into the site analysis process
- Identify the physical and chemical properties of soil
- Identify signs of healthy soil
- Identify soil-borne bacteria and fungi that are beneficial to plants
- Discuss the contribution of organic matter to soil health
- Implement healthy soil management practices

TERMS TO KNOW

Brownfields
Cations
Core aeration
Essential nutrient
Humus
Hyphae
Mineral soils
Mycelia
Nematode
Nitrogen fixation
Nutrient cycling
Organic soils
Soil profile
Soil structure
Subsoil
Topsoil

INTRODUCTION

Soils are important to landscaping in part because the type of soil and its characteristics will determine what plants can be used in a location and how well they will do there. Before landscaping begins, however, construction must be done to build the residence or commercial building, including a driveway or parking lot, walkways, service areas, and possibly other items. Unfortunately for the landscaper, the exact construction history is often an unknown. Yet, it is safe to assume that for recent construction, if heavy equipment was present on site, topsoil was probably removed and not replaced adequately or at all, and grading has been done to direct water away from the building using heavy equipment. Such construction activities compact the soil and can greatly reduce the ability of water to drain through the soil. Some studies have shown infiltration rates have been reduced by as much as 35 percent on altered sites as compared to undisturbed sites.

It is also possible that construction wastes are buried on site, including concrete debris, which can raise alkalinity in surrounding soil. Wood and other buried debris can actually cause excessive drainage from a site.

All of these activities can be added to any previous history of the site, such as whether it was a wooded lot, farmland, or wetland, or used for other purposes. Since a lot of earth is moved during some construction projects, it may not be clear at the end of the construction phase, just exactly what the soil type is compared to native soil that was there originally. A soil profile can reveal the stratifications of varying soil types. Figure 8.1, showing the soil profile from an urban location, indicates that

FIGURE 8.1 Soil profile typical of urban soil. (Photo courtesy Ramona L. Taylor, USDA.)

non-native soils that have been layered over native soils. In native soil, topsoil is found on the top layer, unless there is excessive organic matter above it. Topsoil is important to plant growth because it provides the necessary aeration, and adequate drainage and moisture, and can store and make available to plants essential plant nutrients. Plant root growth is not as restricted in topsoil as in the underlying subsoil. Often in a construction area, topsoil is buried beneath subsoil, or is removed and not replaced.

SOILS AND CONSTRUCTION ACTIVITIES

A characteristic of soils where construction activities have occurred is the great amount of variability, both vertically and horizontally, present. This affects not only plant health due to constraints on root growth but also the ability of water and fertilizer to infiltrate down through the soil profile. Therefore, the landscape manager needs to examine the physical and chemical properties of the soil prior to beginning the design process and selecting plant species to bring into the site. Soil porosity should be examined in order to determine water and fertilizer infiltration rates.

Some of the landscape issues related to construction that are addressed by the Sustainable Sites Initiative are given as follows:

- Sediment runoff rates from construction sites. These can be up to 20 times greater than agricultural sediment loss rates and 1,000–2,000 greater than those of forested lands.
- Compaction, which is caused by the use of heavy machinery during construction, degrades soil structure and reduces infiltration rates, which increase the runoff volume and flooding potential.
- Soils can contain as much as or more carbon than living vegetation. For example, 97 percent of the 335 billion tons (304 billion metric tonnes) of carbon stored in grassland ecosystems is held in the soil.
- In addition to carbon dioxide, disturbed soils also release substantial amounts of methane and nitrous oxide, both gases that trap heat even more effectively than carbon dioxide.

The Sustainable Sites Initiative has several credits and prerequisites related to soil health, including limiting disturbance of prime farmland soils, preserving existing soils and topography, and restoring soils disturbed during construction and by previous development. They also require the development of soil management plan during construction.

There are two top issues of concern for healthy plant growth: whether there is topsoil, and if so, how deep it is, and whether compaction has occurred. The landscape designer needs to ascertain these factors at the outset of the design or installation process. Another concern for landscapers concerning the history of a site is whether existing plants were left on site during construction, and whether the roots were adequately protected from damage. If not, there is a risk that a plant will begin declining, and if it is a very large specimen plant, it may require removal, and this is something to be aware of prior to making additional changes at a site.

Finally, sometimes, in the process of digging a foundation or basement, subsoil is placed over topsoil rather than removed from the site. Subsoil consists of largely fine-textured particles accompanied by small pore spaces and lacks organic matter, proving a poor medium for plant growth.

Case Study: The Building Envelope

Reference: Wasowski, Andy. 1997. The Building Envelope. The American Gardener: 26–33.

Andy and Sally Wasowski were not much different from any other couple when they decided to build a new home for themselves in the Hill Country west of Austin, Texas. Perhaps one difference was that they didn't desire a landscaped yard, but instead preferred the wild and natural look of native vegetation. They developed a plan to protect as much of the native vegetation and soil as possible using an approach they call the building envelope.

The building envelope is derived by placing the house and driveway on the property in such a way as to minimize their impact on the rest of the property. The home and its environs, if any (such as a swimming pool and garden), are designated as the private zone. A band of 5–15 ft is drawn around the private zone, and this is the building envelope. A fence is installed around this, the transition zone. This is the area in which work takes place, equipment is allowed to maneuver, and staging of materials can take place. The fenced-out area is the natural zone. Parking, workers, equipment, and construction activities are prohibited in this area.

The Wasowski's had to work with their contractor to ensure all the workers understood and accepted the parameters of the worksite. Compromises had to be made when there simply was not enough room to access the building site and the fence had to be adjusted. However, the final product was preferable to traditional construction sites where bulldozers have scraped the land clean of topsoil, removed indigenous trees, damaged root systems, compacted soil, and destroyed valuable shrubs, grasses, and wildflowers.

Some municipalities, such as Aspen, Colorado, have included the building envelope into their code. Pinnacle Peak in Scottsdale, Arizona, provides an example of a development whose designer, Gage Davis, embraced the concept of the building envelope.

BROWNFIELDS

Brownfields are sites that have been used for industrial or commercial purposes and have been contaminated or polluted with hazardous substances. They require special understanding and treatment, are an important aspect of sustainability, and should be included in development planning activities. Redevelopment of a brownfield can be accomplished by following guidelines provided by a local, state, or federal agency, or as defined by an environmental assessment using ASTM Standards or a local Voluntary Cleanup Program. The publication "ASTM E1903–97(2002)

Standard Guide for Environmental Site Assessments: Phase II Environmental Site Assessment Process" describes how to conduct a site assessment, including how to determine whether hazardous substances, including petroleum products, have been disposed of there. The website for more information on this is provided in the section "Further Reading".

Case Study: The Augustus F. Hawkins Natural Park, Los Angeles

References: http://www.pps.org/great_public_spaces Viewed Dec. 27, 2011.

Suutari, Amanda. USA – California (Los Angeles) Natural Urban Park. http://www.ecotippingpoints.org/our-stores/region-usa-canada.html Viewed Dec. 27, 2011.

The 8.5-ac Augustus F. Hawkins Natural Park is located in South Central Los Angeles. It is the former brownfield site of municipal detritus such as broken sewer pipes and other relics. It was fenced in with chain link topped with razor wire and bordered four gang territories. With a suggestion from Rita Walters, a Los Angeles City council member in the 1990s, the Santa Monica Mountains Conservancy (SMMC) began the undertaking of providing nature to an impoverished neighborhood in an urban environment.

Landscape architects were involved in the planning, as were community groups. The chief landscape architect was Stephanie Landregan, who faced opposition in the form of disbelief and concern about providing a natural area to poor people who certainly had greater, more important needs. Nevertheless, efforts to involve community groups and neighborhoods residents helped move the plan forward.

In order to address safety concerns, there was a decision to keep the area fenced, although a more artistic wrought iron fence was installed, along with gates on all four sides. Soil was brought in to create hills to create a sense of refuge from the surrounding area. The soil came from an unexpected source when landslides near Malibu released excessive amounts of usable soil. The original brownfield soil was unusable due to contamination.

Existing concrete was crushed to create a parking area. The original ecosystem, an alluvial plain covered with shrubs and grasses, was not recreated. Instead, plants were selected to mimic local plan communities such as riparian forest, oak woodland, freshwater marsh, and chaparral. Native birds rarely seen in urban settings have been observed using the park.

In addition to the environmental aspects, the social aspects are emphasized, too. After-school programs, cultural activities, and other community events take place in the park. There is now a Discovery Center where free classes are held daily for adults and children, and on Saturdays, field trips are organized to local natural areas.

SOIL TESTING

During the sustainable landscape audit, the landscape professional should conduct a soil test. Some onsite tests that are easily done include a ribbon test to determine soil texture and a pH test using a soil probe. Further testing could be done by using kits that are readily available or by sending a sample to a soil testing laboratory in the area.

While sending soil to a lab, samples should be collected from similar areas and combined into a composite sample. If areas in the landscape are to be used differently, such as flower beds, vegetables, fruit trees, or lawn, then separate soil tests should be conducted.

Soil properties that affect plant growth can be broadly grouped into two categories: physical and chemical properties. Living organisms in the soil are a crucial component of soil health.

Physical Properties of Soil

Physical properties of soil include soil texture, soil structure, and compaction. Living organisms such as bacteria, fungus, earthworms, and others play a role in breaking down organic matter. They may also be involved in plant root–soil interactions, as well as contribute to the development of good soil structure.

Most soils fall into the category of mineral soils, as opposed to organic soils, which are not suitable for construction. Organic soils contain at least 30 percent organic matter, but most mineral soils contain less than 5 percent. By comparison, a rich, dark agricultural soil may have 5–10 percent organic matter. Soil texture and soil structure are properties of soil that relate to soil particle sizes and to their ability to form clusters known as aggregates or peds. The particle sizes determine the sizes of the pore spaces between particles. Compaction alters soil properties of soil structure such as infiltration and porosity.

Soil Texture

The soil is made up of soil particles of various sizes, pore spaces between the soil particles, and a large variety of macro- and micro-organisms that inhabit the spaces among soil particles. When water is applied to the soil, most of it moves downward by gravitational pull, while some portion adheres to the soil particles against the pull of gravity. Water in pore spaces is available for uptake by plant roots. Pore spaces are filled with a combination of water and air at all times. Ideally, about 40–50 percent of soil is comprised of solid particles, and another 40–50 percent is air and water. Organic matter varies, but 5 percent is generally considered a healthy amount.

Mineral soils are grouped into soil texture categories based on the percentage of differently sized particles. Sand, silt, and clay are the three particle sizes that make up soil. To conduct a soil test, first collect the soil sample from the desired locations. For small areas, such as a small flower or shrub bed, or for planting a single tree, a single sample should suffice. For larger gardens or turf areas, collect several samples and combine them for a composite sample. To collect the soil, use a soil augur, soil tube, or spade. Remove leaf litter or other organic matter from the surface of the soil. While using a spade, dig a V-shaped hole, remove a half-inch slice from the side of the hole, and remove most of the sample. Allow the soil to air dry. Do not heat the sample in a microwave or conventional oven to dry it. For transport, the soil may be placed into a plastic bag when it is dry.

Soil Health

There are two easy tests that can be done to determine soil texture: the jar test and the ribbon test. The jar test uses a small amount of soil which is placed into a jar. One inch of soil is a convenient amount to use. Water is then added, to the 2-in. mark, and the jar is shaken. When the shaking ends, the sand will settle out in only 30 seconds. The level of the sand should be noted. The silt settles out in 30 minutes, and its level on the jar should also be noted. At this point, the relative amounts of sand and silt can be compared to the starting amount, and the soil texture can be deduced using the soil texture triangle (Figure 8.2). Clay takes up to 24 hours to settle out. If desired, the texture can be deduced after this time, but it is not necessary to wait once the sand and silt fractions are known.

To use the texture triangle, read the percentage of sand across the bottom of the triangle and follow the line upwards and to the left until it intersects with the percentage for clay on the left side of the triangle and the level of silt given on the right side of the triangle. The intersection of these three lines is located in a labeled area in the triangle indicating the texture. This is your soil texture.

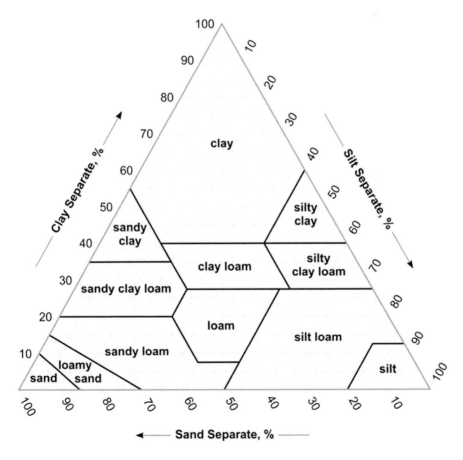

FIGURE 8.2 Soil texture triangle. Read percentages across bottom (sand), left side (clay), and right side (silt) to determine the texture of your soil sample. (Source: USDA.)

The ribbon test is done using a small ball of moistened soil. Place the soil between the thumb and forefinger, and squeeze it gently between them, pushing it out to form a ribbon. Follow the steps below to determine the soil texture of your sample:

1. Place a fistful of moistened soil into the palm of your hand. Moisten it to form it into a ball shape.
2. If it does not form a ball, it is sandy. If it forms a ball shape, then proceed to the next step.
3. Massage the soil and roll it between your thumb and forefinger to make a ribbon.
4. If it will not form a ribbon, it is loamy sand or silty sand.
5. If a ribbon forms of about ½ to 1 in. (1.2–2.4 cm) long before breaking off, it is sandy loam.
6. If a 1 in. (2.5 cm) ribbon forms, it is loam.
7. If it makes flakes instead of a ribbon, it is silt or silt loam.
8. Sandy clay loam makes a ribbon of 1–2 in. (2.5–5.0 cm).
9. Sandy clay makes a ribbon of 2–3 in. (5.0–7.5 cm).
10. If it continues to form a ribbon of more than 3 in. (7.5 cm) long, it is clay.

The best texture for plants varies by plant species, but generally loam or a modified loam is best for most plants. Desert plants will do better in a sandy soil, and water-loving plants will tolerate a heavier silt or clay soil.

Soil Structure

Some plants can grow in subsoil or very poor topsoil. But, in addition to poor plant growth, subsoil is very difficult or impossible to cultivate, either by hand or by machine. Plants that do well in subsoil tend to have shallow root systems.

Soil structure is very important for plant growth, as it determines how well plant roots can grow, as well as whether aeration is adequate, and how well water is held or can drain out of the root zone. The best soil structure is granular, a quality which can be seen by simply picking up a handful of soil and observing it. It should crumble when gently manipulated between the thumb and fingers. It is good for plant growth because it allows for appropriate air and water exchange, drains well, and provides adequate pore spaces for root growth. Granular soil structure is developed by the actions of freeze–thaw cycles, wet–dry cycles, and the activity of earthworms, microbes that break down organic matter, and physical and chemical changes that occur with root growth. The addition of organic matter is a good aid to building soil structure.

In the built environment, good soil structure for plant growth and good soil structure for construction activities are very different. The two activities are thus at odds with one another, a fact that continues past the original building construction into other construction activities that may occur. For example, if an addition is made to an existing building, not only may plants be destroyed or removed in the process, but the soil may also be destroyed with respect to the ability of plants to grow there later on. There are many events in the built environment that create hazards for plants, including building additions, hardscaping installation, sidewalk and roadway repairs, and maintenance or installation of underground plumbing or electrical lines.

Good soil structure results from a few years to hundreds of years of physical and chemical activity, including plant–root interaction with soil particles. Soil structure can be damaged in a matter of minutes. Driving heavy machinery on soil, or even driving a car or truck on wet soil, can cause compaction that will kill turf and allow weeds to invade a site within weeks. Trees often die over a longer period of time when they have been subjected to this treatment, and sometimes, they only partially die, but they still lose their value and may actually become a safety hazard. Anyone who has driven a car on wet soil knows that a rut forms that does not go away on its own. The fact that the rut holds water is clear evidence of the kind of damage that has occurred to the soil structure.

Compaction

Compaction is caused by several activities, including those associated with construction, vehicle and pedestrian traffic, athletic and recreational activities, and landscaping activities performed when the ground is wet. Soil disturbance occurs during construction and installation of hardscaping materials and plants. Fine textured clay or silt soils and wet soils are more prone to compaction than coarser, sandy soils, or dry soils.

Compaction can also occur near roadsides where traffic causes vibration that increases soil settling. Heavy rainfall and flooding also contribute to compaction by allowing smaller particles to fill in larger pore spaces. Athletic fields are subject to compaction during use and must be regularly treated using core aeration to alleviate the situation.

Compaction and soil disturbance can often be identified by the types of weeds that invade an area. For example, plantain (*Plantago* spp.) grows well in compacted soil, and lambsquarters (*Chenopodium album*) grows in alleyways, flower beds, and other disturbed and cultivated areas.

Root structure often changes when plants are growing in compacted soil. They are less branched, shorter, and thicker, allowing them to exert greater pressure as they grow through the soil. Roots must push aside soil to grow, and this is most effective for them when the soil is granular, with adequate sizes and number of pore spaces, such as is found in good quality topsoil. Compaction adversely affects root growth because it decreases both the number and sizes of pores spaces. Smaller particle sizes settle into the pore spaces during compaction. The reduction in pore sizes has a two-pronged effect of reducing the amount of oxygen available, and decreasing the ability of water to move through the spaces. Unfortunately, this water is more tightly held to the small particle sizes and is less available to plant roots.

The decrease in water infiltration down through the soil profile means that compacted soil holds water for a longer period of time, which leads to waterlogged soils, and greater erosion due to increased runoff. Fertilizers applied in the area are also subject to increased runoff, posing a problem to surface water.

To test for poor drainage, a simple percolation test can be conducted. To do this, dig a hole 12 in. (30 cm) across and 12 in. deep (30 cm), and fill it with water. Mark the water line with a stick or other object. Pay attention to the water line, and measure or estimate how long it takes for the water level to drop 1 in. Water will drop 1–2 in. (2.5 – 5 cm) per hour in a well-draining soil.

Manual and digital tools are available that can aid in testing for soil compaction. They consist of a rod-shaped probe that measures how much pressure in pounds per square inch is required to penetrate the soil. Soil is tested to the depth plants require good soil for growth, typically 3–4 ft (1–1.3 m) deep. A steel probe can be used, with pressure gauged by the operator. If hard, compacted soil is encountered, it will be quite noticeable. Several spots should be checked, since soil often lacks uniformity over large areas.

CHEMICAL PROPERTIES OF SOIL

Soil pH and cation exchange capacity (CEC) and salinity are chemical properties that can easily be assessed using soil tests. Soil pH determines nutrient availability to plants. Plants should be selected for the pH of the site, although pH can be adjusted within reason using readily available materials. CEC is not as easily changed, although amendment with organic matter will improve CEC somewhat. Plant nutrients are present in the soil in varying amounts, based on a number of factors. A soil testing lab can test for essential nutrients and may also provide fertilizer recommendations. Sustainable fertilization will be discussed further in Chapter 10. However, nutrient cycling, the process whereby nutrients are broken down from complex, bound forms, to simple, ionic forms which plants can take up, is a function of healthy soil and is addressed in this chapter. Salinity is a detrimental condition that is more likely to occur where groundwater is contaminated with salts, where ice-melting salts are used or where fertilizer build-up in the soil occurs.

Soil pH

Soil pH is a measure of acidity or alkalinity, which determines the availability of plant nutrients. The range of pH is 0 (most acidic) to 14 (most alkaline), with 7 indicating neutral. Table 8.1 shows the pH of various chemical-based products. The range for soil pH is approximately 4–10. At too high or too low a pH, essential nutrients can get bound into compounds which makes the nutrients unavailable to plants. The ideal soil pH for most plants falls in the 6.5–7.5 range. Acid-loving plants such as azaleas and rhododendrons require a lower pH, around 5–5.5. During the site analysis, the soil should be tested for pH to determine whether it is adequately acidic or alkaline. This test can easily be conducted using soil probes that are readily available through horticultural supply companies.

Cation Exchange Capacity

CEC is a reflection of the amount of clay and organic matter present in the soil. Both of these components increase the ability of soil to hold and exchange cations or positively charged ions, the form in which many essential nutrients are taken up by plants (Table 8.2). Soil can hold and exchange cations because of particles having negative charges. Clay soils may have five to greater than thirty times the number of negatively charged sites than sands for a given amount of soil (Table 8.3).

Soil Health

TABLE 8.1
pH Values of Some Common Substances

Substance	pH Value
Battery acid	0
Hydrochloric acid	1
Stomach acid	1.2–1.4
Lemon juice	2.2–2.4
Orange juice	3.0
Vinegar	3.0–4.0
Beer	4.0–5.0
Hydrogen peroxide	4.5
Coffee	5.0
Aspirin	5.0
Tea (black)	6.0
Milk (cow's)	6.5
Human saliva	6.0–7.5
Water	6.5–8.0
Baking soda	8.0
Sea water	8.3
Borax	9.0
Lava soap	10.0
Ammonia	10.0
Milk of magnesia	10.6
Oven cleaner	13
Liquid drain cleaner	14

TABLE 8.2
Essential Plant Nutrients Present as Cations in Soil

Essential Nutrient	Ionic Form in Soil Available to Plants
Hydrogen	H^+
Aluminum	Al^{+++}
Calcium	Ca^{++}
Magnesium	Mg^+
Potassium	K^+
Ammonium	NH_4^+
Sodium	Na^+

Salinity and Deicers

Salinity should be tested in areas where salts in the groundwater are of concern. Salts in irrigation water can build up in soil over time, presenting a toxic environment for plant roots. When there are high levels of salts in the soil, plants may not survive or thrive. Saline soils can burn plant roots and kill the plant.

TABLE 8.3
Cation Exchange Capacity of Different Soil Textures

Texture	CEC
Clay	>30
Clay loam	15–30
Loams and silt loam	5–15
Fine sandy loam	5–10
Sand	1–5

In climates where it snows in winter, salts are applied as deicers to roadways and sidewalks. Sodium chloride is chosen for its low cost and effectiveness. The salt contributes to saline or sodic soils and can damage leaf and stem tissue. Evergreens are particularly vulnerable to salt damage due to desiccation of leaves exposed to salt spray. Salt contamination of soils creates a chemical change that makes it more difficult for plants to absorb water. When sodium is present, it can break down soil aggregates that are the hallmark of good soil structure. When this happens, crust forms on the surface of the soil and seals it. In the winter time, walkways in public areas, and sloped walks are of particular concern, due to the increased liability for falling and injury as well as accessibility concerns. Salt-tolerant plants should be selected for such areas, or alternatives to plants, such as mulch or other decorative items. Excess salts may be leached from the root zone by flushing with water when soil conditions permit.

A number of alternatives to sodium chloride are available. Calcium magnesium acetate (CMA) has been recognized by the Federal Highway Administration as an effective and acceptable road deicer. Magnesium and calcium chloride are available in crystalline form and are more effective than sodium chloride at lower temperatures, are less damaging to concrete, are less corroding, and have a reduced negative impact on soils and vegetation. Brine solutions of magnesium chloride are available as liquid deicers.

In general, alternatives to sodium chloride are more costly on a per unit basis. However, they may require a smaller amount of product to be applied, due to their increased effectiveness. The landscape manager should select the product which is most effective in their climate, while minimizing salt build-up.

ADDRESSING PROBLEMS WITH SOIL CHEMISTRY

The professional landscaper can use a variety of soil amendments to address most problems that occur. In the case of pH, appropriate plants must be selected. A good example of using the wrong plant for soil pH is the thousands of pin oaks that were planted in the Midwest which suffer from iron chlorosis due to incorrect soil pH. Some estimates are that newly planted trees in urban areas last an average of only ten years, whereas many more survive but do not thrive. Most books and websites on landscape plants will provide a pH range for each plant. This is a problem that should be addressed at the design stage of the landscape process.

In the case of low CEC, the landscaper may add organic matter to improve the nutrient-holding ability of the soil. Organic matter is addressed in Chapter 10. It is worth stating here, though: organic amendments can aid in CEC as well as contribute to better soil structure in the future. The landscape manager should try to retain organic matter on site at every opportunity. This includes using organic mulch, leaving grass clippings to decompose in place, and adding shredded leaves to flower beds.

Salinity is a difficult problem to solve if saline irrigation water is the issue, since it may require changing the water source for irrigating plants. Whereas water can be treated to remove excess salts, this is a very expensive solution. A better one would be to grow plants that do not require supplemental irrigation. Salts can be leached out of the root zone, but this is not particularly practical unless the water used does not itself contain high levels of salts. If a fertilizer spill occurs on turf grass, under trees, or in beds, leaching with water is the recommended treatment.

SOIL ORGANIC MATTER

In addition to plant roots, soil organic matter contains living organisms, decomposing matter, and soil humus, the substances left after organic matter has been broken down in biological and chemical processes.

LIVING ORGANISMS IN SOIL

Organisms present in the soil include many species of earthworms, nematodes, fungus, bacteria, algae, protozoa, root-feeding insects, millipedes, and centipedes. Some organisms are referred to by the function they perform, such as decomposers, whereas others are referred to by their relationship to plants, such as symbionts, which are sometimes called mutualists.

Earthworms

Earthworms create large pore spaces in the soil as they move through it, digesting organic matter, and excrete fecal pellets called castings. Some earthworm species live in organic environments, and others live in mineral soil environments. A popular example of organic-dwelling earthworms is the red wiggler. They are good worms to add to a compost pile or bin, and they can usually be found at local bait stores. The common earthworm, or nightcrawler, prefers the mineral soil environment. They build burrows in the soil and leave small mounds of castings near the entrances. They can ingest soil along with decaying organic matter.

Do not confuse the two different types of earthworms. This has been done in some areas, with disastrous environmental effect. In Minnesota, where there are no native ground-dwelling earthworms, discarded fishing worms have invaded the forest areas, wreaking havoc on the ecosystem. Tree seedlings, ferns, and wildflowers are threatened. Native wildflowers are disappearing because the undecayed organic matter that would normally be present to provide natural mulch and slow-release nutrients is no longer available. The introduced earthworms feed on organic matter, and they are consuming the leaf litter and other organic material on the forest floor.

As a result, soil is eroding away, carrying valuable nutrients with it. A better alternative would be to take the fishing worms back home and introduce them to a compost bin or pile. Composting is treated more fully in Chapter 10.

Nematodes

Soil-dwelling nematodes are microscopic, unsegmented worms. They live in soil or inside plant roots. They have a needle-like mouthpart which they use to puncture plant tissue in order to feed from them.

There are hundreds of species of nematodes, some have beneficial effects, and others are detrimental to plants. Those that harm plants are categorized as parasitic nematodes. Nematodes that feed on plants harm them by reducing vigor and creating a wound that is then subject to infestation by pathogenic bacteria and fungi. Omnivorous nematodes feed on decaying organic matter for the most part. Two methods to control detrimental nematodes are the use of predaceous nematodes which prey on other nematodes, and the use of fungus which can strangle harmful nematodes.

Decomposers

The primary decomposers in soil are actinomycetes, fungi, and bacteria. They break down organic residue deriving from fallen leaves, dead stems, grass clippings, and so on. In the process, they produce compounds (humic acid) that help the soil form into aggregates, thus building soil structure. The strands of fungus, or hyphae, also help to bind soil particles together. Often, a mass of hyphae, or mycelia, is visible as a white cottony substance on organic matter.

Waste products from the breakdown of organic matter in the soil provide new organic compounds that, in turn, feed other organisms. Some fungi and bacteria compete with or inhibit pathogenic fungi and bacteria, thus contributing a benefit to plants growing there.

Symbionts

A number of species of soil-dwelling bacteria and fungus form mutually beneficial relationships to plants. Notable among these are the nitrogen-fixing bacteria and mycorrhizae fungus. Nitrogen-fixing bacteria in the genera Rhizobia and Bradyrhizobia can form an association with plants in the Fabaceae family or legumes, and provide much of the nitrogen required for plant growth.

Mycorrhizae

Many species of mycorrhizae have been identified as forming symbiotic relationships with plants. Mycorrhizae are a type of fungus that grow in association with plant roots, some species only grow externally to the plant root (endomycorrhizae), while other species actually invade the plant root and grow internally to the plant (ectomycorrhizae). The host plant provides up to 15 percent of sugars produced in their leaves to the invading organism. But the relationship is mutually beneficial, because the mycorrhizae provide a vast network of hyphae that take up water and nutrients from the soil surrounding the plant's own root system. It is thought that mycorrhizae impart greater drought tolerance and disease and pest resistance to

TABLE 8.4
Mycorrhizae Known to Occur with Plants Used in Landscaping

Name
- *Pisolithus tinctorius*
- *Laccaria bicolor*
- *Suillus bovinus*
- *Cenococcum graniforme*
- *Rhizopogon* spp.
- *Thelephora terrestris*
- *Glomus fasciculatum*

affected plants. Perhaps the single greatest advantage imparted on affected plants is increased phosphorus uptake.

Mycorrhizae occur naturally in soil, but can be adversely affected by stockpiling of soil, as is often done during the construction phase in the built environment. This is also done with mining operations and other soil-disturbing activities, and this soil may eventually be sold to landscapers and others. Eventually, however, mycorrhizae are naturally re-introduced to soils that are depleted of them because they are so prevalent in the environment, and apparently are carried by wind. In some cases, a host plant must be present in order for the mycorrhizae to reproduce.

Endomycorrhizae that grow inside plant tissue form structures that are either balloon-like (vesicles) or dichotomously branching invaginations (arbuscules). For this reason, they are also referred to as vesicular-arbuscular mycorrhizae (VAM). They are found in around 85 percent of all flowering plant families, including many crop species.

Ectomycorrhizae are found in birch, eucalyptus, oak, pine, and rose families. They form a sheath around plant roots, with hyphae extending out in an extensive branching network in the soil. In some cases, the hyphae penetrate the root cells, and they are then called ectendomycorrhizae. Laccaria bicolor is an ectomycorrhizae that has been found to lure springtails, kill them, and provide some of the resulting nitrogen to its host plant. Other mycorrhizae species that occur with landscape plants are listed in Table 8.4.

Symbiotic Nitrogen-Fixing Bacteria

A large amount of nitrogen becomes available to plants through a process known as nitrogen fixation. Nitrogen-fixing bacteria convert the gaseous form of nitrogen (N_2) from inaccessible forms to forms plants can take up and metabolize. Rhizobia and Bradyrhizobia are two genera of symbiotic bacteria that form nodules in the roots of certain plants, where they metabolize, and gaseous nitrogen. The nodules are formed by the plants themselves, which also supply the invading bacteria with essential minerals and sugars from photosynthesis.

Ornamental plants which can form such mutually beneficial relationships with nitrogen-fixing bacteria include many species in the Fabaceae family (legumes),

Lupine, Digitaria, Douglas fir (*Pseudotsuga menziesii*), and white fir (Abies concolor). Commercially available inoculants are added to seeds or can be applied directly to the soil. Under ideal conditions, a legume such as red clover can fix approximately 2.6 pounds of elemental nitrogen per 1,000 ft^2.

ESSENTIAL PLANT NUTRIENTS

Plants are about 90 percent water and about 5 percent carbon. The remaining 5 percent is comprised of other nutrients. Of the more than 100 elements in the periodic table, 16 are considered to be essential for plant growth. The essential plant nutrients include the macronutrients oxygen, hydrogen, carbon, nitrogen, phosphorus, potassium, calcium, magnesium, and sulfur. Micronutrients are needed in smaller amounts and are sometimes referred to as trace elements. They include boron, chlorine, manganese, molybdenum, nickel, and zinc. Some plants may require silicon, sodium, or cobalt. Some other elements that are sometimes found in plants include aluminum, nickel, and selenium. These other components are provided to plants through minerals present in the soil, by the breakdown of organic matter, in the air, or by fertilization.

Carbon, hydrogen, and oxygen are provided by air and water. Most of the nitrogen present on earth is found in the atmosphere, in a tightly bound molecule comprised of two nitrogen atoms, N_2. Unfortunately, plants cannot take up or use nitrogen in this bound form. Rather, they require nitrate (NO_3^-) or ammonium (NH_4^+) forms of nitrogen.

Other than lightning, which can break the chemical bond between two atoms of nitrogen, plants rely primarily on bacteria to convert nitrogen into usable forms. These bacteria also release nitrogen from organic material back into the atmosphere.

Nitrogen is considered to be the primary limiting element to plant growth, assuming plants have all the air and water they need. Ninety to ninety-five percent of the nitrogen in unfertilized soil comes from organic matter. Nitrogen and other nutrients are released during the decomposition process, as living organisms or plant matter in the soil die and decay. Nitrogen and carbon are utilized by the micro-organisms that help break down organic matter. Soil microbes require nitrogen for survival and for population growth; thus, it is often the limiting factor in the decomposition process. Non-symbiotic nitrogen-fixing bacteria, some of which are found in the genera Azotobacter and Clostridium, do not require a symbiotic relationship to a host plant and are capable of fixing gaseous nitrogen (N_2). Thus, nitrogen is cycled through various organisms and through different forms over a period of time.

IMPROVING SOIL HEALTH FOR LANDSCAPING

The first step to soil health in a built environment is avoiding or preventing harm in the first place. Trees should be protected, topsoil should be conserved, and construction debris should be removed from the site. Salt contamination should be avoided, and organic matter should be incorporated when it is feasible to do so. In addition, soil physical and chemical properties should be tested before plants are installed.

Soil Health

Good soil management should be practiced by every landscape manager to avoid unnecessary problems and prevent problems from occurring. This includes avoiding driving heavy trucks or equipment over the landscape, not tilling wet soil, avoiding unnecessary tilling of flower and vegetable beds, and not driving or parking on wet soil.

CORRECTING COMPACTION

Even regular activities, such as mowing lawns and walking or playing on turf, can cause compaction over time. For turf areas, a good activity to maintain soil health and reduce compaction is core aeration using a hollow or solid tine aerator 4–6 in. deep. Under ordinary conditions, this activity can be performed once each year. For high-use areas, it may be conducted twice annually. If solid tine aeration is used, the holes that are created may be backfilled with organic matter such as peat moss, or simply topdressed with good topsoil. A hollow-tine aerator will remove cores, which can then be left as topdressing. Cores can be broken up by dragging chain link fencing over the area. For trees, aeration to 18 in. is required.

Deep tillage will break up compaction, but is recommended if a larger area is being treated. This technique is used in farming to break up a plow-pan or hardpan, a layer of impermeable soil that develops after repeated use of tractors and other equipment on the land. A soil amendment, such as calcined clay, is then applied to amend the soil and prevent further compaction.

PRESERVING AND REPLACING TOPSOIL

In the landscape setting, topsoil should be conserved and replaced following construction.

Prior to construction activities, steps should be taken to preserve the valuable topsoil that provides the optimum growing environment for plant roots. Although the process of removing and storing topsoil will adversely affect soil qualities, such as structure, biological activity, and organic matter, it is still preferable to save it and replace it when construction is completed, due to its soil texture and organic content. In developing codes and specifications for protecting urban quality during construction and landscaping projects, Hanks and Lewandowski (2003) make several suggestions about replacing topsoil after construction. These include ensuring all soil layers are free of compaction, tilling 2–3 in. of topsoil at a time into the underlying soil layer, and amending generously with compost.

REDUCING SUBSOIL AT THE SURFACE

Subsoil, when present, is the soil horizon found below topsoil. It is characterized by a lack of organic matter, and domination of clay particles. It is poorly aerated and does not drain well, all of which make for poor plant growth. Subsoil should be removed from a site whenever possible, or used appropriately, given its greater soil strength or ability to resist movement due to physical pressure.

Providing Adequate Soil Quantity for Root Growth

Plants require varying levels of topsoil for good growth, depending on root depth. For trees, two cubic feet per square foot of mature tree canopy is the recommended amount of topsoil. Trees have soil requirements for both depth and breadth, or width. Since 80 or even 90 percent of a tree's roots are within the top 12 in. (30 cm) of soil, they often spread over a very large area that can span a wider area than even the crown does. Thus, when trees are going to be used, it is important to prevent compaction and provide topsoil over a very large area and not just the planting hole. Annual bedding plants tend to have roots that grow only 6 in. or so deep, whereas herbaceous perennials can have roots 12 in. or more deep. Shrub root spread and depth varies, but they don't tend to be as wide-spreading as tree roots in general.

Amending with Organic Matter

Organic matter, such as compost, can be used as soil amendments to areas which are not yet planted. It may also be used to side-dress or topdress planted areas, although it is more difficult to incorporate into areas where there are plant roots. Organic mulches break down over time, and they can provide some organic matter, particularly near the soil surface. If organic mulch is placed over landscape fabric, as is often done to aid in weed suppression, only permeable fabric should be used. Eventually, the organic mulch on top of the fabric will accumulate, since most of it cannot get through the fabric. This may require removing it and replacing it. Sometimes, plant roots grow into the organic layer on top of landscape fabric, resulting in a precarious situation for the plant, since these roots are more susceptible to drying out. Care should be taken, and in some instances, it may be more desirable to avoid using landscape fabric altogether.

Composting

Composted materials can include plant material from pruning, weed removal, leaves, and dead matter from fall yard clean-up. This valuable organic material helps to build healthy soil by feeding soil organisms and contributing to the nutrient cycle. Recycling this material also reduces yard waste and the disposal costs associated with it. In sandy soils, adding organic matter allows the soil to retain more rainwater, thus decreasing runoff, and increases soil moisture. Nutrients from organic matter are released more slowly than they are from inorganic fertilizers.

Composting is a well-known, tested, and favored practice that not only recycles plant material in the form of leaves, grass clippings, branches, and herbaceous stems, but also is an inexpensive organic fertilizer. It is a renewable resource, a very important factor in sustainability. This topic will be discussed in more detail in Chapter 10.

Using Mulch

When mulch is placed directly over a planted area, plant roots may grow into the mulch. The deeper the mulch is, the more likely this is to happen, especially if adequate moisture is available. Thus, recommendations for 2–4 in. (5–10 cm) of mulch

Soil Health

around plants and in beds should always be followed. Mulch is also useful in providing low-impact walkways in areas that are subject to a lot of foot traffic.

Organic mulches break down over time, and they can provide some organic matter, particularly near the soil surface. If organic mulch is placed over landscape fabric, as is often done to aid in weed suppression, only permeable fabric should be used. Eventually, the organic mulch on top of the fabric will accumulate, since most of it cannot get through the fabric. This may require removing it and replacing it. Sometimes, plant roots grow into the organic layer on top of landscape fabric, resulting in a precarious situation for the plant, since these roots are more susceptible to drying out. Care should be taken to avoid this situation, and in some instances, it may be more desirable to avoid using landscape fabric altogether.

SUSTAINABLE FERTILIZATION

In Chapter 9, we will discuss sustainable fertilization. There are several ways to make fertilization in the landscape a more sustainable practice while building and maintaining healthy soil. A sustainable fertilization program recognizes that healthy, even vigorous plants are the primary reason for using supplemental fertilizer.

Sustainable fertilization practices include using less inorganic fertilizer and more organic fertilizer to build healthy soil, and targeting fertilizer use to actual plant needs.

SUMMARY

Soil health is a major factor in determining plant health, yet soils where construction activities have occurred present major challenges. These soils are often compacted, are notoriously inconsistent, may have construction debris buried in them, and often have an unknown history of activities that can impede plant growth.

The Sustainable Sites Initiative provides credits and prerequisites that pertain to preserving soil during construction activities and maintaining farmland soils.

Brownfields are land areas that have been contaminated with hazardous substances. Special treatment is required to restore them to some level of use. An example of a former brownfield is the Augustus F. Hawkins Natural Park located in South Central Los Angeles.

Landscape designers and managers should be aware of soil properties affecting plant growth, and be able to test for chemical and physical properties. They must also understand how poor soil infiltration affects fertilizer runoff. In addition to physical and chemical properties, healthy soils contain many living organisms, both macroscopic and microscopic.

Remedies for poor soil health include alleviating compaction, amending soils to adjust for incorrect pH and low CEC, and to improve soil structure. A good landscape manager will understand the importance of organic matter to soil health.

REVIEW QUESTIONS

1. What are two chemical properties of soil?
2. What does pH measure?

3. What are three soil particle sizes that determine soil texture?
4. Why is topsoil important for plant growth?
5. Name two ways in which construction activities affect landscape plants in the present or the future.
6. Name two actions that can be taken by landscapers to ameliorate the effects of construction activities on landscape plants.
7. What is a brownfield?
8. What turf management activity is commonly used to alleviate compaction?
9. What is a percolation test? How is it done?
10. What essential nutrient can be "fixed" by soil-borne bacteria so that it can be taken up by plants?

ENRICHMENT ACTIVITIES

1. Identify a brownfield in your area and research its history. Determine what could be done to improve the site with landscaping.
2. Interview a project manager at a construction company in your area and learn how they treat topsoil before, during, and after a project.
3. Conduct a soil test on your local soil and determine its texture, pH, and levels of nitrogen, potassium, and phosphorus. Identify any living organisms that are visible to the unaided eye. Determine the organic matter using the color chart online at https://pubsplus.illinois.edu/moreinformation/AG1941/AG1941_ex1.html.

FURTHER READING

Amthor, J. S. et al. 1998. *Terrestrial Ecosystem Responses to Global Change: A Research Strategy. ORNL Technical Memorandum*, Oak Ridge National Laboratory, Oak Ridge, TN, 37 pp.

Bender, S. F., C. Wagg, M. G. A. van der Heijden. 2016. An underground revolution: biodiversity and soil ecological engineering for agricultural sustainability. *Trends Ecol. Evol.* 31(6): 440–452.

Central Salt. http://centralsalt.com. Retrieved July 3, 2019.

Donahue, R. L., R. W. Miller, and J. C. Shickluna. 1983. *Soils: An Introduction to Soils and Plant Growth*. 5th ed. Prentice-Hall, Englewood Cliffs.

Flannery, T. 2005. *The Weather Makers*. Grove Press, New York, NY, 357 pp.

Hanks, D. and A. Lewandowski. 2003. Protecting urban soil quality: Examples for landscape codes and specifications, USDA-NRCS, Editor, 20 pp.

Ingham, E. R. The soil biology primer. https://www.nrcs.usda.gov/wps/portal/nrcs/main/soils/health/biology/. Retrieved July 3, 2019.

Kelling, K. A. and A. E. Peterson. 1975. Urban lawn infiltration rates and fertilizer runoff losses under simulated rainfall. *Soil Sci. Soc. Am. Proc.* 39(2): 348–352.

Lindsey, P. and N. Bassuk, 1991. Specifying soil volumes to meet the water needs of mature urban street trees and trees in containers. *J. Arboric.* 17(6): 141–49.

Plaster, E. J. 2009. *Soil Science and Management*, 5th ed. Thomson-Delmar Learning, Florence KY. 495 pp.

Nolan, C. et al. 2018. Past and future global transformation of terrestrial ecosystems under climate change. *Science*. 361(6405): 920–923.

Smith, K. A., T. Ball, F. Conen, K. E. Dobbie, J. Massheder, A. Ray. 2003. Exchange of greenhouse gases between soil and atmosphere: Interactions of soil physical factors and biological processes. *Eur. J. Soil Sci.* 54: 779–791.

Standards Worldwide. ASTM E1903–97. 2002. Standard Guide for Environmental Site Assessments: Phase II Environmental Site Assessment Process. http://www.astm.org/Standards/E1903.htm. Retrieved July 3, 2019.

Sustainable Sites Initiative. http://www.sustainablesites.org. Retrieved July 3, 2019.

U.S. Environmental Protection Agency. 2005. Stormwater phase II final rule: Construction site runoff control minimum control measures. Department of the Interior, Editor.

9 Sustainable Fertilization

OBJECTIVES

Upon completion of this chapter, the reader should be able to

- Compare traditional fertilization and sustainable fertilization practices
- Discuss the problems associated with traditional fertilization practices
- Identify sustainable fertilization practices
- Explain the benefits of sustainable fertilization practices

TERMS TO KNOW

Chelates
Fertilizer analysis
Iron chlorosis
Municipal solid waste
Nitrogen fixation

INTRODUCTION

Fertilizers are used to supplement the naturally occurring nutrients in the soil in order to meet the needs of a growing plant. Landscape plants vary in their needs for supplemental fertilization depending on the desired objectives. Many long-lived plants require little to no fertilization if they are planted in healthy soil. On the other hand, high-maintenance plants, such as turf on a golf course, or flowering shrubs that are expected to produce long seasons of showy blooms, will require supplemental fertilization to maintain high standards of quality. Furthermore, managed landscapes usually have plant nutrients removed in the form of grass clippings, fallen leaves, or pruned branches. Such nutrients may be returned to the soil if they are allowed to decompose in place, either intact or after shredding, or if they are composted and returned as a topdressing or soil amendment. However, all too often these organic materials are removed from the landscape and disposed of in remote locations.

Nutrients are added to the landscape in the form of fertilizers, most visibly to high-maintenance turf areas. The first lesson in fertilizing is to conduct a soil test first and then apply fertilizer as needed. However, the formulations of fertilizers and the inconvenience of conducting soil tests often result in misled guesses and one-size-fits-all applications of fertilizers in the landscape. For example, perhaps a soil test reveals that no phosphorus or potassium is required in a given landscape. Yet, all of the readily available fertilizers have a combination of nitrogen, phosphorus, and potassium. The result is overuse of phosphorus and potassium.

Fertilizer applied to lawns and landscapes just prior to a major rainfall event may be most susceptible to runoff. In a study conducted by Kelling and Peterson

(1975), fertilizer applications to a lawn and losses afterwards were looked at. They compared one area where the fertilizer had been watered in as recommended with an area where the fertilizer had not been watered in, to see if there were differences in fertilizer loss. What they found was that after a simulated storm, which was applied immediately after fertilizer application, fertilizer losses from the non-watered area averaged 10.6 percent, and fertilizer losses from the watered area were only 1.7 percent of the amount applied.

Granule fertilizers are the form commonly used on lawns. When all the granules of fertilizer do not land on the lawn, such as when they are broadcast by a rotary spreader, they may instead land on sidewalks and driveways where they may end up in storm sewers, even if it does not rain right away. Because of these and the other problems associated with fertilizer contamination of surface and ground water, some states, such as Illinois, Minnesota, Wisconsin, and New York, are banning the use of fertilizer containing phosphorus except when soil tests indicate it or when seeding a new lawn. Organic, low-phosphorus fertilizer is allowed.

FERTILIZER

Fertilizers are any materials that are applied to the growing environment of plants in order to supply the needed nutrients for healthy plant growth. Some states have legal definitions of products that must meet minimum requirements in order to be labeled as a fertilizer.

The **fertilizer analysis** of a product is a statement of the percentages of nitrogen, phosphorus, and potassium it contains. On a fertilizer label, these three numbers are prominently portrayed. For a fertilizer having equal amounts of 20 percent each of nitrogen, phosphorus (expressed as P_2O_5), and potassium (expressed as K_2O), the fertilizer analysis 20-20-20. Of these three nutrients, nitrogen is required at the highest levels. It is a component of proteins, nucleic acids, chlorophyll, and other plant compounds. After oxygen, hydrogen, and carbon, nitrogen is the most important element for plant growth. Nitrogen is available in its elemental form in the air, but plants cannot take it up in this form. They require either ammonium (NH_4^+) or nitrate (NO_{3-}) forms of nitrogen. Neither of these forms of nitrogen are retained in the soil for long periods. Nitrate is readily leached, and both forms are readily transformed into gaseous (N_2) nitrogen and lost to the atmosphere or else they are consumed by soil micro-organisms.

Nitrogen fixation is a process whereby soil microbes (*Rhizobia and Bradyrhizobia*) convert elemental nitrogen (N_2) into the forms used by plants. When it decomposes, organic matter releases nitrogen in the form of ammonium. This process occurs more rapidly in warmer weather and is also facilitated by good aeration, a neutral pH, and average moisture content.

TOXIC FERTILIZERS

Caution must be exercised in purchasing fertilizer, because legal regulations do not always stipulate exactly what can be sold as a fertilizer. Whereas essential plant nutrients must be listed on a fertilizer label, there are often other chemicals added

Sustainable Fertilization

that are not required by law to be listed or stated explicitly anywhere on a fertilizer package. In some cases, toxic substances have been sold as fertilizers. According to a study by Matt Schaffer at the California Public Interest Research Group (CALPIRG), 29 fertilizers were tested and found positive for 22 toxic metals, including arsenic, cadmium, lead, mercury, and uranium. The authors of the study state that "Between 1990 and 1995, 600 companies from 44 different states sent 270 million pounds of toxic waste to farms and fertilizer companies across the country. The steel industry provided 30 percent of this waste".

PLANT FERTILIZER REQUIREMENTS

Fertilizer needs of individual plants vary depending on the type of plant and the desired effect. For production of flowers and fruits, a different fertilizer analysis is required than for green, leafy, or other vegetative material. The three primary nutrients required by plants and provided in fertilizers are nitrogen, phosphorus, and potassium. Fertilizer products are labeled according to the percentage of N, P, and K they contain. In addition to the three major nutrients in fertilizer: nitrogen, phosphorus, and potassium, plants require thirteen other nutrients in varying amounts. The other macronutrients include calcium, magnesium, and sulfur. Micronutrients are iron, manganese, boron, molybdenum, chlorine, copper, and zinc.

A soil test can reveal available nutrients in the soil, except for nitrogen. Soil nitrogen levels can vary depending on temperature and moisture, and in addition, it leaches readily from the soil, and can be changed into forms unavailable to plants by soil-dwelling micro-organisms. For landscape plants, a soil test should be conducted every 3 to 5 years.

NITROGEN

Nitrogen is used by plants in numerous ways, including as a structural component of amino and nucleic acids, chlorophyll, pigments, and hormones. It is the most limiting nutrient in plants, as it is required in the highest amounts after oxygen, hydrogen, and carbon. These three are usually readily available from the air and water, and are not provided artificially as other essential plant nutrients are. Nitrogen contributes to the leaves and stems of plants. Trees, flowers, and lawns vary in their specific nitrogen needs (Table 9.1). Turf grasses require more nitrogen than either phosphorus or potassium, due to their leaf production, and the continual removal of leaf material during the growing season. Anywhere from 1 to 5 pounds of nitrogen per 1,000 ft^2 is recommended for annual application to turf areas. By leaving grass clippings to decompose in place, nitrogen fertilizer requirements can be reduced by as much as 30 percent.

PHOSPHORUS AND POTASSIUM

Phosphorus and potassium are commonly present at adequate levels for landscape plant growth. However, phosphorus fertilizer is often used as a starter fertilizer for newly seeded turf grass. In the garden, phosphorus aids in flowering and

TABLE 9.1
Typical Nitrogen Requirements of Landscape Plants

Plant Type	Nitrogen Fertilizer
Trees	2–6 pounds per 1,000 s.f. of crown area
Shrubs	2–4 pounds per 1,000 s.f. bed area
Flowers	1–4 pounds per 1,000 s.f. bed area
Turf	2–6 pounds per 1,000 s.f. of turf area

*Depends on species, soil, climate, organic matter

fruit production. It is provided in higher amounts in fertilizers designed to stimulate flowering in bedding and container plants. Phosphate fertilizer is sold on the basis of available phosphate and is expressed as P_2O_5 by longstanding convention. Total phosphate differs from available phosphate in that it is not easily taken up by plants. Rock phosphates, which are often derived from ancient marine deposits, provide only about 1–2 percent available phosphate, even though they contain around 30 percent total phosphate.

Potassium aids in overall plant health and contributes particularly to healthy root growth. A healthy root system provides better tolerance to stresses in the environment, such as drought, excessive heat, and cold. Potassium is included with nitrogen in winterizer fertilizers that are applied in the fall on cool-season grasses, to aid in root growth in fall and spring, which improves spring green-up with warming temperatures. Warm-season grasses do not receive winterizer fertilizer due to the fact that they are dormant in winter, and the warm climates in which they grow are not as challenging to their root development. If soil tests indicate a need for potassium, it should be applied in spring or summer on warm-season grasses.

CALCIUM AND MAGNESIUM

Calcium and magnesium can compete with each other, if both are present in similar amounts. However, calcium is not usually deficient in most landscapes. Both of these macronutrients can be leached from the root zone in high-rainfall areas. One outcome of that is a lower pH. Calcium deficiencies can be easily remedied by the addition of limestone. If a soil test reveals a deficiency in magnesium, it can also be remedied using dolomitic limestone, which contains both calcium and magnesium.

SULFUR

Sulfur is commonly used on alkaline soils to lower the pH. It is also incorporated into soil where low-pH-requiring plants are desired and soil levels are naturally too high. Rhododendrons and azaleas are common landscape plants that require relatively low pH levels of around 5.0–5.5. Aluminum sulfate is commonly used for this purpose. However, sustainable landscape practices do not support artificially amending soil in order to grow plants in areas where they would not otherwise thrive.

Sustainable Fertilization

TABLE 9.2
Plants Susceptible to Iron Chlorosis

Common	Botanical
Pin oak	*Quercus palustris*
Silver maple	*Acer saccharinum*
Azalea	*Rhododendron* spp.
Chrysanthemum	*Chrysanthemum morifolium*
Sweetgum	*Liquidambar styraciflua*
Spirea	*Spiraea japonica*
Birch	*Betula nigra*
Hibiscus	*Hibiscus* spp.
Crabapple	*Malus* spp.
Plum	*Prunus* spp.
Gardenia	*Gardenia* spp.
Bermudagrass	*Cynodon dactylon*
Ixora	*Ixora* spp.
Bougainvillea	*Bougainvillea* spp.

IRON

Of the remaining nutrients, iron is often the second most common nutrient deficiency seen in landscape plants. Some plants, such as pin oaks, do not grow well on soils high in calcium due to increased difficulty taking up the iron that is present. Such plants will have yellowed leaves, a condition known as **iron chlorosis**. The best decision with respect to sustainable practices is to not try growing plants on soils that are not naturally suitable for them. Table 9.2 lists other plants that are susceptible to iron chlorosis.

FORMS OF FERTILIZERS

Fertilizers are available in fast-release or slow-release forms, and may be applied dry or in a liquid solution. They should only be applied in the landscape situation when soil tests indicate a deficiency. Some proponents of sustainability in landscaping suggest that only plants adapted to the soil type should be used in a specific situation. Rather than apply fertilizers at all, which encourages unwanted weed growth, such sources suggest using plants that are adapted to the soil nutrient level. Unfortunately, many urban situations have soils which have been excessively disturbed, or native soils that have been removed and not returned following construction. Thus, in some situations, fertilizer may be needed just to grow acceptable plants (Table 9.3).

FERTILIZER SOURCES

Fertilizers may be composed of crushed rock or minerals, or organic materials, such as manure, or they may be a product of a manufacturing process. In the following

TABLE 9.3
Foliar Levels of Essential Plant Nutrients in Woody and Herbaceous Plants

Macronutrient	Woody (%)	Herbaceous (%)
Nitrogen – evergreens	1.5–3.5	NA
Nitrogen – deciduous	2.0–4.5	1.5–3.5
Phosphorus	0.2–0.6	0.15–0.4
Potassium	1.5–3.5	1.5–3.5
Calcium	0.5–2.5	1.2–2.7
Magnesium – evergreens	0.2–2.0	NA
Magnesium – deciduous	0.3–1.0	0.3–0.65
Micronutrient	**Woody (ppm)**	**Herbaceous (ppm)**
Boron	30–50	15–65
Copper	6–50	4–10
Iron	50–700	60–400
Manganese	30–800	50–400
Molybdenum	0.6–6.0	0.5–7.0
Zinc	30–75	30–75

discussion, fertilizers will be treated in two categories: inorganic or mineral fertilizer, and organic fertilizer. The latter comes from plants or animals.

MINERAL FERTILIZERS

Mineral fertilizers are mined or manufactured. They usually are highly soluble and provide a rapid growth response. Most mineral fertilizers have a high nutrient content. Ammonium sulfate, ammonium nitrate, urea formaldehyde, calcium nitrate, potassium nitrate, and urea are examples of mineral fertilizers. In general, mineral fertilizers are less expensive than organic ones, are more readily available for plant use, and often provide higher levels of nutrients (Table 9.4).

Nitrogen

Mineral fertilizers, also called synthetic fertilizers, are mined or are products of a chemical process. Nitrogen carriers are compounds that supply elemental nitrogen. Most nitrogen carriers are manufactured using the Haber process. This process uses natural gas as a source of hydrogen (H_2) and combines it with nitrogen from the air (N_2), resulting in anhydrous ammonia ($2NH_3$). Anhydrous ammonia can then be further processed into urea, ammonium nitrate, sodium nitrate, calcium nitrate, ammonium sulfate, or ammonium phosphates.

Phosphorus

Phosphorus carriers are mined from rock phosphate in Florida and other states (Figure 9.1). Phosphate rock contains calcium phosphate, also known as apatite. Apatite is found in the western United States. It can be crushed and applied directly

TABLE 9.4
Inorganic Fertilizers that Serve as a Major Source of Nitrogen, Phosphorus, and Potassium

Fertilizer	% N	% P	% K
Ammonium nitrate	33	0	0
Ammonium sulfate	21	0	0
Calcium nitrate	17	0	0
Diammonium phosphate	18	46	0
Gaseous anhydrous ammonia	82	0	0
Monoammonium phosphate	11	48	0
Potassium nitrate	13	0	44
Sodium nitrate	16	0	0
Urea	46	0	0
Urea formaldehyde	37	0	0
Concentrated superphosphate	0	45	0
Phosphoric acid	0	23	0
Rock phosphate	0	11–15	0
Superphosphate	0	20	0
Triple superphosphate	0	20	0
Potassium chloride (muriate of potash)	0	0	60
Potassium magnesium sulfate	0	0	22
Potassium sulfate	0	0	49

FIGURE 9.1 Manure can be properly composted and used as an organic supplement.

to the soil, but is often treated with sulfuric or phosphoric acid to produce fertilizers such as superphosphate, phosphoric acid, ammonium phosphates, and triple superphosphate.

Potassium

Most of the potassium used in the United States is mined in New Mexico, Utah, California, and Canada. Since the mined materials contain impurities, such as salts of potassium, sodium, and magnesium, they have to be purified.

Effect of Inorganic Fertilizers on Soil Health

Most fertilizers are salts. If accidentally spilled on the lawn or planting beds, the salts from fertilizers can kill the grass or other plants. When over-applied, fertilizers "burn" plant roots by drawing water out of them into the soil. Soil structure, which is formed by the aggregation of soil particles, can be destroyed by salts. Fertilizer salts are also detrimental to soil-dwelling organisms that provide beneficial effects in the soil. Thus, the positive effects on soil structure realized with organic fertilizer sources are not seen with inorganic fertilizers.

Nitrogen fertilizers added to soil feed micro-organisms that live there and thus can support microbe populations. They do not have breakdown products such as humus in organic matter. In general, fertilizers primarily provide N, P, and K, although there may also be some iron, sulfur, calcium, or magnesium compounds as ingredients.

Contamination of the Environment

Fertilizer contamination occurs when applications of fertilizer end up in non-target areas, such as bodies of water, where they can cause problems for the environment, wildlife, and human health. Most nitrogen fertilizers are soluble in water, moving readily through the ground in wet conditions, and washing off when rain fall follows nitrogen application in the landscape. Nitrogen moves more readily in coarser, sandy soils than finer, clay soils. When there are too much nitrogen and phosphorus in surface water, it leads to a condition known as eutrophication. An undesired result is that aquatic plants grow excessively. This is part of a cascade of events that leads to hypoxia, or low oxygen levels in the water that can harm fish and shellfish.

Excessive nitrates in drinking water can adversely affect children's health, sometimes causing blue baby syndrome, in which the oxygen-carrying capacity of hemoglobin in the blood is reduced to dangerously low levels. Only about 13 percent of the soluble nitrogen in soil water is derived from fertilizers, however. Soil humus, human and animal manure, and nitrogen fixation by soil-dwelling micro-organisms account for 77 percent, with the remaining amount coming from rain and snow.

Reducing Nutrient Runoff and Leaching

Movement of nitrates from fertilizer can be slowed by plant uptake or by rapidly expanding populations of soil micro-organisms. Rain gardens and bio-swales can be placed between lawn and garden areas that receive fertilizer and storm water drains to remove nitrates and prevent contamination of the surface water. Chapter 7 discusses rain gardens and bio-swales in detail.

Sustainable Fertilization

Slow-release (controlled-release) nitrogen sources and organic sources of nitrogen can help reduce environmental contamination from nitrate fertilizers. Some commercially available slow-release nitrogen fertilizers are urea formaldehyde and related urea-based formulations, isobutylidene diurea, and sulfur coated urea. Plastic-coated fertilizers, such as Osmocote™, Nutricote™, MagAmp™, and Duration CR, are controlled-release products. In the case of the latter, Duration CR, the polymer coating allows soil moisture to dissolve the encapsulated urea nitrogen. Then, the liquid nitrogen is released over a predictable period of time, its release being activated by temperature. As a result, only one application per year is required, lasting 1–7 months, depending on the specific product. Nutrient usage, runoff, and leaching may all be reduced. In addition to these inorganic sources, many organic fertilizers and materials are released slowly, as well.

ORGANIC FERTILIZERS

Organic fertilizers come from organic materials that may otherwise be considered waste materials. The primary organic fertilizer materials are animal manure and sewage sludge. Other organic materials that can be used for fertilizer are mushroom compost, oyster shells, reed or sedge peat, rice hulls, and wood ashes. Cottonseed meal, blood meal, and bone meal are organically derived byproducts. Of course, organic matter, such as leaves, shredded wood and wood chips, grass clippings, and composted yard waste, may also be used as fertilizer, although they are typically used as a soil amendment to improve soil structure. Even these materials will release nutrients over time. Chapter 10 discusses how to improve the soil with organic matter in more detail.

Organic fertilizers usually release their nutrients slowly, except in the case of incompletely cured manure, which can burn plants. This slow-release action means that less nutrients are lost through leaching. Table 9.5 shows the range of nutrient analysis for various types of organic materials that may be used as fertilizer.

In a survey research on mulch, Chalker-Scott (2007) reported that living mulches, such as turf or other groundcovers, and organic mulches, such as shredded wood, may increase, decrease, or have no effect upon nutrient levels. The important factors were type of mulch, soil chemistry, and which nutrients are concerned. Both living and organic mulches release nutrients into the soil as they decompose. Green manure (growing plants) and animal manures supply nutrients at higher rates than other mulch choices (straw, bark, and wood chips) and often perform better than organic fertilizers. It is important to keep in mind, however, that organic matter helps build soil structure and improves soils quality. Soil with organic components retains water better than soil lacking organic components, while also providing better draining in clayey soils.

Animal Manure

Manure can come from many sources, including cattle, poultry, horse, sheep, swine, and even rabbits. Commercially produced manure compost is most cost-effective when it is processed close to both its source and its end-use. Manure should always be "cured" before it is applied to soil around plants because of extremely high levels

TABLE 9.5
Organic Fertilizers and Their Nutrient Analyses

Fertilizer	Analysis (%)			Comments
	N	P	K	May contribute excessive salts; may also contribute micronutrients.
Manure				
Cattle	0.5–8	0.3–1	0.5–3	6–15 percent TSS
Chicken	0.9–8	0.5–2	0.8–2	2–5 percent TSS
Horse	0.6	0.3	0.6	
Sheep	0.9–5	0.5–3	0.8–3	1–2 percent TSS
Swine	0.6–5	0.5–1	0.4–2	1–2 percent TSS
Mushroom	1	1	1	5 lbs./ft^2 recommended application rate
Sludge				
Sewage	2	1	1	5 lbs./ft^2 recommended application rate
Activated sewage	6	5	0	3–4 lbs./ft^2 recommended application rate
Tankage	4	1.5	2	High in potash. 5 lbs./ft^2 recommended application rate

of nitrogen that can burn plant roots which will kill plants. Curing is another name for composting or decomposing. Manure is heaped into piles or mounded rows for composting (Figure 9.1). As it is decomposed by organisms, the pile heats up in the center. After a period of time, the temperature begins to drop, and the compost must be turned because material located towards the outside of the pile is cooler and will not decompose as readily. During this time, the decomposing organisms are increasing in number through reproduction. The nitrogen and carbon in the manure serve as food sources. Adequate moisture must be available in order for the process to continue. Sometimes, rows of manure may be covered with plastic tarp to maintain even temperatures within the piles.

Depending on the origin of the manure, the temperature and moisture of the manure pile, methods of managing the pile, and other factors, various periods of time are required for the adequate decomposition of the manure. A well-managed composting process may be ready in as little as a month.

The fertilizer analysis of manure varies due to the number of factors involved. Any given lot of processed manure requires a separate analysis, so even if the same source is used, the exact amount of nutrients may vary. Cured, or composted, manure typically has low levels of nutrients that are released slowly over a period of time (Table 9.5). They also may contain higher levels of micronutrients than inorganic, or synthetic, fertilizers. Due to the organic matter content of manure, it can help build soil structure as it breaks down after being applied to the landscape. The organic matter serves as a site for cation exchange and also provides **chelates**, which hold the nutrient in a form that makes it available to plant roots, rather than getting tied up in unavailable compounds, and so provides benefits to landscape plants that inorganic fertilizer does not.

Case Study: Organic Soil Amendment

Reference: Loper, Shawna, Amy L. Shober, Christine Wiese, Geoffrey C. Denny, Craig D. Stanley, and Edward F. Gilman. 2010. Organic soil amendment and tillage affect soil quality and plant performance in simulated residential landscapes. *HortScience* Vol. 45(10) October: 1522–1528.

Citing the poor soil conditions in the urban environment, a group of researchers wanted to find out whether application of compost with or without tillage or aeration could aid plant growth in residential landscapes. To test their idea, they simulated twenty-four different landscape plots using 4 in. of loamy sand fill material over a compacted base of native soil (sand). The research plots were set up in Florida, so the landscape plants used were common to that area. Part of each plot was planted with turfgrass (*Stenotaphrum secundatum*). The organic soil amendment used was composted dairy manure solids.

All the plants, including the turfgrass, but excluding *Rhaphiolepis indica*, showed improved growth when compost was applied. Tillage or aeration had little effect on soil properties, but adding the organic amendment contributed to a better moisture-holding capacity. Whereas the control plot had a slightly alkaline pH of 7.46, composted plots had a slightly lower pH initially, but it rose again after a couple of months. Plots receiving only tillage or aeration did not see this decrease in pH. Unfortunately, the pH levels never reached the ideal level for some of the plants.

Phosphorus levels were higher in the composted plots, but even in the control plots, phosphorus levels were adequate for normal growth. The researchers concluded that composted dairy manure solids can be incorporated into the soil or used as a topdressing for similar growth responses in selected landscape plants. Shallow tillage and aeration, on the other hand, did not positively affect landscape plant growth. Nor did these practices show the improved soil physical properties that were seen with the addition of composted manure solids.

Disadvantages of Manure as Fertilizer

Animals are often fed salt (sodium chloride) as part of their diet, and this comes out in the manure. Total soluble salts (TSS) in manure can range from 6 to 15 percent for beef or dairy cows, to 1–2 percent in sheep and swine manure. Excess manure should be avoided due to the potentially toxic accumulation of salts, especially in arid situations where irrigation is not used. In a moist environment, salts can leach out of the root zone. Nevertheless, they should not be allowed to leach into groundwater that serves as a source of drinking water.

Applying Manure Fertilizer

Recommendations for manure application in landscape situations may be extrapolated from those given for agricultural use. Typically, fertilizer use for landscape plants is given in pounds per 1,000 ft^2, whereas, for agriculture, it is given in pounds (or tons) per acre. Donahue et al. 1983, using several sources, estimate optimal rates of manure to be about 18–22 tons/acre for agricultural production. This comes out to about one-half to one pound per square foot, or one-quarter to one-half ton per 1,000 ft². Higher rates will likely result in excessive salts that can be harmful to plants or leached into groundwater. About 45 percent of the nitrogen in the manure is released in the first year, and 25 percent is released the second year.

Alayne Blickle, Pierce Conservation District in Puyallup, Washington, recommends spreading a half-inch layer at a time, for a total of 3–4 in. per year of horse manure on flower beds during the growing season. Koenig and Johnson offer the following formula for calculating fertilizer applications:

$$\text{Fertilizer needed} \times \text{lb of nutrient} \times 1 \text{ lb fertilizer} \times Z \text{ ft}^2 \text{ area}$$
$$= 1000 \text{ ft}^2 Y \text{ lb nutrient,}$$

where X is the required or recommended amount of a nutrient, Y is the nutrient percentage in the fertilizer, and Z is the square footage of the area to be fertilized.

Cowsmo brand compost is made from cow manure from their dairy farm operation. It is composted using a windrow turning system, and the compost is monitored regularly to ensure it maintains a minimum temperature of 131°F (55°C) for 15 days. The manure is turned at least five times using a compost turner that ensures uniformity of the finished compost. It complies with standards set for organic products set by the USDA (United States Department of Agriculture). The product is recommended for landscape mulch or planting bed establishment at a rate of 2–3 in. (5–8 cm), or for topdressing on lawns, and for incorporation into gardens at a rate of 1 in. (2.5 cm), applied annually.

If animal manure is not properly composted, it may contain weed seeds and/or disease pathogens. Be sure to obtain composted manure from a reputable source. Commercial operators of composting facilities should provide details of their operations and are usually amenable to giving tours to interested parties.

Municipal Solid Waste

Municipal solid waste is receiving increased attention as a horticultural product due to increasing waste disposal costs and environmental concerns with some disposal methods. Municipal solid waste derives from human effluent, among other things. Table 9.6 shows the nutrient value of human effluent. Compost made from municipal solid waste has been successfully used as a substitute for peat moss in production of container plants and as a seedbed for turfgrass production (Rosen et al., 1993).

In sewage treatment in the United States, there are commonly three processes, or stages, known as the primary, secondary, and tertiary stages. During the primary treatment process, large items and wastes that float or sink, such as wood and rocks, are removed. Next, the wastewater is pumped, if necessary, to aeration tanks, which are often placed higher than the rest of the system. This allows gravity flow to be used for the remaining stages of the process. Aeration allows dissolved gases, such as hydrogen sulfide, to be released. This is the gas that gives wastewater treatment plants the bad smell, similar to rotten eggs.

TABLE 9.6
Human Effluents and Their Major Plant Nutrient Content

	Major Plant Nutrients (Dry-Weight Basis), %		
Effluent Type	Nitrogen	Phosphorus	Potassium
Sewage sludge	4	2	0.4
Septage	4.4	1.6	0.4

Source: Donahue, Miller, and Shickluna, 1983.

Oxygen is pumped through tanks that hold the wastewater, facilitating the decay of the organic matter in the solution. During aeration, small particles such as sand can settle out and be removed.

In the secondary stage, wastewater goes into sedimentation tanks where the organic portion, or sludge, settles out. It is then pumped out of the tanks, and some of the water is removed. The sludge is then processed in digesters, large tanks containing naturally occurring bacteria. Left floating on the surface of the water is oil, grease, plastics, and soap. This residue, or scum, is removed, thickened, and pumped to digesters along with the sludge.

Some municipalities use a third stage, filtration. The liquid portion is passed through tanks of sand to remove almost all bacteria and odors, and removes most other solid particles that still remain. Water is sometimes filtered through carbon particles, which removes organic particles.

In a final stage, chlorine is added to kill bacteria. Usually, the chlorine is eliminated in this process, but sometimes, it must be neutralized using other chemicals. This treated water is called effluent. In many situations, the effluent is discharged to a local river or into the ocean.

The solid-waste material that is left from wastewater treatment is kept for 20–30 days in digesters, where it is heated so that bacteria can break it down. The water content is decreased, and odors are eliminated while disease-causing organisms are killed.

When the finished product has 20 percent or more solids, it can be handled with a shovel. It may be dried down and used as a fertilizer or soil conditioner. When sold as a soil conditioner, guaranteed nutrient analyses are not required as they are for fertilizers. Sludges are variable in their nutrient composition, but in general, they contain 3–5 percent nitrogen, around 2 percent phosphorus, and 0–0.4 percent potassium on a dry-weight basis. In addition to beneficial nutrients for plants, they also contain micronutrients as well as toxic heavy metals such as arsenic, boron, cadmium, chromium, copper, lead, mercury, nickel, selenium, and zinc. They could also contain pesticides or other chemical compounds that are toxic to plants. Milwaukee, Wisconsin, sells their municipal sludge under the product name "Milorganite", which has a fertilizer analysis of 5-2-0. Sewage sludge that is used in agriculture is regulated under the Clean Water Act and is currently subject to concentration limits for the metals arsenic, cadmium, copper, lead, mercury, molybdenum, nickel, selenium, and zinc.

COMPOST SOLUTIONS

Compost tea and compost extract are liquid solutions made from compost. They are distinguished from one another in that compost tea is brewed under aerobic conditions and compost extract is made in an anaerobic process. A common method of obtaining compost extract is to suspend the compost in a burlap sack in a barrel of water for 7–14 days. Compost extract is primarily used as liquid fertilizer which supplies soluble nutrients.

Compost tea is obtained from an aerobic process, in which the compost is placed into a vat or barrel. A food source for the micro-organisms present is provided in order to increase the microbial populations. Some food sources that are commonly used are kelp, molasses, and fish powder. Humic–fulvic acid, yucca extract, and rock dust are used as microbial catalysts. An aerator such as a sump pump is used to provide oxygen to the solution during the brewing process, which lasts 24–36 hours. Plans to build a 25-gallon (95-l) compost tea brewer are available online from the Oregon State University Extension Service. Greater Earth Organics is a company that specializes in compost tea brewing systems. Their website, http://greaterearthorganics.com/, has instructional videos and schematics for making a compost tea brewing system.

Research results on the efficacy of compost tea vary. There is a lot of variability in the qualities of the tea due to the ingredients that go into it. In a study conducted by Carbollet and colleagues (2009), manure source, presence or absence of aeration, and temperature during brewing were all factors that altered the final product. The researchers did find phytotoxicity could be a problem at higher rates of application.

Case Study: Municipal Solid-Waste Compost and Compost Tea

Reference: Hargreaves, J.C., M. S. Adl, and P. R. Warman. 2009. The effects of Municipal Solid Waste Compost and Compost Tea on mineral uptake and fruit quality of strawberries. *Compost Science and Utilization*. Volume 17 (2):85–94.

Municipal solid-waste compost (MSWC) represents a recycling opportunity for many communities. However, some MSWC has high levels of salts and metals that are of concern. Compost tea can be produced from MSCW, which presents income potential.

Compost tea is popular among organic growers and is the subject of many claims, including that it confers disease resistance, aids beneficial soil-dwelling organisms, and supplies plant nutrients. Compost tea is used as a foliar spray and thus claims are that the nutrients provided are available more quickly than soil-applied nutrients.

In this study, the researchers used three levels of MSWC and two levels of compost tea made from the MSWC. The compost was applied to the soil surface, whereas the compost tea was applied as a foliar spray. They applied treatments for two consecutive years, with the first year being a vegetative-only year for the strawberry plants. In the second year, foliage and fruit nutrient levels were determined, as well as vitamin C and anti-oxidant levels in the fruit.

Results from the study indicated that yields were much lower than expected for all treatments, and nitrogen levels in the plants indicated insufficient nitrogen.

Fruits had suffered from a condition called leather rot, which may have contributed to the low yields. Sugar content of the fruit and anti-oxidant levels were comparable to those observed in other studies. The full study can be read at http://soilecology.biology.dal.ca/Reading%20Picks%20Assets/Hargreaves%20et%20al%202009%20compost%20teas.pdf.

OTHER ORGANIC FERTILIZERS

Bone meal is readily available commercially. It contains about 27 percent available phosphorus, which is nearly all the phosphorus present. Greensand, or glauconite, is a source of potassium. Total potassium content of greensand is around 7 percent, all of which is locked up in the mineral and only available for plant use over a long period of time. Greensand is said to have a beneficial effect on soil structure.

Daniels® Plant Food is manufactured and marketed by Ball DPF LLC in formulations for the greenhouse, organic, landscape, and retail marketplaces. For ornamental landscape plants, Daniels Ornamental Landscape 10-4-4 is available and comes in 2.5-gallon twin packs, 55-gallon drums, and 275-gallon totes and tankers. Daniels Plant Food is derived from oilseed extract. Since there are an estimated 325,000 metric tons of oilseed extract available in the United States annually, transforming it into fertilizer is not only a sustainable practice, but also reduces mining of potassium and phosphate, and natural gas demand in making nitrogen fertilizers. Since it comes from seeds, there are fewer salts as compared to mineral-based water-soluble fertilizer. The recommended feed rate for Daniels 10-4-4 is 0.10 lb. of actual nitrogen per 1,000 ft^2 which should be applied every 7–10 days. One 2.5 gallon container of Daniels 10-4-4 will feed up to 25,000 ft^2. Many landscape companies are using this fertilizer, which is readily available from Daniels Plant Food distributors throughout North America.

Griffin Industries LLC produces Nature Safe® Natural and Organic Fertilizers, which are formulated from animal proteins. Their products are approved by the Organic Materials Research Institute and are used on turf and ornamental areas, in addition to farming and gardening.

GREEN MANURE AND INTER-PLANTING

Green manure is the practice of planting a cover crop that gets disked in prior to seasonal planting. Nitrogen-fixing legumes have been used as green manure. Inter-planting a nitrogen fixing plant is another way to get the benefits of nitrogen fixation while maintaining a perennial crop, such as turf grass. To this end, it was once common practice to inter-plant clover in turf areas. Clover is a legume, meaning that it fixes nitrogen. Since turf requires relatively high levels of nitrogen, it is sensible to consider growing clover with it to meet its nitrogen demands. When nitrogen fertility is low, turf areas may become have a lot of clover in them.

In discussing low-maintenance turf, Tom Cook, professor at Oregon State University, has tested several mixes combining turf grasses with broadleaved plants in order to develop an ecologically stable lawn that would reduce inputs, such as mowing, irrigation, fertilizer, and pesticides. He found that strawberry clover (*Trifolium fragiferum* "Fresa") was not as invasive as white clover (*T. repens*), although it still attracted bees when in bloom.

SUMMARY

Fertilizers are applied by landscapers as needed to permit healthy growth during the appropriate times. Although many plants require little fertilization, if any, others such as turf, do. Turf is constantly being mowed during the growing season and thus requires replacement nutrients to sustain healthy growth. Furthermore, managed landscapes often have plant nutrients removed in the form of plant materials – nutrients which may be returned to the soil if the organic matter is allowed to decompose in place, or if they are composted and returned as a topdressing or soil amendment.

Other than nitrogen, fertilizer should be applied to the landscape after a soil test is conducted to determine the nutrient status of the plants. Application of fertilizer should be informed by the nutrient requirements of the plants. Trees, shrubs, flowers, and turf have differing nutrient requirements. Overuse of fertilizer can occur when a "one-size-fits-all" approach is taken.

Fertilizer contamination from landscapes often occurs when heavy rains follow fertilizer application. This is more common in the spring, but can occur at other times, too. Fertilizer granules can be lost when they are spread on non-target areas, such as streets, driveways, and sidewalks, where they are easily washed into storm drains. Some states are banning the use of phosphorus fertilizer on landscaped areas in order to reduce phosphorus contamination of surface waters.

Toxic fertilizers are a problem in many states, because labeling only requires percentages of nutrients to be stated. Other chemicals, including toxic metals, are not required to be identified.

The major nutrients required by plants that are provided in fertilizers are nitrogen, phosphorus, and potassium. Of these, nitrogen is required in the greatest amounts and is not stored in the soil to any great extent. Phosphorus and potassium are usually present at adequate levels in landscape soils. Phosphorus fertilizer is used as a starter fertilizer for newly seeded turf grass and also aids in flowering and fruit production. Potassium contributes to healthy root growth, leading to drought, heat, and cold tolerance. Potassium is included in winterizer fertilizers for turf. Calcium, magnesium, sulfur, and iron are the remaining macronutrients required by plants.

Fertilizers are available in various forms and come from a variety of sources. Two general groupings of fertilizer sources are those that are inorganic, including from mineral sources, and those that are organic, meaning they originate with plant or animal material. Fertilizers may be fast-acting or slow-release. Slow-release fertilizers are generally better at reducing runoff and contamination.

Organic fertilizers include animal manure and sewage sludge. Organic matters, such as leaves, shredded wood and wood chips, grass clippings, and composted yard waste, are often used as soil amendments rather than fertilizer. Organic fertilizers usually release their nutrients slowly, except in the case of incompletely cured manure, which can burn plants. Green manure involves inter-planting a nitrogen-fixing plant, such as clover in turf.

Sustainable Fertilization

REVIEW QUESTIONS

1. What information is provided by the fertilizer analysis?
2. Why are some fertilizers considered toxic?
3. Where do toxic fertilizers come from?
4. Compare the nutrient needs of turf, flowers, and woody plants.
5. What are the main nutrients in winterizer fertilizer and why are they used?
6. What are the two main forms of nitrogen taken up by plants.
7. What is nitrogen fixation and how does it occur?
8. What are some of the benefits of potassium to plants?
9. Compare the differing amounts of nitrogen fertilizer required for cool season turf grasses if clippings are removed or not.
10. What is green manure and how would it work in a landscaped environment?

ENRICHMENT ACTIVITIES

1. Using the formula provided by Koenig and Johnson, estimate the amount of various mineral and organic fertilizers and compare the results. Check the prices for each type of fertilizer for a price-per-nutrient comparison.
2. Interview the manager of wastewater treatment facility that recycles their sewage waste into compost. What did it take to switch from their previous method of treating waste? What are the pros and cons?
3. Identify potential sources of organic fertilizer in your area. Interview the owners or managers of facilities to find out whether they have considered recycling the waste rather than disposing of it. Conduct a feasibility study to discover what the logistics and expenses would be for such an endeavor.

FURTHER READING

Carballo, T., M. V. Gil, L. F. Calvo, and A. Moran. 2009. The influence of aeration system, temperature, and compost origin on the phytotoxicity of compost tea. *Compost. Sci. Util.* 17(2): 127–139.

Chalker-Scott, L. 2007. Impact of mulches on landscape plants and the environment—a review. *J. Environ. Hort.* 25(4): 239–249.

Donahue, R. L., R. W. Miller, and J. C. Shickluna. 1983. Soils: an introduction to soils and plant growth. 5th ed. Prentice-Hall, Englewood Cliffs. 667 pp.

Gilman, E. F., I. A. Leone, and F. B. Flower. 1987. Effect of soil compaction and oxygen on vertical and horizontal root distribution. *J. Environ. Hort.* 5(1): 33–36.

Hensley, D. L. 2010. *Professional Landscape Management.* Stipes Publishing, Champaign, IL. 354 pp.

Ingham, E. R. 2011. *The Compost Tea Brewing Manual.* 5th ed. Soil Foodweb Inc., Corvallis, OR, 88 pp.

Kelling, K. A. and A. E. Peterson. 1975. Urban lawn infiltration rates and fertilizer runoff losses under simulated rainfall. *Soil Sci. Soc. Am. Proc.* 39(2): 348–352.

Koenig, R., and M. Johnson. 2011. *Selecting and using Organic Fertilizer.* Utah State University, Logan, UT. http://extension.usu.edu/files/publications/factsheet/HG-510.pdf. Retrieved July 5, 2019.

Nelson, P., C. Niedziela, and D. Pitchay. 2009. Daniels plant food: Does it work? *GrowerTalks*, February.

Oregon State University Extension Service. Plans for a homemade 25-gallon compost tea brewer. https://www.swcd.net/wp-content/uploads/2011/06/Compost-Tea.pdf. Retrieved July 5, 2019.

Pelczar, R. 2008. Comparing natural and synthetic fertilizers. *The American Gardener*, March/April: 51–53.

Plaster, E. J. 2003 *Soil Science and Management*. Delmar Learning, Clifton Park, NY, 384 pp.

Quarles, W. 2001. Compost tea for organic farming and gardening. *IPM Pract.* XXIII(9): 1–8.

Rosen, C. J., T. R. Halbach, and B. T. Swanson. Horticultural uses of municipal solid waste composts. *HortTechnol.* 3(2): 167–173.

Rosenow, J. 2011. Cowsmo compost pamphlet. Cowsmo, Inc., Cochrane, WI. www.cowsmocompost.com. Retrieved February 25, 2011.

Shaffer M. Waste lands: the threat of toxic fertilizer. http://www.pirg.org/toxics/reports/wastelands/. Retrieved July 5, 2019.

University of Hawaii Sustainable Agriculture Research and Education (SARE). Compost tea. https://www.sare.org/content/download/66749/944806/Compost_Tea_Manual.pdf. Retrieved July 5, 2019.

USGS. A visit to a wastewater-treatment plant: Primary treatment of wastewater. In: *Water Science for Schools*. https://www.usgs.gov/special-topic/water-science-school/science/a-visit-a-wastewater-treatment-plant?qt-science_center_objects=0#qt-science_center_objects. Retrieved June 9, 2020.

Vitousek, P. M., J. D. Aber, R. W. Howarth, G. E. Likens, P. A. Matson, D. W. Schindler, W. H. Schlesinger, and D. G. Tilman. 1997. Human alteration of the global nitrogen cycle: sources and consequences. *Ecol. Appl.* 7(3): 737–750.

10 Improving Landscape Soils with Organic Matter

OBJECTIVES

Upon completion of this chapter, the reader should be able to

- Explain the role of organic matter in landscapes
- Discuss the benefits of organic matter to soil
- Describe the composition of organic matter
- List the major components of organic matter
- Identify types of organic amendments and their properties
- Identify effects of organic amendments in landscaped environments

TERMS TO KNOW

Amendments
Dripline
Fulvic acid
Humic acid
Humin
Humus
Mulch
Vermicompost

INTRODUCTION

Organic matter is important to soil for a number of reasons:

- The addition of organic matter is a good aid to building healthy soil structure.
- Cation exchange capacity, and thus nutrient holding ability, is increased by organic matter present in the soil.
- As it breaks down, it releases nutrients that may become available to plant roots.
- Organic matter in the soil can help retain moisture that is readily available to plant roots.

How is organic matter added to landscape soils? The landscaper should first try to retain organic matter on site. To accomplish this, mulch grass clippings in place and add shredded leaves to flower beds. Also, recycle both woody and herbaceous prunings by shredding and/or composting them. Next, incorporate organic matter into planting sites, prior to planting. Use composted yard waste, peat moss, or other readily available organic materials.

Organic matter is already used to varying degrees in different parts of the country and in different landscape applications. However, current landscape practices often involve removing large amounts of organic matter from the landscape and disposing of it in a remote location rather than using it to renew soil fertility. In doing so, an opportunity to recycle carbon, nitrogen, and other essential nutrients back into the landscape is lost. Nutrients that are recycled through decomposition may be taken up by landscape plants on site. The improvement to soil structure can have long-lasting effects that are evident years later.

Due to size restrictions, many municipalities no longer allow landscape waste to be disposed of in landfills. Some have initiated yard waste disposal sites, and some even shred the yard waste and provide free mulch from it. Some landscape companies have begun their own composting operations as a response to the need to dispose of landscape waste such as tree prunings, leaves, and grass clippings. Others are purchasing composted manure from animal production facilities and dairies, which have a need to dispose of animal wastes.

In this chapter, we will look at the effects of organic amendments to soil health. We will look at ways to incorporate more organic matter in the landscape in order to reduce landscape waste while reaping additional benefits from it. Chapter 9 discussed sustainable fertilization, including organic fertilizers, whereas this chapter will focus on adding organic matter into the landscape and leaving organic matter in the landscape rather than removing it.

ORGANIC MATTER IN THE LANDSCAPE

In a traditional American landscape consisting of lawn, trees, shrubs, and flower beds, there is a cycling of organic matter throughout the growing season. This traditional landscape can be found throughout the Northeast, Midwest, and Eastward to the Atlantic Ocean, and other areas that receive adequate rainfall and a growing season that permits a fair level of organic matter to develop in spring and summer. Even in more arid and subtropical climates, lawns are mowed, and shrubs and trees are pruned, creating organic matter that is removed from the landscape. This is contrasted to natural areas, such as woodlands, meadows or prairies, wetlands, and deserts, where plant materials that are shed due to natural life cycles or during dormant seasons remain in place where they eventually decompose.

Simply put, organic matter derives from plants and animals, including tissues and waste products. In the soil, organic matter may be present as living organisms (see Chapter 8). Plant roots could also be included in this category of living organisms in soil. Soil organic matter may also be present as undecayed, partially decayed, or fully decayed plant and animal materials. In this chapter, we are concerned with the latter materials that may be used as soil amendments with the objective of improving soil health.

FATE OF ORGANIC MATTER

Organic matter that has fully broken down into stable residues is referred to as **humus**. Humus is comprised of many individual substances:

- Carbohydrates
- Hydrocarbons
- Alcohols
- Auxins
- Aldehydes
- Resins
- Amino acids
- Aliphatic acids
- Aromatic acids

Included in humus there may also be gases such as ethylene and hydrogen sulfide. However, the humic substances listed above can be very stable and may last hundreds to over a 1,000 years in the soil.

Three classes of humic substances are distinguished:

1. Fulvic acid
2. Humic acid
3. Humin

Humic substances are produced through biological and chemical degradation, and the activities of soil microbes. Fulvic acid is smaller and less stable than humic acid or humin. According to Vaughan and Malcolm, when plant debris is added to the soil:

> Initially there is rapid decomposition, with a flush of CO_2 as a result of enhanced microbial activity...and some of the more resistant materials such as cellulose and lignins, are decomposed.

The remaining cellulose and lignin are utilized, leaving the more stable "humic substances, which may persist for thousands of years".

ORGANIC MATTER AND SOIL HEALTH

Organic matter has generally positive effects on soil health, assuming it is not too high in salts, doesn't have chemical contaminants, and doesn't harbor disease pathogens or deleterious pests. Humic acid forms a glue-like substance that holds soil particles together, forming aggregates. This can help to build good soil structure and provide large pore spaces in the soil, which aids both aeration and drainage. These are important features that are often lacking in clay soils. The increased porosity also aids root growth in heavier soils.

Another benefit of organic matter is that it tends to be present in relatively large sizes, especially in comparison with soil particles. The larger surface area results in greater water-holding capacity, as well as improved cation exchange capacity. These are beneficial aspects in any soil, but are more pronounced in sandy soils, since they are typically low in colloids, unlike clay soils. Nevertheless, well-decomposed humus can have 10–100 times the cation exchange capacity of clays typically found in soil.

Humic substances have been shown to influence the uptake of ions by plant roots. They have also been shown to have positive effects on plant growth, including increases in length, fresh and dry weights, of both roots and shoots, and increases in number of lateral roots and of flowers.

Two other notable effects of organic matter are given as follows:

1. It interacts with pesticides.
2. It forms stable complexes with some metals in the soil.

Pesticides, such as DDT, heptachlor and 2,4-D have been shown to degrade in the presence of organic matter in the soil. Aluminum toxicity, which is a major problem in many acid soils, can be ameliorated by the addition of organic amendments. Fulvic acid forms stable complexes with aluminum ions, making them unavailable to plants. Unfortunately, humic acid forms complexes with copper in the soil, also rendering this essential plant nutrient unavailable for plant uptake.

TYPES OF ORGANIC MATTER FOR LANDSCAPED AREAS

Organic matter in landscapes includes both materials that are incorporated into the soil, which are grouped under the heading of **amendments**, and those that are applied to the surface, or **mulch**.

Organic Amendments

Whereas a good body of research has been conducted on organic matter and crop production, a smaller amount of research has been published specifically on managed landscapes. Nevertheless, some of the former research may be applicable to landscapes. Organic amendments that have been studied specifically for landscape applications include composted dairy manure solids, poultry manure (pelletized, aged, dewatered, and composted), municipal yard waste compost, undefined compost, municipal compost, milled pine bark, peat, cotton gin compost, paper sludge, undigested wood material, poultry meal, granular humate (from household solid waste, rabbit manure, horse manure, and poultry manure), and **vermicompost**. Vermicompost is a composting system that uses earthworms in a specially designed bin system. References at the end of the chapter provide a good start on studies about compost usage in the landscape. There are also many extension articles on the Internet that are specific to compost usage, availability, and recommendations for individual states (Figure 10.1).

Mulch

Organic matter in the form of decorative mulch is already widely used in the landscaping industry. Mulch helps retain moisture, thus reducing the need for watering plants in flower beds, and under trees and shrubs. Another benefit of mulch is that

FIGURE 10.1 Composting bins can be purchased or easily crafted from readily available materials.

it can smother out weeds that compete for water and nutrients in the soil and for the sunshine that fuels the growth of all plants. It can keep soil cooler in summer keeping roots cooler, too, and mulch can keep soil warmer in winter, reducing soil upheaval due to the freeze-thaw cycle.

Good mulch will decompose slowly over time, releasing valuable nutrients and humic acid that helps to bind soil particles together and build good soil structure. The best soil structure for plant growth is crumbly, and often reveals earthworms and other organisms. Soil having a deep, dark brown to black color indicates that it has a healthy dose of organic matter, something that it can get from the use of mulch.

Case Study: Mulch on Crape Myrtle

References: Zajicek, Jayne M., and J. L. Heilman. 1991. Transpiration by crape myrtle cultivars surrounded by mulch, soil, and turfgrass species. *HortScience* Vol. 26(9) September: 1207–1210.

Heilman, J. L., C. L. Brittin, and J. M. Zajicek. 1989. Water use of shrubs as affected by energy exchange with building walls. Agr. For. Meteorol. 48:345–357.

Water use on landscape plants is often seen as a luxury, especially in times of drought and in regions with drier climates. Mulch is often touted as one method that can reduce water loss from soil, thus reducing water requirements of landscape plants. Bare soil was compared to pine bark mulch and turfgrass to see how various crape myrtle cultivars would respond. Instead of looking at soil moisture, transpiration was measured on a daily basis. This technique would give researchers a better idea of the actual water used by the plant.

Unexpectedly, plants that were surrounded by pine bark mulch lost more water through transpiration than trees grown on bare soil or those surrounded by turfgrass. Researchers think the reason is because temperatures of the bark mulch were higher than those of either the bare ground or the turfgrass. In a related study (Heilman et al., 1989), shrubs were found to lose more water when they were adjacent to east- and west-facing walls. This was due to the heat emitted from the building walls. Other parameters that could influence surface temperatures of mulch in the landscape are sun exposure and albedo.

Leaf size and canopy architecture were cited in the study as influencing transpiration rates. Since transpiration, and not evaporation from the soil surface, was measured, it is not possible to state whether more or less water was lost in mulched areas. Similar results may be found in other locations, but this study was done in College Station, Texas, in July and August. So, one should be careful when drawing conclusions for other climates or seasons.

Organic and living mulches can be effective at removing heavy metals from landscape and garden soils. Lead, cadmium, and copper were all found to be removed or tied up into less toxic forms by a variety of mulch materials (Chalker-Scott, 2007).

Mulch Materials

Some materials make better mulch than others. The best mulch is one that has a blend of carbon and nitrogen. Carbon is more predominant than nitrogen in plant materials, but the ratio is much higher in woody materials than green, leafy materials. Green materials, such as grass clippings, decompose relatively rapidly. So, while they make good mulch while they last, they have to be replenished often in order to keep doing their job. Wood chips, on the other hand, have a very high carbon-to-nitrogen ration, and therefore, they take a long time to break down and may actually rob nitrogen from the plants they are supposed to be protecting from competition from weeds.

Cypress mulch is a good choice because it has a nice color, breaks down moderately slowly, yet doesn't compete for nitrogen with the plants it is supposed to protect. Cypress also resists rot and insects. The bad news about cypress mulch is that mature cypress trees in the Gulf Coast are harvested for their mulch, even though the cypress swamp forests that cover hundreds of thousands of acres in coastal Louisiana play a vital role in protecting the coastline from flooding. Cypress swamps prevent floods by collecting storm waters and clean water by filtering out pollution. In addition, they provide habitat for migratory songbirds and many other animals.

Cocoa, cottonseed, and buckwheat hulls are sold as mulch in various parts of the country. They are decorative and lightweight, which causes them to be easily lifted by the wind. Cocoa hulls have a chocolate scent and so add another aesthetic dimension to the landscape. However, they may be toxic to dogs if they are ingested in large enough amounts. Cocoa hulls have a high potassium content and may be toxic to some plants.

In the fall, when deciduous trees and shrubs drop their leaves, they provide a rich source of organic matter. Whereas property owners desiring a well-manicured

Improving Landscape Soils

landscape may not approve of using fallen leaves as mulch, this is a more sustainable practice than sending them to the landfill. In areas where fall leaves are abundant, every effort should be made to gain the advantage of the nutrients they store for the benefit of the trees, shrubs, and flowers in the landscape. Shredding the leaves, while speeding the decomposition process, may prove to be more aesthetically acceptable to some.

Trees and shrubs aren't the only plants that provide organic matter. Flowers and vegetable gardens can also be a rich supply of this valuable material. Once again, plant stems may be unsightly if placed directly under trees and in beds, but shredding them is more attractive and also allows the organic matter to become available much sooner. While winter cold will slow or stop the decomposition process, in spring and summer, the remaining material will likely be completely gone. In addition to adding organic matter to the soil, birds will gather stems, leaves, vines, and other pieces during nest-making activities.

How to Apply Mulch

Recommendations for depth of mulch to apply are in the 2–4 in. (5–10 cm) range. This is because deeper mulch can reduce oxygen availability to plant roots that are under the mulch. As a result, some plants will send small, fine roots into the mulch layer. Since the mulch is much more porous than soil, it will dry out during periods of drought, and these many small roots will die. Two inches (5 cm) is the minimum amount recommended for suppression of weed seed germination and for moisture conservation. If there is already a layer of mulch from previous applications, then only enough mulch to bring the total depth to 4 in. is required.

Under trees, the area of mulch that should be applied is the circumference of the canopy. This is sometimes referred to as the **dripline** of the tree, because it indicates where rainfall would drip directly from above. For many larger, older trees, this is quite an expansive area. When there is a large area planted to numerous trees, the dripline of one tree merges with the dripline of another tree. To cover the entire area with mulch would result in large expanses of mulch with no turf or other groundcover. This is appropriate, yet in some situations, it would not be acceptable, since activities that could be conducted on turf may not be as easily done on mulch. Therefore, it is possible to compromise by growing some turf in the area and limiting the areas of mulch.

If the area under the trees is extremely shady, it may not be possible to grow turf or other groundcovers under them. The best turf for growing in shady areas is fine fescue for cool-season areas and St. Augustine grass for warm-season areas. Tall fescue tolerates shade fairly well, as does zoysiagrass. For moderately shady areas in cool season zones, Kentucky bluegrass and perennial ryegrass will perform all right. For warm-season zones, centipedegrass and carpetgrass will perform fair to moderate.

To calculate the amount of mulch to apply to an area, one must figure the volume needed. Volume is a measure of width × length × depth and may be calculated in cubic inches, cubic feet, or cubic yards. However, some suppliers sell mulch in tons, so the conversion from cubic feet or cubic yards to tons must be performed. There is 27 cubic feet in a cubic yard, and although the weight of mulch varies by material type, there are about 3.5 cubic yards in a ton of wood-based mulch. A cubic yard of mulch will cover approximately 81 ft² to a 4 in. depth or 162 ft² to a 2 in. depth.

Mulches should be applied in early spring before weeds become a problem. It is possible to smother weeds with mulch while they are still small, and there will be some success at eradicating them. However, larger weeds will be quite difficult to kill this way, and other types of control will be needed.

Problems with Mulch

If mulch is stored improperly or for too long, certain problems may arise. For example, if mulch remains overly wet, anaerobic conditions cause "sour" mulch, which gives off a foul odor. If the mulch contains leaves that don't readily decompose, such as gingko or maple leaves, they will contribute to the overall wetness of the mulch pile. Shredded leaves may mat together, forming a barrier to water and oxygen, if they are not shredded finely enough. Care should be taken to avoid this problem by allowing the leaves to partially rot before using them as mulch.

Mulch that is stored in a pile will begin to decompose, and during this process, it can heat up to rather high temperatures. That is why some mulch piles have steam coming from them. Eventually, this mulch will turn into compost, and while it still can be used as a soil amendment, it will decompose much more rapidly after application than mulch that is still in larger pieces. Pine needles are acidic and when used as mulch will contribute an acidifying effect to the soil. Due to the nature of mulch, there are many different types of fungal spores that germinate and grow there. Some common ones are artillery fungus and slime mold. It is impossible to rid the mulch of these various fungi, as conditions that support them are optimum: adequate food sources, moisture, and proper temperatures during the growing season. For the most part, they will not harm the plants around them, nor are they likely to spread to other areas, since fungi live on specific hosts. To avoid such problems, make sure the mulch you use has been adequately composted prior to its use. Heating a mulch pile to adequate temperatures will kill many fungal organisms as well as weed seeds that may be present.

Many studies have been conducted to determine whether woody mulch materials actually rob the plants around them of nitrogen. Some claims have been made to this effect, due to the high nitrogen needs during decomposition. However, since the mulch is lying on top of the soil rather than interacting with activity in the root zone, this is not generally thought to be an issue of concern.

ORGANIC SOIL AMENDMENTS

Organic matter is added to planting beds to improve moisture-holding ability of sandy soils, increase nutrient retention ability of clay soils, and reduce the pH for low-pH-requiring plants. Leaves, spent flowers and vegetable stems, and other garden and landscape plant matter can all become part of the nutrient cycle.

Some organic materials that may be used as soil amendments in landscapes are compost, vegetable meals such as soybean meal, corn gluten meal, cottonseed meal, and peanut meal, blood meal fish emulsion or meal, greensand, and bone meal.

Improving Landscape Soils

PEAT MOSS

Peat moss is harvested from peat bogs after draining the area and allowing the undecomposed plant material to dry down. In the United States, most of the peat used in landscaping is harvested from bogs in Canada. Peat moss is somewhat acidic, ranging between approximately 4 and 5 on the pH scale. It may be tilled into beds prior to planting for best effect. Addition of lime may be required to adjust for pH. In spite of its popularity, there are concerns about the sustainability of continual harvesting of peat moss due to its long regeneration time. Furthermore, this is not a local material for most people. Thus, composted materials from local sources are a considered to be a more sustainable resource.

COMPOST

Landscape companies can contribute to the development and use of compost in two ways: they can recycle plant material from their own operations, and they can build or install compost bins for clients.

How to Build a Compost Bin System

Clients may want their own compost bin so that they can recycle plant material from kitchen and yard waste. The ideal size for a compost bin is one cubic yard. However, on a commercial scale, this is not adequate.

What to Add to the Compost

Many different plants and plant byproducts, including kitchen scraps, can be used in compost (see Table 10.1 for compost ingredients). Avoid using meat scraps in compost piles to avoid scavengers, raccoons, or other pests. Some non-plant materials that are acceptable in compost are included in Table 10.1. Under ideal conditions, and depending on the material used, compost may require only 3–4 weeks to develop, or it may take as long as 6 months.

TABLE 10.1
Compost Ingredients

Type of Material	Carbon/Nitrogen	Details
Algae, seaweed, and lake moss	N	Good nutrient source.
Cornstalks, corn cobs	C	Shred and balance with nitrogen-rich materials.
Grass clippings	N	Decompose rapidly.
Oak leaves	C	Shredded leaves break down faster.
Pine needles and cones	C	Acidic and decomposes slowly. Use sparingly.
Sawdust and wood shavings (untreated wood)	C	Slow to decompose; add nitrogen-rich material to speed up. Avoid treated woods.
Weeds	N	Remove seed heads first.
Prunings – woody	C	Best if shredded, use sparingly, treat like wood shavings.

Carbon and Nitrogen: Finding the Right Balance

The micro-organisms involved in the decomposition process are present in native soils, but spores can be purchased and mixed into the compost. Since they require adequate nutrition for reproduction, plant materials containing differing levels of carbon and nitrogen should be included in such a way as to achieve a balance of 30 parts carbon to every 1 part nitrogen. This 30:1 ratio can be achieved if plenty of green vegetative plant parts are included in higher numbers than woody plant parts (Table 10.2) because fresher, greener materials are relatively higher in nitrogen than dry materials or well-rotted manure. Wood is difficult to break down due to the lignins present in its cellular structure. Even sawdust is not recommended for compost piles because it is relatively low in nitrogen. If it must be used, then add plenty of blood meal or some other high-nitrogen-containing material so that decomposition is not unduly impeded. A higher nitrogen content than the recommended amount will result in a speedier breakdown of organic material.

Proper Conditions

Soil or compost thermometers can be used to monitor temperatures in the compost pile. The optimal temperature range for decomposition is between 104°F and 130°F (40°C and 54.5°C). The higher end of this range is adequate for killing many weed seeds and plant pathogens. If the temperature drops below this range, or exceeds it, it is a good time to turn the pile. If weed seeds appear in the pile, or if they have been a problem in the past, allow the pile to remain at 140°F–150°F (60°C–65.5°C) for half an hour to kill them.

TABLE 10.2
Average Carbon:Nitrogen Ratio of Some Compost Ingredients

Material	C:N Ratio
Blood	3:1
Cardboard	400:1
Coffee grounds	25:1
Dry leaves	60:1
Fruit waste	16:1
Grass clippings	15:1
Manure, fresh	8:1
Manure, well-rotted	20:1
Sawdust	225:1
Small sticks and twigs	375:1
Weeds (vegetative parts)	10:1

Note: Actual ratios vary, depending on the source of the material. A 30:1 ratio can be achieved by mixing ingredients in the appropriate proportions.

Improving Landscape Soils

Moisture

Watering is not necessary with adequate rain, but there will be reduced microbial activity in a dry compost pile. On the other hand, an excessively wet pile is anaerobic and will smell bad. If the pile is too wet, add dry materials or turn it over to help aerate it. Some people insert aeration tubing consisting of plastic tubes or pipes with holes drilled in them. The holes may be covered with screen to prevent pipes from clogging.

Turning

Turn the compost pile regularly if you want to speed up the process. This will distribute the heat evenly and allow different materials to be in the center of the pile, where heat builds up most. Turning increases aeration in the pile. New materials should be turned in whenever they are added to an existing pile.

A rapid way to obtain compost is to shred the material before it is added and to add a nitrogen supplement such as cottonseed meal, dried manure, or blood meal at 1 part for every 2 parts of shredded plant material.

COMPOST SOLUTIONS

Compost solutions such as compost tea and compost extract derive from compost. They are not particularly useful as organic amendments, but as organic fertilizers. There are claims of disease-suppressing qualities as well. So, these topics are covered in more detail in Chapters 9 and 12.

GRASS CLIPPINGS

Mulching mowers are designed to leave grass clippings on the lawn. The blade deck is designed to circulate the leaves until they have been chopped to smaller bits than a non-mulching mower can do. This allows the clippings to settle down into the lawn, reducing their ability to stick to shoes and get tracked into areas where they are not wanted. When grass clippings are bagged rather than left in place after cutting, they must be disposed of, and a common place for that has been the municipal landfill. However, many municipalities have outlawed grass clippings, necessitating another solution.

Grass clippings should be left on the soil during mowing in order to return the organic contents back into the soil. This practice can return up to 25 percent of nitrogen needed to grow healthy turf. The analysis of grass clippings is approximately 4N:2P:1K.

SUMMARY

Organic matter, such as peat and compost, is used to amend soil; organic mulches are used around trees and in planting beds; grass clippings are often left to decompose on lawns and other turf areas. Current landscape practices that involve removing large amounts of organic matter from the landscape are currently under renewed examination. Studies show it can be beneficial to soil health and plant growth.

Organic matter, also called humus, is comprised of many individual substances, including carbohydrates, hydrocarbons, amino acids, and other acids, and gases. Humic substances can be very stable, lasting many years in the soil. The three classes of humic substances most often discussed and studied are fulvic acid, humic acid, and humin. They are degraded through biological and chemical processes and the activities of soil microbes.

In general, organic matter has positive effects on soil health. But it should not contain high levels of salts, chemical contaminants, disease pathogens, or deleterious pests. Humic acid holds soil particles together, forming aggregates that are important to good soil structure, which results in good aeration and drainage. It also helps hold mineral nutrients in the soil very well. This feature is especially important in lighter, sandy soils, but can also be beneficial in clay soils.

Humic substances appear to aid in nutrient uptake at the root surface. They also interact with pesticides, helping them to degrade rather than persist in the soil, a trait that is of concern due to the ability of pesticides to contaminate both ground and surface water. Furthermore, organic substances form stable complexes with some metals in the soil, aiding in toxicity problems in the plants growing there.

Organic amendments that have been studied specifically for landscape applications include composted animal waste, landscape waste, sewage sludge, and other materials. Mulch is usually composed of larger particles which will degrade slowly over time, providing a decorative effect in the landscape, alleviating competition between turf areas and tree root areas, for informal walking paths, and in shrub and flower beds to aid in moisture retention and weed control.

Both hardwoods and softwoods are used as mulch. Materials having a high carbon:nitrogen ration tend to take much longer to degrade, and so they last longer in the landscape setting. Cypress is a soft wood which breaks down moderately slowly and resists rot and insects. Cypress harvested in the Gulf Coast should be avoided.

Leaves, tree prunings, flower stems, and other garden waste may be shredded and applied to areas where wood mulch would be used. They are not as decorative as wood mulches, but can degrade in a shorter period of time, improving soil structure and providing nutrients to plant roots.

REVIEW QUESTIONS

1. Why is organic matter important in the managed landscape?
2. Organic matter is composed of what three substances?
3. Name three benefits of organic matter to soil health.
4. Name three benefits of organic matter to plants.
5. Mulch has what benefits?
6. Name three plant materials that are used as compost.
7. Name three animal materials that are used as compost.
8. Distinguish between organic amendments and mulch
9. What is the optimal temperature range for decomposition in a compost pile?
10. How long do grass clippings take to decompose?

ENRICHMENT ACTIVITIES

1. Conduct a survey in your local area to find out which municipalities allow (1) leaf burning and (2) landscape waste at the landfill. Research alternatives that are offered or required, and interview an official associated with the city to learn the recent history of yard waste disposal in your area.
2. Construct a vermicomposting bin for your home or school. Record what ingredients are placed into the bin, how long it takes to fill the bin, and how long it takes to convert the organic matter into compost.
3. Investigate sources of local organic materials that could be used for mulch. Research the pros and cons of each material, including how long they last, the nutrients they release as they break down, any detrimental aspects, and their beneficial qualities.

FURTHER READING

Bellows, B. 2003. *Sustainable Turf Care*. ATTRA, Fayetteville, AR. https://attra.ncat.org/publications/attra-publications/page/38/?s. Retrieved July 5, 2019.

Chalker-Scott, L. 2007. Impact of mulches on landscape plants and the environment—a review. *J. Environ. Hortic.* 25(4): 239–249.

Cogger, C., R. Hummel, J. Hart, and A. Bary. 2008. Soil and redosier dogwood response to incorporated and surface applied compost. *Hortic. Sci.* 43(7): 2143–2150.

Cornell Gardening Resources. Mulches for landscaping. http://chemung.cce.cornell.edu/resources/mulch-in-the-landscape. Retrieved July 5, 2019.

Donahue, R. L., R. W. Miller, and J. C. Shickluna. 1983. *Soils: An Introduction to Soils and Plant Growth*. 5th ed. Prentice-Hall, Englewood Cliffs, NJ.

Ferreras, L., E. Gomez, S. Toresani, I. Firpo, and R. Rotondo. 2006. Effect of organic amendments on some physical, chemical, and biological properties in a horticultural soil. *Bioresour. Technol.* 97(4): 635–640.

Gonzalez, M., E. Gomez, R. Comese, M. Quesada, and M. Conti. 2010. Influence of organic amendments on soil quality potential indicators in an urban horticultural system. *Bioresour. Technol.* 101(22): 8897–8901.

Guckenberger Price, J., A. N. Wright, and R. L. Boyd. 2009. Organic matter application improves post-transplant root growth of three native woody shrubs. *Hortic. Sci.* 44: 377–383.

Hartz, T. K., J. P. Mitchell, and C. Giannini. 2000. Nitrogen and carbon mineralization dynamics of manures and composts. *Hortic. Sci.* 35(2): 209–212.

Kaminski, J. E. and P. Dernoeden. 2004. Soil amendments and fertilizer source effects on creeping bentgrass establishment, soil microbial activity, thatch and disease. *Hortic. Sci.* 39: 620–626.

Kluepfel, M. and B. Polomski. All about mulch. http://www.savvygardener.com/Features/mulch.html. Retrieved July 13, 2010.

Loper, S., Shober, A. L., C. Wiese, G. C. Denny, C. D. Stanley, and E. F. Gilman. 2010. Organic soil amendment and tillage affect soil quality and plant performance in simulated residential landscapes. *Hortic. Sci.* 45(10): 1522–1528.

Maggard, A. O., Will, R. E., Hennessey, T. C., McKinley, C. R., and J. C. Cole. 2012. Tree-based mulches influence soil properties and plant growth. *Hortic. Tech.* 22(3): 353–361.

Norrie, J., C. J. Beauchamp, and A. Gosselin. 1994. Use of paper sludges as soil amendments in landscape horticulture. *Hortic. Sci.* 29: 246.

Scheiber, S. M., R. C. Beeson, Jr., and S. Vyapari. 2007. Pentas water use and growth in simulated landscapes as affected by municipal compost and mined field clay soil amendments. *Hortic. Sci.* 42: 1744–1747.

Tipton, J. L., E. Davidson, and J. Barba. Effect of hole size, organic amendments, and surface mulches on tree establishment in southwestern soils. *Hortic. Sci.* 31: 606.

Tiquia, S. M., Lloyd J., D. A. Herms, H. A. J. Hoitink, F. C. Michel Jr. 2002. Effects of mulching and fertilization on soil nutrients, microbial activity and rhizosphere bacterial community structure determined by analysis of TRFLPs of PCR-amplified 16S rRNA genes. *Appl. Soil Ecol.* 21: 31–48.

Tukey, R.B., Schoff, E.L., 1963. Influence of different mulching materials upon the soil environment. *Proc. Am. Soc. Hortic. Sci.* 82: 68–76.

Vaughan, D. and R. E. Malcolm. 1985. *Soil Organic Matter and Biological Activity.* Kluwer Academic Publishers, Hingham, MA, 469 pp.

Volk, T. Tom Volk's fungi. http://tomvolkfungi.net/. Retrieved July 31, 2010.

11 Pesticides in the Landscape

OBJECTIVES

Upon completion of this chapter, the reader should be able to

- Discuss the types of pesticides used in the United States
- Discuss the history of pesticide regulation in the United States
- Identify the major categories of pesticides used in the landscape
- Identify the chemical classes of pesticides used in landscaping
- Explain concerns of human health associated with pesticides
- Explain environmental problems associated with pesticides
- Name the federal statutes that regulate pesticide use in the United States

TERMS TO KNOW

Acetylcholinesterase inhibitors
Active ingredient
Acute effects
Adjuvant
Bactericide
Carbamate
Cholinesterase
Chronic effects
Endocrine disruptor
Erythrocyte
Formulation
Fungicide
Herbicide
Hydrocarbon poisoning
Insecticides
Integrated pest management
LD_{50}
MSDS
Mutagen
Personal protective equipment (PPE)
Pesticide
Restricted use pesticide (RUP)

Spreader
Synthetic pesticide
Teratogenic

INTRODUCTION

Pesticides are used in the landscape industry to destroy plant pests, whether they be insects, disease pathogens, or weeds. They are an important component of landscape plant management. The Turf & Ornamental Greenbook lists over 500 chemical protection products. Although sustainable landscaping practices seek to reduce or eliminate the use of pesticides in landscaping, they still play a major role, both within the profession, and by homeowners, gardeners, and others. They also contribute to environmental problems, as evidenced in the results of water quality studies discussed in Chapter 5.

Some consider pesticides a "necessary evil" because insect pests and diseases are an inevitable part of a natural ecosystem. Yet, due to an aversion to damaged and diseased plants, or weedy invasions of managed landscapes, people resort to the use of chemicals that may be hazardous to humans, pets, and wildlife. The hazard that pesticides present is a combination of toxicity levels and exposure time. Pesticides affect landscape sustainability adversely when they contaminate the environment and threaten the health and well-being of its inhabitants.

Ultimately, sustainable landscaping practices dictate taking a multi-pronged approach to reduce dependence on pesticides in the landscape. Such approaches include using pesticides with the least toxicity, using them as little as often to still be effective, and using and disposing of them properly. Ecological landscape practices can help to change the perspective of what constitutes a "beautiful" landscape. Other approaches to reducing pesticide use in the landscape are addressed in the next chapter. The Sustainable Sites Initiative has the same goals as those discussed throughout this text. Specifically, they require reducing plant stress, decreasing negative effects on human health, and effecting the lowest possible negative ecological and environmental impacts by promoting practices based on observation and planning to minimize or eliminate synthetic pesticide and fertilizer use.

In addition to proper use and disposal of pesticides to minimize adverse effects, an appreciation of pesticides and their toxicity will aid the landscape manager in making informed decisions concerning their use. This chapter will cover basic information about pesticides and their use in the landscape. It provides a discussion of the types of pesticides used in the landscape industry, and their regulation in the United States, and presents evidence of human health and environmental problems associated with pesticides. It discusses the handling, storage, and disposal of pesticides as mandated by federal law. In Chapter 5, there is a thorough discussion of contamination of water by pesticides and pesticide breakdown products. Chapter 12 presents the many alternatives to chemical pesticides.

A pesticide is any compound that is used to control pests. This usually indicates killing them in an attempt to eradicate them. However, total eradication is often not possible. The landscape industry is mainly concerned with ornamental plants. Common pests of ornamental plants may be insect pests, fungal infections, or bacterial infections. Weeds are also considered a landscape pest. Plant growth regulators (PGRs) are included in the group of pesticides and are regulated by the same laws. In the landscape industry, PGRs are used to cause flowers to fall off of trees or shrubs in order to avoid nuisance fruit that can cause a mess. They are also used to slow growth in turfgrasses as a means of reducing mowing.

PESTICIDE USE IN THE LANDSCAPE

Sixty-seven million pounds of pesticides are applied to lawns every year in the United States, according to the EPA. Millions more are applied to the rest of the ornamental plants in the landscape to combat insects, disease pathogens, and weeds. According to a Minnesota study published in 2010, weed control represented the highest lawn and garden pesticide usage in that state.

In the landscape industry, and in individual residences and businesses, a large number of plant species are present. As a result, the number of insect, pathogen, and weed pests is enormous. The complexity of the landscaped environment presents a challenge to pest management strategies. Table 11.1 lists some of the commonly used pesticides in the landscape.

Many pesticides are used in various activities, including manufacturing, forestry and agriculture, inside and outside the home and other buildings, for industrial purposes, water treatment, dry cleaning, and others. Of these, the largest amount is used by the agriculture sector, representing approximately 80 percent of all pesticides used in the United States in 2010 (Figure 11.1).

Roughly 20 percent of all pesticides used in the United States are for non-agricultural purposes. In the landscape, many people are directly exposed to pesticides as compared to on-farm use. Furthermore, pesticides used in the landscape are often not used according to label directions, and they are often not disposed of properly. According to Behe and Beckett, more than 80 percent of the 93.3 million households in the US participate in gardening activities. The 2011 EPA study found that $2.5 billion was spent on non-agricultural pesticide uses.

Pesticides used in the landscape are targeted mainly for ornamental plants, including turfgrass. By contrast, pesticides used around the home, and other buildings include those targeting termites and other threats to structures, as well as rodents and other nuisance pests. The landscape pesticides fall into categories of **bactericides**, **fungicides**, **insecticides**, and **herbicides**, in increasing order of importance or use.

Whereas landscape professionals are required to hold a current pesticide applicators license, many homeowners apply their own pesticides. As a result, mis-use of pesticides in the landscape may occur.

Sometimes, the use of a pesticide fails to give the desired outcome. There are several reasons for this, and understanding them can aid in better decision-making in the future:

- Pest resistance to the pesticide. Some weeds are resistant to certain herbicides, and some insects are resistant to Bt, and some are resistant to malathion.

TABLE 11.1
Selected List of Pesticides Used in the Turf and Ornamental Industry

Common Name	Product Name	Landscape Use	Mode of Action
2,4-D Dimethylamine salt	Weedestroy	Herbicide	Synthetic auxin (PGR)
Abamectin, avermectin	Avid	Insecticide/Miticide	Avermectin
Acephate	Orthene	Insecticide	Organophosphate
Atrazine	Atrazine	Herbicide	Triazine
Azadirachtin	Azatin XL	Insecticide	Insect growth hormone
Bacillus thurengiensis	Gnatrol	Insecticide	Biological
Bentazon sodium salt	Basagran T/O	Herbicide	Photosynthesis inhibitor
Carbaryl	Sevin	Insecticide	Carbamate
Chlorothalonil	Daconil, Manicure Ultrex	Fungicide	Substituted benzene
Chlorpyrifos	Dursban Pro	Insecticide	Organophosphate
Cryolite	Prokill Cryolite 96	Insecticide	Mineral
Daminozide	B-Nine WSG	PGR	PGR
Dicamba	Vanquish	Herbicide	Synthetic auxin (PGR)
Ethephon	Proxy	Flower/fruit drop enhancer	PGR
Fenoxycarb	Acclaim extra	Insect growth regulator	Carbamate
Fenpropathrin	Tame	Insecticide	Synthetic pyrethroid
Fipronil	Chipco Choice	Fungicide	Neurotoxin
Gibberellic acid	ProGibb	PGR	Cellular growth
Glufosinate	Finale	Herbicide	Phosphonic acid
Glyphosate diammonium salt	Touchdown Pro	Herbicide	Glycine, EPSP synthase-inhibitor
Glyphosate isopropylamine salt	Accord, Rodeo, RoundUp	Herbicide	Glycine, EPSP synthase-inhibitor
Glyphosate monoammonium salt	RoundUp Pro Dry	Herbicide	Glycine, EPSP synthase-inhibitor

(*Continued*)

TABLE 11.1 (Continued)
Selected List of Pesticides Used in the Turf and Ornamental Industry

Common Name	Product Name	Landscape Use	Mode of Action
Imidacloprid	Marathon	Insecticide	Blocks acetylcholine receptors
Mancozeb	Dithane, Fore	Fungicide	Ethlenebisdithio-carbamate (EBDC)
Mecoprop	Triamine	Herbicide	
Methiocarb	Mesurol 75-W, Mesurol Pro	Insecticide	Carbamate
Metolachlor	Pennant liquid	Herbicide	Chloroacetamide
Monosodium acid methanearsonate	MSMA soluble granules	Herbicide	Organic Arsenical
Neem oil	Triact 70	Insecticide	Botanical
Oryzalin	Surflan AS T/O	Herbicide	Dinitroaniline
Paclobutrazol	Turf Enhancer	PGR	PGR
Permethrin	Astro, Dragnet	Insecticide	Pyrethroid
Petroleum cistillate, refined	Horticultural oil	Fungicide, larvacide, oocide	Petroleum oil
Potassium salts of fatty acids	Insecticidal soap	Insecticide	Soap
Simazine	Princep liquid	Herbicide	Photosynthesis inhibitor
Spinosad	Conserve SC	Insecticide	Acetylcholine-binding inhibitor
Steinernema feltiae	Nemasys	Nematicide	Nematode
Steinernema scapterisci	Nematac S	Nematicide	Nematode
Triclopyr	Garlon, Momentum, Battleship, Eliminate	Herbicide	Synthetic auxin
Uniconazole-p	Sumagic	PGR	PGR

Source: T&OR – Turf and ornamental reference for plant protection products. C&PO Press. For updates, see www.bluebooktor.com.

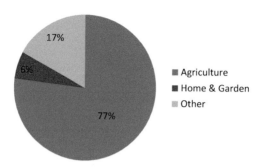

FIGURE 11.1 US pesticide use by sector. (Illustration by the author.)

- Misapplication of the pesticide. Pesticide applied to the wrong part of a plant or pesticide that is insufficiently applied are two examples of pesticide misapplication.
- Applying pesticide too early or too late in the life cycle of the pest. Crabgrass and other weeds controlled using a pre-emergent herbicide will not be controlled if the herbicide is applied too early or too late. Many insects emerge at specific tomes only once or twice in the growing season, so insecticides must be carefully timed to be effective.
- Incorrect pesticide. Using an insecticide on mites will not control them. Bacterial infections require antibiotics, such as streptomycin or oxytetracycline.

TYPES OF PESTICIDES

Early records of pesticide use date to Greece and China, where arsenic and sulfur were used for fumigation of homes and eradication of garden pests. In Bordeaux, France, a fungicide mixture was concocted using copper sulfate and lime. This became known as Bordeaux Mixture and has been widely used on grapes, apples, and other food crops. These naturally occurring compounds are still in use today. However, developments in chemistry have led to synthetic compounds that have pesticidal qualities. Many of these developments occurred during the 1940s and 1950s. Naturally occurring compounds and biological organisms are also used for their pesticidal qualities.

The **active ingredient** in a pesticide is the ingredient having the pertinent control characteristic of that pesticide. A pesticide may have more than one active ingredient.

SYNTHETIC PESTICIDES

Synthetic pesticides are manufactured and have come into widespread usage following World War II. They fall into three general categories based on their chemical composition: organochlorides, organophosphates, and carbamates. Organochlorides are also known as chlorinated hydrocarbons. They are often persistent in the environment, due to their tendency to degrade or break down,

Pesticides in the Landscape 209

very slowly. Organophosphates are also known as organic phosphates. They are highly toxic to humans, but are not persistent in the environment. Carbamates are considered highly toxic to humans.

NATURALLY OCCURRING PESTICIDES

Naturally occurring plant compounds have been used at various times in history, including the present day. These include the nicotine, neem, and pyrethrum, all of which have been derived from plants. A newer generation of pesticides focused on the target pest, using more tailored methods to eradicate it while not harming other organisms. Among this group of non-toxic pesticides are insect growth regulators (IGRs), soaps, and horticultural oils. See Chapter 12 for further discussion of these pesticides.

The use of non-synthetic pesticides has been coupled with better targeting of pesticide application, as well as other methods, such as the use of biological organisms to control pests. Biological controls include predators, parasites, and pathogens that attack plant pests. Together, these are treated under the practice known as **integrated pest management (IPM)**. In some disciplines, other terms are used to denote a similar development concerning pest control, such as total plant care or integrated plant management. Chapter 12 discusses this topic.

PESTICIDE REGULATION

In the United States, the Environmental Protection Agency (EPA) has been charged with the responsibility of regulating pesticide use at the national level. The history of the EPA began with publication of the book *Silent Spring* by Rachel Carson in 1962. EPA historian Jack Lewis compares the role of Carson's book on environmentalism to the role *Uncle Tom's Cabin* played in abolishing slavery. The book asserts that chemical pesticides were not only polluting the environment, but that they were persistent in the environment, creating a hazard that would permeate our food, water, and air.

Through a series of events, actions, and policy decisions, a bill known as the National Environmental Policy Act (NEPA) was signed into law on New Year's Day in 1970. In April of that same year, the first Earth day was held. Some have referred to 1970 as "the year of the environment". The mission of the NEPA was to

- Establish and enforce environmental protection standards
- Conduct environmental research
- Provide assistance to others combating environmental pollution
- Assist the CEQ (Council on Environmental Quality) in developing and recommending to the President new policies for environmental protection

Two Congressional Acts provide authority to the EPA to regulate pesticide use in the United States. They are the Federal Insecticide, Fungicide, and Rodenticide Act (FIFRA) and Federal food, Drug, and Cosmetics Act (FFDCA).

FIFRA

The Federal Insecticide, Fungicide, and Rodenticide Act (FIFRA) originated in 1910 with the first pesticide law passed in the United States. In 1947, when FIFRA was first passed, it established registration procedures and labeling provisions for pesticides. FIFRA underwent several other changes over the decades, but the 1988 amendments to the act strengthened EPA's authority in pesticide re-registration, and of storage and disposal of cancelled pesticides. The EPA is in charge of pesticide labeling, in which they are authorized to "establish pesticide packaging standards, regulate pesticide and container disposal, issue experimental permits, conduct research on pesticides, and alternatives, and monitor pesticide use and presence in the environment". Both labels and Manufacturer's Safety Data Sheets (**MSDS**) must be easily accessible at a work site. They are provided by manufacturers upon purchase of their product, as well as on websites maintained by each company. The MSDS provides technical information on the chemical components of the pesticides, plus handling, storage, and emergency procedures. Labels and MSDS sheets are available online at manufacturer's websites.

Under the new law, pesticide manufacturers were required to obtain EPA registration prior to manufacture, transport, or sale of their products. Every pesticide label includes this number on it. The EPA requires manufacturers to conduct research and to provide adequate data demonstrating the product safety. Manufacturers are required to pay fees to cover the costs of registration and other administrative costs.

One major change that was ushered in with these amendments was a certification process for pesticide applicators. Such programs are now approved by the EPA administrator and enforced within each state.

FFDCA

FFDCA is the Federal Food, Drug, and Cosmetics Act. It provides the EPA authority to require testing of all pesticides and gives them authority to require testing for additional substances found or used in pesticides if they may have a cumulative effect similar to a pesticide ingredient.

The EPA and DDT

An early act taken by the EPA was to ban the chemical DDT (dichloro-diphenyl-trichloroethane), which Carson had demonstrated was present throughout the ecosystem, and became concentrated exponentially as it moved up the food chain. Effects were seen in the eggs of bald eagles, for example, even though pesticide was intended for insect pests on crops. In writing about the danger of synthetic pesticides, she said:

> They have been recovered from most of the major river systems and even from streams of groundwater flowing unseen through the earth. Residues of these chemicals linger in soil to which they may have been applied a dozen years before. They have entered and lodged in bodies of fish, birds, reptiles, and domestic and wild animals so universally that scientists carrying on scientific experiments find it almost impossible to locate subjects free from contamination. They have been found in fish in remote mountain lakes, in earthworms burrowing in soil, in the eggs of birds – and in man himself.

She goes on to state that active ingredients from pesticides are found in mother's milk and even in the tissues of unborn children.

This information is echoed in Sandra Steingraber's writings, including a chapter in a book by Johnson and Hill called *Ecology and Design*, in which she wrote:

> We know with certainty that a whole kaleidoscope of chemicals linked to cancer exists inside all of us. Pesticide residues, industrial solvents, electrical fluids such as PCBs, and the unintentional by-products of garbage incineration – namely dioxins and furans – are now detectable in breast milk, body fat, blood serum, semen, umbilical cords, placentas, and even in the fluid surrounding human eggs extracted from women who are undergoing *in vitro* fertilization. Residues of household pesticides and wood preservatives are found in the urine of American schoolchildren.

RESTRICTED USE PESTICIDES

Pesticides are categorized as either general use or **restricted use pesticides (RUPs)**. Whereas the former may be used by the general public and do not require special permits or certification, the latter do have those additional restrictions on their use. RUPs are those pesticides which have been deemed to pose a high risk to human health or the environment. They may only be applied by an applicator that has passed a test and been awarded certification, or by one who is under the supervision of a certified applicator. Each state has regulatory jurisdiction over pesticide matters, as long as they meet the minimum requirements set forth by the EPA. Therefore, each state has its own testing procedure and certification requirements. Often, states have reciprocal agreements with one another regarding certification.

The EPA maintains an RUP report that lists active and canceled RUP. In addition, a 6-month summary is published which lists pesticide active ingredients having undergone RUP changes, such as banned pesticides or pesticides with cancelled registrations. The EPA is required to report such information to other governments concerning pesticide exports from the United States.

HUMAN HEALTH HAZARDS

Mutagens cause genetic mutations that may lead to cancer. Compounds that are known to cause cancer are **carcinogens,** and those that cause birth defects when a fetus is exposed during pregnancy are **teratogenic**. Some pesticide ingredients have been shown to be one or both (Table 11.2). Others are **endocrine disruptors**, meaning that they interfere with the body's hormones. Organophosphates and carbamates are **acetylcholinesterase inhibitors** and have neurological effects.

Carbamates can also cause or contribute to oxidative stress of red blood cells (**erythrocytes**). Many pesticides can cause skin rashes, difficulty breathing or swallowing.

Pesticides are tested for their toxicity using animals such as rabbits, mice, and rats. These animals are given varying levels of pesticide active ingredients until 50 percent or more of the population is killed. The amount of active ingredient that kills 50 percent of the test animals is said to be the LD_{50}, where LD = lethal dose. The amount of pesticide is then extrapolated to humans, and the lethal dose is presented

TABLE 11.2
Human Health Issues Related to Pesticides

Pesticide	Landscape Use	Health Effect
2,4-dichlorophenoxyacetic acid	Herbicide	Lymphatic cancer
Acephate	Insecticide	Cholinesterase inhibitor
Barricade (active ingredient Prodiamine)	Herbicide	Liver alteration and enlargement, thyroid effect (hormone imbalances)
DDT, vinclozolin, endosulfan, toxaphene, dieldrin, and DBCP	Insecticide	Endocrine disruption
Ethalfluralin	Herbicide	Teratogenic
Fluazifop-butyl, Fluazifop-P-butyl	Herbicide	Teratogenic
Flumioxazin	Herbicide	Teratogenic
Imazethapyr	Herbicide	Bladder cancer
Malathion	Insecticide	Acetylcholinesterase inhibitor, oxidative stress of erythrocytes
MANY		Asthma, emphysema, respiratory problems
MCPA (4-chloro-2-methylphenoxyacetic acid)	Herbicide	Brain cancer
Pendimethalin, EPTC	Herbicide	Pancreatic cancer
Propanil	Herbicide	Methemoglobinenemia, immunotoxicity, nephrotoxicity

Sources: http://www.fluoridealert.org/pesticides/effects.teratogenic.htm; Steingraber, http://www.chem-tox.com/pesticides/#breastcancer; www.chem-tox.com/bbb; Kross et al. (1996); Fleming et al. (1999).

in milligrams of pesticide to kilograms of body weight. The higher the number of milligrams of active ingredient per kilogram of body weight, the safer the active ingredient is. Pesticides are grouped into categories based on their toxicity. The most highly toxic group is labeled "Danger – Poison", moderately toxic chemicals are labeled "Warning", and the least toxic group is labeled "Caution" (Table 11.3).

Children (and pets) may be directly exposed to pesticides used around the home, and since exposure rates are based on body weight, children and most pets are at higher risk for toxic effects than adults. There are other issues concerning the toxicity of pesticides to children as compared to adults. For example, children are in developmental stages that may be more sensitive to toxic chemical exposure. They also tend to be active in lawns and on the ground, and are more likely to put dirt or dirty objects into their mouths. All pesticide labels are required to state "Keep Out of Reach of Children". They must be stored safely away from places where children or pets could accidentally come into contact with them.

EPIDEMIOLOGY

Epidemiological studies look at links between health problems and risk factors, or those factors that are associated with the health problem in question.

TABLE 11.3
Pesticide Chemical Categories and Associated Toxicity Levels for Five Health Categories

Signal Word (Category)	Oral Toxicity (mg/kg)	Dermal Toxicity (mg/kg)	Inhalation Toxicity (mg/l)	Primary Eye Irritation	Primary Skin Irritation
DANGER (Category I)	0–50	0–200	0–0.5	Corrosive (irreversible destruction of ocular tissue) or corneal involvement or irritation persisting for more than 21 days	Corrosive (tissue destruction into the dermis and/or scarring)
WARNING (Category II)	>50–500	>200–2,000	>0.05–0.5	Corneal involvement or other eye irritation clearing in 8–21 days	Severe irritation at 72 hours (severe erythema or edema)
CAUTION (Category III)	>500–5,000	>2,000–5,000	>0.5–2	Corneal involvement or other eye irritation clearing in less than 7 days	Moderate irritation at 72 hours (moderate erythema)
(NONE) (Category IV)	>5,000	>5,000	>2	Minimal effects clearing in less than 24 hours	Mild or slight irritation at 72 hours (no irritation or slight erythema)

Source: Hensley, David L. 2010. *Professional Landscape Management.* 3rd ed. Stipes Pub.

Epidemiological studies do not prove a causal relationship between the risk factor and the health problem. Only direct observation in a controlled study can do this. For example, smoking is a risk factor for lung cancer, a fact which has been widely accepted since at least 1964, when the US Surgeon General issued a warning stating as much. However, the causal mechanism was not known until 1996, when both the mutagen found in tobacco smoke (benzo-a-pyrene) was identified, as damaging a specific gene, p53.

ACUTE AND CHRONIC EFFECTS

Acute effects of pesticide poisoning are immediate, usually appearing within 12 hours of exposure. They may be caused by accidentally spilling it on the skin, ingesting it, or inhalation. Immediate action must be taken to counteract the poisoning. Instructions for treatment can be found on the pesticide label or the MSDS. Every landscape operation that uses pesticides should have a well-stocked first-aid kit to address such incidents, as well as employees trained in their use.

Chronic effects of pesticide poisoning are illnesses or injuries that appear only after much time has passed, years, or even decades. Some examples of chronic effects are tumor formation, cancer, birth defects, infertility, sterility, anemia, skin discoloration, respiratory problems, and kidney failure.

CARCINOGENS

A Canadian government report suggests that non-Hodgkins lymphoma (NHL) and reproductive organ tumors result from exposure to pesticides. The widely used herbicide 2,4-D has been implicated in numerous studies linking it to cancer; however, some studies suggest that dioxin, a contaminant and breakdown product of 2,4-D, is the actual problem. 2,4-D is used to kill broadleaf weeds in turf. It sells under a variety of trade names (Weedestroy, Weed-b-gone, Trillion, Killex), either alone or in combination with other herbicides, typically mecoprop and dicamba. It is also available in combination with fertilizer in "Weed and Feed" products. Another 2,4-D breakdown product 2,4-dichlorophenol is a suspected endocrine disrupter and possible carcinogen. Stiengraber has found that farmers, Vietnam War veterans exposed to Agent Orange, and golf course superintendents have been shown to contract NHL at higher rates than the general public. She also has found that dogs whose owners regularly apply herbicides to their lawns suffer from double the rates of canine NHL when compared to dogs whose owners did not use herbicides on their lawns. Dr. Vincent Garry at the University of Minnesota has demonstrated that pesticide applicators have a particular genetic mutation at high frequency – a mutation that is also found at high frequency in NHL patients. Even though no one of these pieces of evidence proves a connection between the pesticide and NHL, further studies are required before a possible connection can be ruled out.

Golf course superintendents have higher rates of lymphoma, and cancers of the brain and prostate than the general population. Furthermore, they die more often from cancer than the general population.

Teratogenic Effects

Pregnant women exposed to toxic pesticides may unintentionally expose their unborn fetuses to higher risks than the adult population. In Iowa, low birth weights have been attributed to herbicides. Pesticide residues are detectable in breast milk, body fat, blood serum, semen, umbilical cords, placentas, and the fluid surrounding eggs extracted from women who are undergoing in vitro fertilization.

Endocrine Disrupters

Endocrine systems are also known as hormone systems. Hormones are chemical messengers that perform special functions in the body. They are carried in the blood from their point of origin in the glands to organs or tissues where the appropriate receptor cells are located. When the hormone reaches the appropriate receptor, the two molecules bind, causing a molecular response that turns on a gene or alters existing proteins in the cell.

Humans and other mammals, birds, fish, and many other types of living organisms have endocrine systems. They are composed of

- Glands that are located throughout the body
- Hormones that are made and secreted into the blood by the glands
- Receptors in various organs and tissues

In the human body, all of the physiological processes are regulated by hormones. Sexual reproduction, development of organs and the nervous system, cell division and growth, metabolism, and blood sugar levels are all regulated by the endocrine system. Table 11.4 lists hormones that are found in the human body.

There are several ways in which the endocrine system can be disrupted. For example, chemicals that mimic a hormone may stimulate an over-production of a

TABLE 11.4
Hormones in the Human Body

Hormone	Gland Responsible	Action
Glycogen and insulin	Pancreas	Glucose metabolism
Thyrozine, triiodothyronine	Thyroid	Growth, mental development
Parathyroid hormone (parathormone)	Parathyroid	Calcium, phosphorus regulation
Melanocyte-stimulating hormone, somatotropin, thyrotropin, corticotropin, prolactin, gonadotropins, oxytocin, and vasopressin	Pituitary	Pigmentation; growth; thyroid secretion; milk production; sex cells and hormones; birthing contractions; blood vessel constriction
Thyrotropin-, Gonadotropin-, and Growth hormone-releasing hormones; Somatostatin; Dopamine	Hypothalamus	Pituitary gland modulation; neurological functions
Catecholamines, corticosteroids	Suprarenal/adrenal	Various functions

protein or even of other hormones. Hormones may be prompted to respond at inappropriate times, as well.

Research on endocrine disruption of pesticides has only recently been explored. New studies are beginning to provide evidence of this link, as well as the health consequences. In addition to human health concerns, fish, frogs and other amphibians, birds, and other wildlife are all threatened. Reproductive systems appear to be particularly susceptible. Reproductive problems, as well as birth defects, have been observed. Endocrine disruptors often have an effect on the offspring of the individual contaminated with the chemical even though the individual may not show ill effects. According to the Tulane/Xavier Center for Biomedical Research, "prebirth exposure, in some cases, can lead to permanent alterations and adult diseases".

In 1996, the Food Quality Protection Act (FQPA) and Amendments to the Safe Drinking Water Act (SDWA) required that the EPA

> Develop a screening program, using appropriate validated test systems and other scientifically relevant information, to determine whether certain substances may have an effect in humans that is similar to an effect produced by a naturally occurring estrogen, or other such endocrine effect as the Administrator may designate.

Furthermore, two other acts that provide testing authority for endocrine disruption to the EPA are the Toxic Substances Control Act and FIFRA.

The EPA endocrine disruptor screening program is designed to evaluate and determine the risks posed by pesticides, as well as commercial chemicals and environmental contaminants. In addition to assessing risk by identifying potential endocrine disruptors, and characterizing the endocrine disruption activity of such chemicals, it will also determine dose–response information, which will allow EPA to set safe and unsafe levels of exposure.

CHOLINESTERASE INHIBITORS

Cholinesterase is an enzyme involved in the proper functioning of the nervous system. It is present in humans and insects, as well as other animals. Carbamates and organophosphate pesticides are toxic to insect pests due to their interference with, or inhibition of, cholinesterase. They can cause certain nerves to malfunction, causing many organs to become overactive and eventually to stop functioning. Table 11.5 lists pesticides that are cholinesterase inhibitors.

TABLE 11.5
Pesticides That Are Cholinesterase Inhibitors

Chemical Name	Product Name
Acephate	Orthene
Aspon	
Azinphos-methyl	Guthion
Carbofuran	Furadan, F formulation
	(*Continued*)

TABLE 11.5 (*Continued*)
Pesticides That Are Cholinesterase Inhibitors

Chemical Name	Product Name
Carbophenothion	Trithion
Chlorfenvinphos	Birlane
Chlorpyrifos	Dursban, Lorsban
Coumaphos	Co-Ral
Crotoxyphos	Ciodrin, Ciovap
Crufomate	Ruelene
Demeton	Systox
Diazinon	Spectracide
Dichlorvos	DDVP, Vapona
Dicrotophos	Bidrin
Dimethoate	Cygon, De-Fend
Dioxathion	Delnav
Disulfoton	Di-Syston
EPN	
Ethion	
Ethoprop	Mocap
Famphur	
Fenamiphos	Nemacur
Fenitrothion	Sumithion
Fensulfothion	Dasanit
Fenthion	Baytex, Tiguvon
Fonofos	Dyfonate
Isofenfos	Oftanol, Amaze
Malathion	Cythion
Methamidophos	Monitor
Methidathion	Supracide
Methyl parathion	
Mevinphos	Phosdrin
Monocrotophos	
Naled	Dibrom
Oxydemeton-methyl	Metasystox-R
Parathion	Niran, Phoskill
Phorate	Thimet
Phosalone	Zolonc
Phosmet	Irnidan, Prolate
Phosphamidon	Dimecron
Temephos	Abate
Tepp	
Terbufos	Counter
Tetrachlorvinphos	Rabon, Ravap
Trichlorfon	Dylox, Neguvon

In humans, repeated exposure of carbamate or organophosphate pesticides over a period of time may result in a mild, slow poisoning. Symptoms of such poisoning are difficult to detect as they often do not cause the pronounced reactions seen in acute cases of poisoning. However, blood tests are available for such poisoning, because blood cholinesterase levels can be measured as a response to carbamates or organophosphates. A baseline measurement must be conducted at a time of year when the applicator is not using pesticides regularly. Subsequent testing during periods of pesticide use will then reveal any effect. If a sufficient drop in cholinesterase levels is seen, pesticide exposure should be avoided until they return to normal.

Repeated exposure to cholinesterase inhibitors can be as dangerous as acute exposure that occurs when they are spilled on the skin or accidentally swallowed. Daily exposure to pesticides should be treated with care, including working around the pesticide storage area, or working in areas recently treated with pesticides. People who are exposed repeatedly to pesticides may not be able to handle acute exposure as effectively because of the toxic effects already present in their body. Symptoms that may be present in persons over-exposed to cholinesterase-inhibiting pesticides range from weakness, dizziness, and blurred vision in mild cases to abdominal cramps, muscular tremors, slow heartbeat, and breathing difficulty.

OTHER HEALTH EFFECTS

Small doses of 2,4-D are toxic to the liver. Increases in liver function tests, jaundice, and acute hepatitis, as well as permanent liver damage leading to cirrhosis, have been reported in exposed golfers. Most pesticides use oil-based chemicals as carriers. These oils consist of hydrocarbons derived from petroleum. They are easily absorbed through the skin, and once in the bloodstream, these hydrocarbons can be fatal. This is known as **hydrocarbon poisoning** and also refers to poisoning with other petroleum products, such as gas or diesel, fuel oil, kerosene, and tar.

Allergic reactions to pesticides can range from skin irritations and rashes, to respiration problems, such as emphysema and asthma, to eye and nose irritation. Oftentimes, a person does not have an allergic reaction upon first exposure, but as they become sensitized to the chemical, an allergic reaction eventually manifests itself. Utmost care should be taken to minimize or avoid this problem by handling pesticides safely and always wearing the proper apparel and equipment.

ENVIRONMENTAL HAZARDS

Water and soil pollution are the primary sources of use or overuse of pesticides. The ability of the pesticide to move through soil and leach into groundwater, or to be transported by downstream and adversely affect wildlife is an important matter of concern. Another issue of concern is the ability of chemicals to deteriorate into breakdown products that may be more toxic than the original active ingredient. A third issue of concern is the persistence of a toxic chemical in the environment.

There are many factors and interactions that occur that leads to pesticide contamination of the environment. For example, in order for pesticides to leach into groundwater four conditions are required:

Pesticides in the Landscape

1. The properties of the chemical are highly mobile in soil.
2. The soil properties do not allow chemicals to adsorb to soil particles.
3. Moderate to high rates of application over large areas.
4. They have moderate to long-lived half-lives in the environment.

Weather conditions can also affect pesticide contamination of water, particularly if they are applied shortly before a rainfall event. This subject is discussed in greater detail in Chapter 5.

Persistence in the environment is a further concern of pesticides due to their direct effect on wildlife. Zoologist Rebecca van Beneden found that clams grown in aquariums contaminated with 2,4-D under experimental conditions had gonadal cancers induced in them. Environmental persistence may also have an effect on soil-dwelling microorganisms that would normally contribute to good soil health.

PESTICIDE HANDLING

For landscape professionals who handle pesticides, care and attention must be taken at every stage of the process: opening and mixing the pesticides, application, storage, and disposal. In mixing pesticides, the method used depends on the pesticide **formulation**, or the form in which the pesticide is carried. The pesticide label shows this abbreviation right after the product name. Some pesticides may be safely combined, and some may have an **adjuvant** or **spreader** added to enhance the application or the effectiveness of the pesticide.

In addition to safety concerns for the applicator, proper cleaning of pesticide equipment and mixing containers must also be observed. Special clothing, including shoes and gloves, must also be handled appropriately. The specific issues for each pesticide are clearly spelled on each pesticide label, a legal document that is regulated by the EPA and is included with each product. Pesticide labels can be found online at a variety of websites, two of which are listed at the end of this chapter: EXTOXNET (The Extension Toxicology Network) and PAN (The Pesticide Action Network). Individual chemical companies also maintain files of labels for their products. EXTOXNET is no longer being updated, but the large amount of information posted there is thorough and easy to read, and covers many areas of concern with respect to pesticides and toxicity.

SAFETY ISSUES

Landscape professionals who are certified pesticide applicators are trained in the proper use, storage, and disposal of pesticides. However, homeowners are often not aware of safe use of pesticides. In surveys conducted in 1986 by McEwen and Madder, and later by Whitmore (1993), less than 50 percent of homeowners read pesticide labels and many ignore common safety procedures. Mixing and loading pesticides are two of the most hazardous activities associated with pesticide use, in part because pesticides are present in their most concentrated form at this time in the process. Protective equipment and clothing must be worn at this time.

Pesticides spilled on clothing or the skin can be hazardous, as can inhalation of granules or powder forms of pesticides. Applicators and mixers have to be particularly careful not to place hands contaminated with pesticides to their mouths when eating or smoking. Approximately 110,000 people are sickened annually by pesticides in the United States. Wearing the proper apparel during pesticide use is the first step in preventing health hazards. **Personal protective equipment**, or **PPE**, such as coveralls, chemical-resistant gloves, boots, and even a chemical mask may be required for handling of some pesticides.

Studies have shown that the dermal route of pesticide contamination is the one that occurs with greatest frequency among pesticide applicators. Failure to wear gloves, long pants, long-sleeve shirt, or shoes and socks contributes to this situation. Regardless of the toxicity of the pesticide, this is the minimal clothing that should be worn. A coverall may be used in lieu of long pants, and a long-sleeved shirt if it covers the arms fully. Additional garments that may prove useful is a hat to keep hair out of the way and a chemical-resistant apron, which will protect the scrotal area, which is the most vulnerable area for dermal absorption.

Pesticide applicators should change clothes and shower at the end of the day after using pesticides. Contaminated clothing should be laundered separately from regular laundry. Some employers provide laundry facilities for this purpose, and to eliminate exposure to pesticides outside the work place. When contaminated clothing is laundered, the person doing the laundry should wear neoprene gloves and then wash them thoroughly before removing them. Hot water (150°) and the recommended amount of detergent should be used. If laundering at home or at a laundromat, after washing the clothes, the machine should be run without a load in it, allowing the soap and hot water to remove chemical residues. Contaminated clothing should be properly discarded.

STORAGE AND DISPOSAL

Pesticides are required by law to be stored in locked areas that are labeled as such. This may mean storing pesticides in a locked area within a building, or even in their own designated building. Usually signs labeling such a site as dangerous or hazardous are required, and they may need to be in more than one language. Training of employees should include a tour of such areas and clarification of which personnel are permitted entry. The many specific instructions regarding storage of pesticides are provided by each state as part of their pesticide training and certification program.

Pesticides can only be properly disposed of following federal guidelines. In no case may pesticides, pesticide containers, or pesticide-related waste be disposed of by open burning, open dumping, or dumping into water. Accepted methods of disposal are generally fairly sophisticated and are often costly. Because of this, it is a good business practice to buy only the amount of pesticide that will be used in a reasonable amount of time, such as within one growing season. It is also a good idea to be aware of possible pesticide registration cancellations and avoid purchasing it. State EPAs can provide information on facilities that can accept hazardous waste products.

SUMMARY

In this chapter, we learned that pesticides are any of a number of compounds that are used to control pests. In the landscape industry, and particularly with ornamental plants, the common pests are insect pests, fungal infections, bacterial infections, and weeds.

Sixty-seven million pounds of pesticides are applied to lawns each year. Whereas the agricultural sector accounts for around 75 percent of the pesticides used, only about 7 percent are used in the home and garden sector. This relatively small amount can have more significant effects on people due to their close proximity to living situations, people's interaction with lawns, gardens, and landscaped areas, and increased exposure to children and pets as compared to agricultural uses.

Landscape pesticides fall into categories of bactericides, fungicides, insecticides, and herbicides. The presence of a large number of plant species in the landscape result in a far greater number of insect, pathogen, and weed pests.

Synthetic pesticides fall into three general categories based on their mode of action: organochlorides, organophosphates, and carbamates. Naturally occurring plant compounds have been used at various times in history, including the present day. These include the active ingredients nicotine, neem, and pyrethrum. A newer generation of pesticides focused on the target pest, using more targeted methods to eradicate it while not harming other organisms.

EPA regulates pesticide use in the United States. Its mission included mandates to establish and enforce environmental protection standards, conduct environmental research, provide assistance to others combating environmental pollution, and assist in developing new policies for environmental protection.

Two Acts, the Federal Food, Drug, and Cosmetics Act, and the Federal Insecticide, Fungicide, and Rodenticide Act, provide the necessary authority to the EPA to require testing, collect fees, institute a registration process, and remove dangerous pesticides from the market, as it deems necessary. An early act taken by the EPA was to ban the chemical DDT, which had become present throughout the ecosystem, resulting in far-reaching consequences to wildlife and the environment. Although it was banned in 1972, it continues to be found in the environment.

Another action that occurred under the EPA is a certification process for pesticide applicators that is administered by states and required for anyone applying or supervising the application of RUP. RUP are those pesticides which have been deemed to pose a high risk to human health or the environment.

Pesticides known to cause cancer are carcinogenic and those that cause birth defects are teratogenic. Other pesticides are endocrine disruptors, meaning that they interfere with the body's hormones. Pesticides are tested for their toxicity on animals that are given varying levels of active ingredients to determine the LD_{50}, or lethal dose of 50 percent of them. This amount is then extrapolated to humans and presented in milligrams of pesticide to kilograms of body weight.

The FQPA and Amendments to the SDWA established a mandate for the EPA to develop a screening program to examine chemicals that could be endocrine disruptors. Results from these studies will allow EPA to set safe and unsafe levels of exposure.

Use, overuse, and mis-use of pesticides lead to environmental problems that include water and soil pollution, and adverse effects on fish and wildlife. Persistence in the environment of both pesticide active ingredients and breakdown products is of concern. Factors and interactions that can lead to environment contamination include soil mobility of a compound, failure to adsorb to soil particles, moderate to high application rates over large areas, and moderate to long-lived half-lives in the environment. Weather conditions, particularly severe rainfall events, can also allow pesticides to run off into surface water and be carried away from their intended target.

Mixing and loading pesticides are two of the most hazardous activities associated with pesticide use. Protective equipment and clothing must be worn at this time. In addition to spilling pesticides on clothing or the skin, inhalation of granules or powder forms of pesticides is also potentially dangerous. Applicators and mixers should not eat or smoke unless they have thoroughly cleaned their hands after handling pesticides. Personal protective equipment, such as coveralls, gloves, boots, and even a chemical mask, may be required for handling of some pesticides. Pesticides are required by law to be stored in locked areas that are labeled as such.

REVIEW QUESTIONS

1. What sector uses the greatest amount of pesticides in the United States? What is the percentage use of that sector?
2. List three reasons why landscape pesticides are an important health concern, although the overall percent usage is not great.
3. What are the four types of pesticides used in the landscape?
4. What are the three types of chemical pesticides used and what are their attributes with respect to toxicity and persistence?
5. What agency was established in 1970, the "year of the environment"?
6. In 1988, Amendments to which Act of Congress were responsible for regulating licensing and registration of pesticides?
7. Why are children more susceptible to the toxic effects of pesticides?
8. Name two conditions that affect pesticide leaching into the groundwater.
9. What does PPE stand for? Name three pieces of PPE that may require for pesticide handling.
10. Explain the conditions required for proper pesticide storage.

ENRICHMENT ACTIVITIES

1. Visit a local landscape company, and tour their pesticide storage area. Request a list of the pesticides they commonly use and research their toxicity levels.
2. Use the list from Activity 1 and look for less toxic alternative pesticides. Complement the pesticide choices with other practices that could reduce or eliminate the need for the pesticide.
3. Look for weedy areas in a local park or other public space. Identify underlying problems that may be contributing to the problem. Suggest solutions for remedying the underlying problem.

FURTHER READING

Aker, W. G., X. Hu, P. Wang, and H-M. Hwang. 2008. Comparing the relative toxicity of malathion and malaoxon in blue catfish *Ictalurus furcatus*. *Environ. Toxicol.* 23(4): 548–554. doi: 10.1002/tox.20371.

Carson, R. 1962. *Silent Spring*. Houghton Mifflin, New York, NY, 368 pp.

C&P Press. 2010. *Turf and Ornamental Reference for Plant Protection Products*. Chemical and Pharmaceutical Press, Inc., New York, NY.

Chemical Pesticides Health Effects Research. http://www.chem-tox.com/pesticides/#breastcancer. Retrieved July 5, 2019.

Denissenko, M. F., A. Pao, M.-S. Tang and G. P. Pfeifer. 1996. Preferential formation of benzo[a]pyrene adducts at lung cancer mutational hotspots in P53. *Science.* 274(5286): 430–432.

Durak, D., F. G. Uzun, S. Kalender, A. Ogutcu, M. Uzunhisarcikli, and Y. Kalender. 2008. Malathion-induced oxidative stress in human erythrocytes and the protective effects of vitamins C and E in vitro. *Environ. Toxicol.* doi: 10.1002/tox.20423. www.interscience.wiley.com.

Edwards, C. A. 2010. Pollution issues: Pesticides. http://www.pollutionissues.com/Na-Ph/Pesticides.html. Retrieved July 5, 2019.

EPA. Pesticides. http://www.epa.gov/pesticides/. Retrieved July 5, 2019.

EPA. Sustainable landscaping. http://www.epa.gov/greenacres/smithsonian.pdf. Retrieved November 29, 2010.

Fleming, L.E., J. A. Bean, M. Rudolph, K. Hamilton. 1999. Mortality in a cohort of licensed pesticide applicators in Florida. *Occup. Environ. Med.* 56(1): 14–21.

Hensley, David L. 2010. *Professional Landscape Management*. 3rd ed. Stipes Publishing, Champaign, IL.

Kavlock, R. J., G. P. Daston, C. DeRosa, P. Fenner-Crisp, L. E. Gray, S. Kaattari, G. Lucier, M. Luster, M. J. Mac, C. Maczka, R. Miller, J. Moore, R. Rolland, G. Scott, D. M. Sheehan, T. Sinks, and H. A. Tilson. 1996. Research needs for the risk assessment of health and environmental effects of endocrine disruptors: a report of the U.S. EPA-sponsored workshop. *Environ. Health Perspect.* 104(Suppl 4): 715–740. http://www.ncbi.nlm.nih.gov/pmc/articles/PMC1469675/. Retrieved July 5, 2019.

Kross, B. C., L. F. Burneister, L. K. Ogilvie, and L. J. Fuortes. 1996. Golf course superintendents face higher cancer rates. *Am. J. Ind. Med.* 29(5): 501–506.

Latimer, J.G., R. B. Beverly, C. D. Robacker, O. M. Lindstrom, S. K. Braman, R. D. Oetting, D. L. Olson, P.A. Thomas, J. R. Allison, W. Florkowski, J. M. Ruter, J. T. Walker, M. P. Garber, and W. G. Hudson. 1996. Reducing the pollution potential of pesticides and fertilizers in the environmental horticulture industry I: Greenhouse, nursery and sod production. *HortTech*. 6: 96–140.

Latimer, J.G., S. K. Braman, R. B. Beverly, P.A. Thomas, J. T. Walker, B. Sparks, R. D. Oetting, J. M. Ruter, W. Florkowski, D. L. Olson, C. D. Robacker, M. P. Garber, O. M. Lindstrom, and W. G. Hudson. 1996. Reducing the pollution potential of pesticides and fertilizers in the environmental horticulture industry II: Lawn care and landscape management. *HortTech*. 6(3): 222–232.

McEwen, F. L. and D. J. Madder. 1986. The use of horticultural oils and insecticidal soaps for control of insect pests on amenity plants. *J. Arboricult.* 15: 257–262.

Miller, T. L. EXTOXNET: The Extension Toxicology Network. University of California-Davis, Oregon State University, Michigan State University, Cornell University, and the University of Idaho. http://extoxnet.orst.edu/. Retrieved July 5, 2019.

Nett, M. 2008. *The Fate of Nutrients and Pesticides in the Urban Environment*. ACS Symposium Series 997. Oxford University Press, Oxford, 277 pp.

Nixon, P. L., C. D. Anderson, N. R. Pataky, R. E. Wolf, R. J. Ferree, L. E. Bode. 1995. *Illinois Pesticide Applicator Training Manual: General Standards.* University of Illinois, Special Pub. 39, Urbana-Champaign, 90 pp.

Pesticide Action Network. http://www.pesticideinfo.org/. Retrieved July 5, 2019.

Racke, K. and A.R. Leslie. 1993. *Pesticides in Urban Environments: Fate and Significance.* ACS Symposium Series 522. 378 pp.

Raloff, J. 2009. Pancreatic cancer linked to herbicides. http://www.sciencenews.org/view/generic/id/44163/title/Pancreatic cancer linked to herbicides/. May 28, 2009 web edition. Retrieved July 5, 2019.

Steingraber, S. 1998. *Living Downstream: A Scientist's Personal Investigation of Cancer and the Environment.* Random House, Inc., New York, NY, 374 pp.

Steingraber, S. 2001. *Having Faith.* Persues Publishing, Cambridge, MA, 342 pp.

Steingraber, S. 2002. Exquisite communion: The body, landscape, and toxic exposures, In: *Ecology and Design*, B. R. Johnson, and K. Hill, Eds., Island Press, Washington, DC, 192–202.

Steingraber, S. http://steingraber.com/. Retrieved July 5, 2019.

Steingraber, S. 2005. Breast cancer linked to home pesticide chlordane. *Breast Cancer Res. Treat.* 90: 55–64.

Templeton, S. R., D. Zilberman, S. J. Yoo. 1998. An economic perspective on outdoor residential pesticide use. *Environ. Sci. Tech.* 32(17): 416A–423A.

Tulane University ehormone. Endocrine disruption tutorial: Human effects. http://e.hormone.tulane.edu/learning/human-effects.html. Retrieved July 5, 2019.

USGS. 1999. Pesticides detected in urban streams during rainstorms and relations to retail sales of pesticides in King County, Washington. USGS Fact Sheet 097–99. 4 pp.

USGS. 2008. Pesticide national synthesis project. http://water.usgs.gov/nawqa/pnsp. Retrieved July 5, 2019.

White, R. S. 1971. *Pesticides in the Environment.* M. Dekker, Inc., New York, NY.

Whitmore, W. H., J. E. Kelly, P. L. Reading, E. Brandt, and T. Harris. 1993. National home and garden pesticide use survey. In: *Pesticides in Urban Environments: Fate and Significance.* K. D. Racke, and A. R. Leslie, Eds., The American Chemical Society, Washington, DC, 18–36.

12 Integrated Pest Management

OBJECTIVES

Upon completion of this chapter, the reader should be able to

- Explain integrated pest management
- Describe the methods that comprise IPM
- Discuss the reasons for using IPM
- Discuss thresholds and injury levels in landscaping
- Describe the importance of pest life cycles
- Identify pest symptoms and signs
- Explain the role of sticky traps in pest monitoring
- Develop an understanding of the concept of degree days and how it applies to pest management
- Identify the various alternatives to chemical pesticides
- Explain the role of genetics in pest management
- Discuss the various cultural practices that support pest management

TERMS TO KNOW

Action threshold
Aesthetic injury level
Biological control
Botanical
Branch beating
Degree days
Economic threshold
Insect growth regulator (IGR)
Integrated pest management
Monitor
Phenology
Pheromone
Physiological time
Trap

INTRODUCTION

Integrated pest management is a term that has come to represent a pest management strategy that integrates numerous methods and practices in order to suppress plant pests while minimizing harmful effects on humans and other animals, non-target insects, and the environment (Figure 12.1). At one time IPM, as it has come to be known, represented an alternative to traditional pest control, which consisted mainly of establishing a calendar-based spray schedule that would ensue throughout the growing season. Currently, however, many programs implement at least some of the methods of IPM as standard practice.

Pests of landscape plants include biotic factors, such as insect pests, disease pathogens, and weeds. Plants can suffer stresses from abiotic factors, too. Air pollution, poor cultural practices, and root injury due to compacted soil are some examples of abiotic factors. Plant damage can appear similar regardless of whether it is caused by pests or from other factors. When assessing problems with landscape plants, the landscape professional must keep all of these factors in mind.

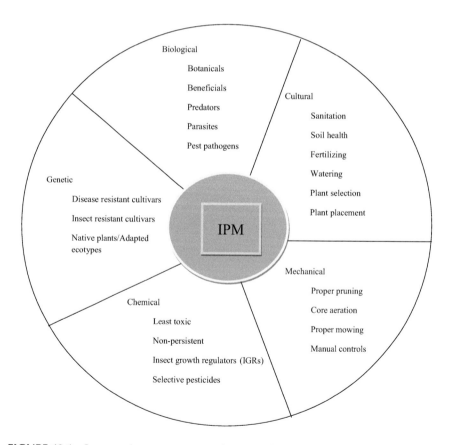

FIGURE 12.1 Integrated pest management has many interactive components. (Illustration by the author.)

There are two main aspects of pest management in sustainable landscaping practices. The first is avoidance, and the second is treatment. Avoidance includes using cultural and mechanical practices, using genetically superior plants. Treatment includes using less toxic pesticides, using pesticides that do not persist in the environment, and use of **biological controls**. In IPM, all applicable avoidance strategies should be implemented before treatment measures are taken.

Case Study: Comparing an IPM Pilot Program to a Traditional Spray Program

Reference: Stewart, C. D., S. K. Braman, B. L. Sparks, J. L. Williams-Woodward, G. L. Wadde, and J. G. Latimer. 2002. Comparing an IPM pilot program to a traditional cover spray program in commercial landscapes. *Journal of Economic Entomology.* 95(4): 789–796.

Five different sites were included in this study to compare a traditional spray program with an IPM program on landscape plants over a 2-year period. The sites were two commercial sites, two residential sites, and one institutional property managed by landscape professionals. The traditional program was calendar-based. The IPM program included the spot treatment of pests using horticultural soaps and oils, and pruning-affected material.

In the first year of the program, pesticide volume was reduced a total of 86.3 percent at the four sites receiving only IPM. In the second year, all five sites saw a reduction of pesticide use by 85 percent. Initially, mean cost per site increased from $703.40 pre-program to $788.26 in the first year of the IPM program. However, in the second year, costs decreased to $582.22.

In addition to landscape plant pests such as spider mites, aphids, lace bugs, scales, whiteflies, and Japanese beetles, beneficial insects were also seen. Spiders, ants, green lacewings, and lady beetles, all predatory arthropods, were well represented.

AVOIDANCE

Cultural, mechanical, and genetic practices comprise avoidance strategies that are the front line of defense against pest problems in the landscape. Among the mechanical practices, proper pruning, aeration of turfgrass, and proper mowing are addressed in Chapter 2.

CULTURAL PRACTICES

There are a number of cultural practices that can reduce pest populations or avoid them altogether. Perhaps first and foremost is the concept of "right plant, right place". This practice would appear to be common sense, yet it is often overlooked. One reason is that conditions in the landscape change over time. For example, plants in a sunny place may eventually become shaded out by maturing trees and shrubs. Powdery mildew is notorious for appearing on landscape plants in cool, shaded areas.

Conversely, shade-loving plants may become exposed to full sun when a nearby tree or shrub is removed by weather conditions, age, disease, or other reasons. Shade-loving plants may become sunburned when exposed to too much sun. Proper plant selection and placement should receive the utmost attention during the design phase.

Plants should be selected for

- Appropriate size for the available space
- Correct sun/shade requirements
- Correct moisture and temperature requirements
- Desirable attributes, such as shade-producing
- Disease and pest resistance
- Good growth habit/good structural form
- Avoiding monoculture

Proper site preparation and proper planting are two other components of cultural practices that can help avoid potential problems. Trees that are planted too shallow can suffer root damage from equipment or foot traffic. The roots may experience drying if they become exposed due to erosion of the shallow layer of soil. Trees that are planted too deep are more susceptible to root or crown diseases. All plants require at least several inches of topsoil for healthy root growth.

SANITATION

Sanitation is an important cultural control practice, especially when disease organisms are present. Diseased branches and leaves should be removed and properly disposed of before spores or other infectious agents are allowed to spread. When pruning trees for diseases that travel through the vascular system, cuts should be made at least 6 in. away from the visibly infected area. Pruners must be disinfected between each cut to avoid spreading the disease organism throughout the plant. When pests appear on landscape plants, they should be dealt with in a timely manner. This is particularly true for pests that use numerous different species as host plants, such as Japanese beetles.

PROPER IRRIGATION/WATERING

A well-watered and fertilized plant growing in the right place is less stressed and therefore less prone to infection or pest invasion. Trees and shrubs that experience drought stress 1 year may show an increase in pests the following year. Due to this delay in problems, it is important to be diligent about watering plants during periods of prolonged drought. It is also important to ensure that trees and shrubs are well-watered and fertilized during the first and possibly the second years in the landscape. The intention is to establish a well-developed root system that will allow the plants to survive succeeding periods of drought.

Turf and trees and shrubs have different watering requirements. In areas of low rainfall or places that experience seasonal drought, drought-tolerant turf species should be selected. Other cool-season turf grasses should be allowed to go dormant during this period. When turf is irrigated in order to keep it green and lush, nearby trees are watered inappropriately. The shallow root systems of turf species such as Kentucky bluegrass and perennial ryegrass require regular, shallow watering, but this can create anaerobic conditions around the trees feeder roots.

Water quality can also be an issue when irrigation is used. Mineral salts may be present at high enough levels to burn plant roots or change soil pH. Magnesium dissolved in water can interfere with calcium uptake by roots. When irrigation is used, salt levels should be monitored and measures taken to avoid such problems.

SOIL HEALTH AND COMPACTION

There are several components of plant health and soil conditions. Fertility, pH, aeration, and drainage are among the soil-related problems that can contribute to pest problems. Plants should be selected based on soil qualities rather than trying to amend the soil for unsuitable plants. Soil health, fertilization practices, and amending the soil with organic matter are addressed in Chapters 8, 9, and 10.

The importance of drainage and plant selection cannot be over-emphasized. Many diseases thrive and spread in soils that remain wet for long periods. Some pathogens that cause root rot include *Fusarium*, *Phytophthora*, and *Pythium*. Trees that have sustained root injury are more susceptible to root rot.

A common symptom of soil-related problems is yellowing or chlorosis of leaves. This may be caused by inadequate nitrogen, iron, or improper pH (too acid or too alkaline). High pH levels can lead to iron deficiency in plants because the iron bonds with other elements in the soil at pH levels above 7.4. Plant pH requirements and soil pH test results are a critical part of landscape design and plant selection.

Compaction is often an issue in the landscape due to construction activities that date to when the house was built. Compaction also occurs with repeated walking, and with the use of a mower or other equipment. Parking vehicles under trees for shade in summer is a common cause of compaction. Compaction problems lead to poor plant performance and thus should be addressed properly right away.

GENETICALLY IMPROVED PLANTS

Landscape plants have been bred and selected for improved traits for many years. Pest and disease resistance are important traits. Breeding programs can be found in university and extension programs, are operated by private companies, and are also conducted by private individuals. The genetic improvement that is made is usually due to specific goals of a breeding program.

The use of ornamental plants in the landscape encompasses hundreds of species, both woody and herbaceous. Whereas it important to avoid using pest-prone plant species, use of improved cultivars that have been tested for resistance to pests and diseases can still allow certain species to be used in the landscape. Table 12.1 lists landscape plants that are particularly prone to pests and diseases. Table 12.2 lists some examples of cultivars of landscape plants that are more pest-resistant than the species or unimproved cultivars.

TABLE 12.1
Select Landscape Plants That Are Susceptible to Pests and Diseases

Plant	Pests
Acacia, *Acacia* spp.	Acacia psyllid
Alder, *Aldus* spp.	Flathead alder borer
Ash, *Fraxinus* spp.	Emerald ash borer
Box elder, *Acer negundo*	Boxelder bug
Crabapple, *Malus* spp.	Fireblight, cedar-apple rust
Crape myrtle, *Lagerstroemia indica*	Powdery mildew
Cypress, arborvitae, juniper, *Cupressus* spp., *Thuja* spp., *Juniperus* spp.	Cypress tip miner
Elm, *Ulmus* spp.	Dutch elm disease, European elm scale, elm leaf beetle
Euonymus, *Euonymus* spp.	Euonymus scale
Lilac, *S. vulgaris*	Lilac borer, powdery mildew
Pear (flowering), *Pyrus* spp., flowering quince, *Chanomeles speciose*	Fireblight
Poplar, *Populus* spp.	Cytospora canker
R. Catawba, *R. catawbiense*	Root weevils
Rose, *Rosa* spp.	Powdery mildew, black spot
Sycamore, *P. occidentalis*	Anthracnose

Using genetically resistant landscape plants can help reduce maintenance levels and chemical (fertilizer and pesticide) inputs. Genetically improved plants may also survive better under adverse environmental conditions (heat, cold, drought, saline soils) than unimproved types.

TREATMENT

If a pest problem exists in the landscape and all avoidance measures have been implemented, treatment may be necessary. Some criteria for deciding whether treatment is necessary are provided in the sections on action thresholds, economic injury level, and aesthetic injury level (AIL). For these levels to be decided, one must first identify the pest. In integrated pest management, observations and record-keeping about possible pest threats are important components. Predictive models for pest emergence are useful in monitoring pest presence. In IPM, pest control is targeted to a particular stage in a pest life cycle.

Whereas IPM does not exclude the use of chemical pesticides, least toxic pest control methods are preferred. Natural enemies are preserved whenever possible and are sometimes introduced into the landscape. Least-toxic chemicals may help to preserve desirable micro-organisms and insect species. They are also more desirable for wildlife, pets, and humans. Furthermore, some non-toxic pesticides will not lead to resistance in the pest population.

Due to the vast number of species used throughout the landscape, each playing as host to one or many pest species, the total number of landscape pests can

TABLE 12.2
Select Landscape Plant Species and Cultivars with Improved Resistance to Pests

Plant	Resistant Alternative
Acacia, *Acacia* spp.	Bailey acacia
Alder, *Aldus* spp.	Italian alder
Ash, *Fraxinus* spp.	Non-ash species
Box elder, *Acer negundo*	Use male trees, or avoid this species altogether
Crabapple, *Malus* spp.	"Adams", "Adirondack", "Donald Wyman", "Prairiefire"
Crape myrtle, *Lagerstroemia indica*	Japanese crape myrtle, *L. faurriei*
Cypress, arborvitae, juniper, *Cupressus* spp., *Thuja* spp., *Juniperus* spp.	*J. chinensis* var. *sargentii* "Glauca", *J. scopularum* "Erecta Glauca", *J. chinensis* "Kaizuka", western red cedar
Elm, *Ulmus* spp.	"Valley Forge", "New Harmony"
Euonymus, *Euonymus* spp.	*E. kiautschovicus* (*patens*) "Sieboldiana"
Lilac, *S. vulgaris*	Meyer lilac, *S. meyeri* "Palibin", Tree lilac, *S. reticulate*
Pear (flowering), *Pyrus* spp., flowering quince, *C. speciose*	"Capitol", *P. ussuriensis* "Prairie Gem", *C. speciosa* "Contorta"
Poplar, *Populus* spp.	"Noreaster"
Rhododendron Catawba, *Rhododendron catawbiense*	"Cilpinense", "Jock", "Dora Amateis", "Moonstone", "Lady Clementine Mitford"
Rose, *Rosa* spp.	"Meidiland" cultivars, *Rosa rugosa* "Alboplena", *R. rugosa* "Frau Dagmar Hastrup", "Knockout" cultivars
Sycamore, *Platanus occidentalis*	London plane, *P.* × *acerifolia*, (*P.* × *orientalis*, *P. hybrida*) "Bloodgood", "Columbia", "Liberty"

be enormous. A good reference book on landscape plants and their pests, or other access to reliable information and identification is necessary. Refer to the sources at the end of this chapter for some suggestions. A local or regional extension service office, botanical garden, or arboretum may be able to provide identification guidance.

A successful pest control strategy includes

1. Knowledge of the expected pests on a given plant species
2. Proper identification of the pests that occur on each landscape plant
3. Knowledge of the life cycle of each pest
4. Knowledge of the damaging stage(s) of the pest
5. Knowledge of the vulnerable stage(s) during the life cycle of the pest

DETERMINING PEST OR DISEASE PRESENCE

In order to apply integrated pest management solutions, landscape plants must be **monitored** for problems. Observations are made and records are kept so that patterns can be identified. Pest emergence patterns can be related to environmental conditions or other activities in the landscape. When these patterns are noted, problems can be more effectively avoided or better treated due to predictability of a problem.

When a new phenomenon occurs, it may have been caused by the introduction of new plants in the landscape, neighboring activity, such as a change in drainage patterns, or other causes. Historical events, such as drought in a previous year, can lead to pest problems. Good record-keeping provides a better understanding of landscape problems that arise in response to such events.

Monitoring and record-keeping also reveal whether a particular insect pest or pathogen is present in the landscape and the relative severity of the problem. Monitoring for pests is a matter of looking for specific pests of each landscape species. Insect pests are counted, or the amount of damage is estimated. That information is then recorded.

In order for integrated pest management to work, both pest species and beneficial species must be correctly identified. Damage to plants must also be attributed to the correct cause. It is helpful to have knowledge of insect and disease organisms that are most likely to affect each landscape plant species. There are generalists, such as aphids, and there are pests and diseases that are specific to a single or small number of plants. Some pests have a life cycle that requires two host species, such as cedar apple rust, *Gymnosporangium juniperi-virginianae*.

Monitoring pests and pest damage is usually done by direct observation of a plant or by the use of a **trap** that allows population estimates to be made (Figure 12.2). Sticky traps, pheromone traps, and branch beating are three common ways to accomplish this goal.

Sticky Traps

Sticky traps are effective at catching adult whiteflies, fungus gnats, leafhoppers, leafminers, psyllids, thrips, and winged aphids. They are usually yellow or another color that is attractive to the pest being monitored. Sticky traps may be hung in the

FIGURE 12.2 Sticky purple emerald ash borer traps are placed in ash trees at least five feet above the ground in the lower to middle canopy. (Photo by the author.)

Integrated Pest Management

canopy of the plant being monitored. They are also used in greenhouses, where they are usually placed on a metal stake in or just above the plant canopy. Sticky apples have been designed for monitoring coddling moth on apple trees.

To count the insects on a sticky card, insects located in a one-inch-wide strip on each side of the card give a good representative sample. For example, when using a 3×5 in. card, count the insects in a 1 by 5 in. strip and multiply by 3 to get the total results for one trap.

Sticky traps should be monitored on a weekly basis and disposed of when they are full. Sticky paste can be painted on wood, metal, or plastic for a more efficient uses of resources. Pestick is a product that has been developed for this purpose. Yellow is attractive to many pests, blue is attractive to thrips, and purple is used to attract emerald ash borers (Figure 12.3).

Pheromone Traps

Pheromone traps use chemical communication signals to attract and trap insect pests. A pheromone is a chemical signal sent by insects to communicate. Pheromones are not detectable by humans and are specific to a particular insect. Pheromone traps indicate when an insect is present, but do not necessarily determine whether it is damaging a particular plant. Because pheromone traps are very sensitive, attracting pests even at very low densities, they are used to determine the first presence of a pest in an area. A pheromone trap has a lure that is impregnated with the pheromone and placed within a conventional trap. Disadvantages of pheromone traps are that they may bring in insects from other areas, they are usually only targeted to one gender (usually male), and they usually only attract adults, whereas the juvenile stage may be the one requiring control.

FIGURE 12.3 Sticky cards can be purchased or made using pre-mixed or homemade product applied to yellow or blue cardboard or plastic. The color depends on the type of insect to be trapped. (Photo by the author.)

Branch Beating

For insect pests that are easily dislodged from trees and shrubs, **branch beating** is an effective method to remove them. This applies to leaf beetles, leafhoppers, mites, psyllids, thrips, true bugs, and non-webbing caterpillars. To accomplish this affectively, the pests should be caught onto a sheet of white paper or some other surface where they will show up clearly. A sheet or a special beating tray may be used. Shake or hit the branch with a stick two or three times. Repeat this procedure using the same technique, including number of hits, over multiple sampling times, and record the results. The best time to perform this monitoring activity is in the morning when temperatures are cooler and insects are less active.

Identify both pest species and beneficial ones, counting and recording each. Sampling will be more accurate if multiple branches and multiple plants are sampled each time. Grid paper can be used to simplify the counting. If insect populations are high, count those in a representative number of grids rather than the entire number collected.

PHENOLOGY AND DEGREE DAYS

Phenology is the study of the relationship between weather, particularly temperature, and development of an organism. Insects are cold-blooded, and therefore, their development is related to accumulation of heat in the environment. For example, insects may emerge from pupation or hatch from eggs, more quickly as temperatures increase. The amount of time required for a pest to complete its development is known as **physiological time** because it is associated with temperatures and not solar time or dates on the calendar. A similar response to temperature is also expressed in many plants. Flowering is one of the easily observed expressions of a plant response to temperature. For example, spring-blooming plants bloom earlier or later in the spring depending on the temperatures that season. Physiological time is expressed in units called **degree days**.

The use of degree days is a method for predicting insect emergence that has been developed in order to better time pest management activities. The degree day system is based on the physiological time of an insect and its association with temperature. Degree days are calculated using a base temperature, which is selected based on an organism's metabolism. Below the base temperature, no significant metabolic activity occurs, but above it, biochemical reactions within the organism are set in motion. When the biochemical reactions reach a threshold level, insects emerge from dormancy, insect eggs hatch, plants bloom, and so on, depending upon the organism. In order to predict insect emergence and therefore more accurately time pest management activities, a base temperature of 50°F (10°C) has been established for general purposes. Other base temperatures are sometimes used, as revealed by studies of specific organisms (Table 12.3).

When the appropriate number of degree days has accumulated, the predicted event occurs. The simplest formula used to calculate degree days is

$$(\text{Max temperature} + \text{Min temperature})/2 - \text{Base temperature} = \text{Daily DD}$$

TABLE 12.3
Degree Day Requirements of Some Landscape Pests

Insect or Disease	Lower Threshold Development (°F)[a]
Aphids	7; 135
Apple maggot	43.5
Codling moth	50
Cooley spruce gall adelgid	120; 1500
Dogwood borer	148
Eastern tent caterpillar	90
Elm bark beetles	7
Euonymus scale	35; 533
European red mite	51.1
Gypsy moth	90
Honeylocust plant bug	58
Juniper scale	22; 707
Lilac borer	148
Northern pine weevil	7
Oriental fruit moth	45
Oystershell scale	7; 363
Peach tree borer	50
Pine bark adelgid	22; 58
San Jose scale	50
Spruce bud scale	22; 912
Spruce spider mite	7; 192; 2375
Taxus mealy bug	7; 246
Tufted apple bud moth	45
Tuliptree scale	12; 2032
White pine weevil	7
Woolly elm aphid	121

Sources: UC-Davis http://www.ipm.ucdavis.edu/index.html, University of New Hampshire http://extension.unh.edu/agric/GDDays/Docs/growch.pdf; and Dept. Entomology, Virginia Tech. http://www.virginiafruit.ento.vt.edu/Understanding_Degree_Days.html.

[a] More than one number indicates multiple emergence times.

For example, if the maximum temperature is 56°F and the minimum temperature is 52°F, the daily degree (DD) day will be

$$(56+52)/2 - 50 = 4 \text{ DD}$$

More complex formulas are used by a variety of groups, including a computer program called Forecaster, developed by the Department of Entomology at the University of Minnesota. Electronic temperature monitoring devices record the temperature multiple times each day and average them out, and are thus able to more accurately calculate degree days than the simple method of using the high and low temperatures of a given day. Many state agricultural extension services provide degree day information for various locales which are posted on their corresponding websites.

Degree days for pest control may be expressed in a range of numbers, from emergence, to first evidence of damage to a plant, and continuing through to the end of the pest's plant injury cycle. They may refer to timing for optimum control or multiple generations of a pest within one season. For example, degree days of 22–92 and 1500–1775 for the Cooley spruce gall adelgid indicate pest activity beginning around 22 degree days and continuing through around 92 degree days, followed by a second period of activity between 1500 and 1775 degree days when control must be administered and will be effective.

The start date for accumulating degree days varies by locale, but several extension offices collect temperature data from March 1 through September 30 for their calculations. In order for proper use of degree day information, monitoring should be used to determine whether a pest problem does indeed exist, so that plans for control measures may be made. Aesthetic and economic threshold information may then be employed.

Phenological studies have been able to correlate bloom times of plants with certain events in pest life cycles. In the book *Coincide* by Don Orton and Tom Green, 58 insect pests are described, and their vulnerable stages are correlated with life cycle events on major landscape plants. This allows the landscape manager to use visual cues in the landscape to determine when to scout for pests and, if they are present, when to treat them. For example, the euonymus scale is most vulnerable in the first stage, a newly hatched crawler. It can be controlled with contact pesticide at that time. These crawlers should be treated when Japanese tree lilac (*Syringa reticulata*) and Catalpa tree (*Catalpa speciosa*) are in early bloom, with four repeat applications at 10–12-day intervals. Alternative indicators are provided for each pest, as is a list of host plants. In separate chapters, common landscape plants are listed along with their major pests, and landscape plants that are used as indicator plants are listed along with the pests for which they are indicators.

ACTION THRESHOLDS

The concept of an **action threshold** is fundamental to the practice of integrated pest management: its purpose is to determine when action against a pest is needed in order to avoid undesirable damage, or economic or aesthetic loss. Commercial crops are grown for profit, so action thresholds are based on economic factors, comparing the costs of taking action to the costs of crop loss if action is not taken. The same criteria do not apply to landscape plants, even though the loss of a landscape plant certainly represents an economic loss to its owner. In the case of a golf course, resort, tourist attraction, or other site, aesthetic factors play the most important role in determining when to take action against a pest. An acceptable level of injury in such cases is often at or near zero.

ECONOMIC THRESHOLDS

In commercial crop production, **economic thresholds** compare the cost of applying pesticides versus the cost of not doing so. When the losses of a crop due to pest

Integrated Pest Management

damage exceed the cost of pesticide application, action is taken. Economic thresholds have been set for many commercial crops, including apple, cotton, corn, soybeans, and alfalfa.

AESTHETIC INJURY LEVEL

For ornamental plants in a landscape setting, economic thresholds play a different role, since loss of a plant has a different economic value for the homeowner than it does for the commercial grower. Appearance, rather than economic loss, is sometimes the deciding factor for taking pest control measures.

The creation and use of the term AIL originated with Olkowski in 1974 and is defined as the lowest level of a pest population that causes aesthetic injury. It forms the basis for making decisions about when to apply pesticides on ornamental plants in non-economic situations. It has been very valuable as a decision-making tool for public and private landscape settings. In 1983, Koehler and Moore quantified the relationship between insect populations and aesthetic quality, and illustrated that it was a simple linear relationship.

For many plants in landscape settings, very low levels of aesthetic injury are acceptable, and therefore, an emphasis on prevention is a key component of the pest management program. For example, Raupp (1989, 1992) and her colleagues reported that the economic injury level for bagworm on American arborvitae was very similar to the AIL as perceived by the average individual.

Some plants can tolerate levels of damage reaching as high as 40 percent, or more, from insect feeding, and it will not kill them or visibly harm them. Furthermore, many ornamental plants have fungal infestations late in the season, but the cost of spraying with a fungicide at that time in the season, when plants will go dormant in another month or so, is usually not important enough to warrant the expenditure and labor.

ALTERNATIVE PEST CONTROLS

There are numerous types of pesticide alternatives. They can be grouped into general categories of biological controls, **botanicals**, non-toxics, and insect growth regulators. Biological controls include predators (Figure 12.4), parasites, and pathogens that attack plants pests. Botanicals are derived from plants. Non-toxics are derived from other sources, and insect growth regulators act by disrupting the life cycle of the pest at some specific point.

BIOLOGICAL CONTROLS

Biological controls are a group of pest control organisms that include predators of insect pests, disease organisms that infect pests, and insects that parasitize other insects. Table 12.4 lists insect and mite predators and parasitoids used to control plant pests. *Bacillus thuringiensis* is commonly known as Bt. It is a naturally occurring bacterial organism that disrupts the digestive system of many species of

FIGURE 12.4 Lady bugs are predators of aphids.

TABLE 12.4
Insects/Mites and Parasitoids for Controlling Plant Pests

PREDATORS	Species	Plant Pest
Ant beetle	*Thanasimus formicarius*	Pine bark beetle
Chinese praying mantis	*Tenodera ridifollia sinensis*	Wide variety of insects
Convergent lady beetle	*Hippodamia convergens*	Aphids, soft-bodied insects
Green lacewing	*Chrysoperla carnea, C. rufilabris*	Aphids, soft-bodied insects, mites, insect eggs
Ground beetles	Many species in the family Carabidae	Wide range of insects
Lady beetle	*Coccinella septemounctata*	Aphids
Mealybug destroyer	*Cryptolaemus montrouzieri*	Mealybugs
Pirate bug	*Orius insidiosus*	Caterpillars
Predatory mite	*Phytoseiulus persimilis, Amblyseius californicus*	Two-spotted spider mites
Predatory mite	*A. cucumeris, A. mckenziei*	Thrips
Rove beetles	Many species in the family Staphylinidae	Wide range of insects
Spined soldier bug	*Podisus maculiventris*	Soft-bodied insects and larvae
Syrphid flies	Family Syrphidae	Aphids
PARASITOIDS		
Encarsia wasp	*Encarsia formosa*	Greenhouse whitefly
Trichogramma wasp	*Trichogramma minutum*	Orchard and forest caterpillars

caterpillar. It is applied as a powder or spray applied to leaves where the caterpillars feed. Another bacterial organism that kills insect pests is miky spore. It is used to kill Japanese beetle grubs in the soil. Avermectin is derived from the streptomyces organism and can kill beetles, mites, loopers, and thrips.

Integrated Pest Management

BOTANICALS

Naturally occurring compounds that are extracted from flowers or other plants are called botanicals. They are not necessarily less toxic than synthetic pesticides.

The two primary botanical pesticides used in landscape pest management are neem extract and pyrethrum. Neem (azadirachtin, azatin) comes from a tree of the same name, and pyrethrum is derived from a particular species of chrysanthemum, *Chrysanthemum coccinea*. Compounds related to pyrethrum have been synthesized and are called pyrethroids. Neem controls aphids, bagworms, beetles, budworms, caterpillars, leafhoppers, leafminers, thrips, and whiteflies. Pyrethroids control many plant pests and home pests, including outdoor pests around the home such as yellow jackets and ticks. Nicotine is also extracted from plants and has been used as a fumigant in greenhouses.

A variety of plant-derived oils have been formulated into pesticides. Caution should be used, however, because some plants will suffer damage from some of these products. New foliage is most susceptible. It is best to test a small area of the target plant first, to ensure no damage is done. Clove, citrus, peppermint, and rosemary oils are some of the plant oils that are used in plant-derived pesticide products. Holy Moley mole repellent repels moles by two modes of action, scent, and taste. It contains castor oil and Fuller's Earth from St. Gabriel organics (Table 12.5).

NON-TOXIC PESTICIDES

Numerous non-toxic substances are touted for use as insecticides. Insecticidal soaps are potassium salts derived from fatty acids. They penetrate insect body cells and break down cell membranes. It is most effective on soft-bodied insects, such as aphids, mites, mealybugs, earwigs, whiteflies, slugs, and others.

A variety of oils can be used against insects, mites, and fungal pathogens. Organocide controls rust mites, spider mites, scales, mealybugs, whiteflies, aphids, leafrollers, leafminers, and thrips. Saf-T-Side™ is petroleum oil for use on a broad range of plants against a variety of insect pests and mites. It may be used in combination with other pesticides and fungicides to increase effectiveness.

Hot pepper wax is used as an insect repellent that also has insecticidal qualities. It works against soft-bodied insects such as aphids, whiteflies, mites, thrips, mealybugs, coddling moth, leafhoppers, scales, lace bugs, and armyworms. It kills pests by smothering them and disrupting the nervous system. It forms a thin layer on leaf surfaces, serving as a fungicide.

Diatomaceous earth is sold under a variety of brand names, including Concern (Hummert International). It disrupts insect digestive systems. It may be used on ants, centipedes, cockroaches, earwigs, millipedes, silverfish, and slugs. Corn gluten meal is sold under the trade name Premerge from St. Gabriel organics and is used as a pre-emergent herbicide to control crabgrass.

Bordeaux mixture is a combination of copper sulfate and lime that is used for control of fungi. It is used on landscape trees and shrubs, fruit trees, and grapes. Another fungicide, Camelot, is also a bactericide and contains the copper salts of fatty and rosin acids. It is labeled for use on azalea, begonia, chrysanthemum, ivy,

TABLE 12.5
Select Botanical (Plant-Derived) Pesticides

Source	Uses	Target Pests and Activity
Anise	Ornamental plants, lawns	Repels dogs and cats
Bergamot	Ornamental plants, homes, garbage cans	Repels dogs and cats
Canola	Food crops, ornamental plants, houseplants	Kills insects
Castor oil	Ornamental plants, lawns, garbage cans	Repels dogs, cats, wildlife such as moles, deer, rabbits, squirrels
Citronella	Humans, outdoor areas ornamental plants	Repels insects and ticks; repels dogs and cats
Eucalyptus	Cats, dogs, humans and their clothing, homes	Repels mites; repels specified insects, including fleas and mosquitoes
Jojoba	All crops	Kills/repels whiteflies on all crops. Kills powdery mildew on grapes and ornamentals
Lemongrass	Ornamental plants, garbage dumps	Repels dogs and cats
Methyl salicylate *[Notes: Also called oil of wintergreen; may be toxic in large quantities]*	Ornamental plants, indoor and outdoor residential sites (including clothing), garbage dumps.	Repels dogs, cats, moths, beetles
Mint	Ornamental plants in ponds with or without fish	Kills aphids on plants (used with thyme herb)
Mustard *[Note: Also known as allylisothio-cyanate]*	Homes, ornamental plants, garbage cans	Repels dogs, cats, wildlife such as deer and raccoons; Repels and kills insects, spiders, centipedes, etc.
Orange oil (*d*-limonene)	Ornamental plants, homes, garbage dumps	Fleas, mites, mosquitos; repels dogs and cats
Pyrethrins (*Pyrethrum*)	Ornamental plants	Insects
Soybean oil	Food and feed crops, ornamental plants, indoor and outdoor sites	Kills mites Kills beetles and other insect pests

Source: https://www3.epa.gov/pesticides/chem_search/reg_actions/registration/fs_G-114_01-Jul-01.pdf

juniper, pine, philodendron, rose, and others. Trichoderma harzianum Strain T-22 is sold under the trade name "Plantshield/Rootshield".

INSECT GROWTH REGULATORS

Insect growth regulators, or **IGRs**, are naturally occurring hormones produced by the insects themselves. By understanding their role in insect development, these chemicals may be synthesized and used to disrupt normal development. A number of such chemicals have been developed and are commercially available.

Kinoprene (Enstar II) is a synthetic analog of a naturally occurring insect growth regulator that is effective against soft-bodied insects such as aphids, mealybugs, scales, thrips, and whiteflies. Prestrike (S-Methoprene) kills mosquitos by interrupting pupae development. Diflubenzuron (Dimilin) and fenoxycarb (Precision) interfere with synthesis of chitin, a component of insect exoskeletons. Halofenzamide (MACH-2) and tebufenozide (Confirm) adversely accelerate insect molting. In addition to being very targeted in their activity and selective for certain species, insect growth hormones have the additional benefit of not harming natural predators and other non-target organisms.

SUMMARY

Integrated pest management is a complex strategy that relies on multiple methods and practices to suppress plant pests while minimizing harm to the environment, humans, other animals, and non-target insects. There are two main aspects of integrated pest management: prevention and treatment. Prevention begins with proper plant selection and placement, followed by using appropriate cultural practices. Avoid using plants with difficult pest problems. Select genetically improved cultivars for pest resistance. Treatment includes determining action thresholds, monitoring, using less toxic pesticides, less-persistent pesticides, and biological controls.

An action threshold guides pest management activities to avoid undesirable damage or economic or aesthetic loss. Monitoring and scouting incorporate the use of record-keeping, traps, or branch-beating.

Degree days have been determined for many pests so that predictions about the timing of pest emergence may be accurate, providing better information than calendar dates alone could. Phenological studies correlate pest emergence with bloom times of common landscape plants, simplifying access to degree day information.

Many pesticide alternatives are available, including biological controls, botanicals, non-toxics, and insect growth regulators. The majority of these alternatives draw from the natural world for pest management resources. The use of genetically improved cultivars should be the first line of defense in the landscape.

REVIEW QUESTIONS

1. What is integrated pest management?
2. What role do plant breeding programs play in integrated pest management?
3. What are five major components of a successful pest control strategy?
4. What is phenology?
5. What are degree days and how are they used?
6. What is the purpose of an action threshold?
7. What is an AIL and how is it used?
8. Name three types of alternative pest controls. What are the two primary botanical pesticides used in landscape pest management?
9. What are two advantages of managing pests using insect growth hormones?

ACTIVITIES

1. Make a list of the top ten insect pests of landscape in your location. Using a web-based degree day monitor for your location, estimate emergence dates for those insects.
2. For the top landscape pests in your location, research least-toxic control methods, including non-pesticide approaches.
3. Conduct a web search to find out what emerging pest problems are a concern for your location.

FURTHER READING

Adams, N. Using growing degree days for insect management. University of New Hampshire Cooperative Extension. http://ccetompkins.org/resources/using-growing-degree-days-for-insect-management. Retrieved July 12, 2019.

Barbercheck, M. E. and E. Zaborski. Insect pest management: Differences between conventional and organic farming systems. https://articles.extension.org/pages/19915/insect-pest-management:-differences-between-conventional-and-organic-farming-systems. Retrieved July 12, 2019.

Cloyd, R., P. Nixon, B. Paulsrud, and M. Wiesbrook. 2000. IPM and pesticide safety, In: *Illinois Master Gardener Manual*. D. Schrock, Ed., University of Illinois Extension, Champaign, IL, H-1–H-32.

Dreistadt, S. H., J. H. Clark, and M. L. Flint. 1994. *Pests of Landscape Trees and Shrubs: An Integrated Pest Management Guide*. University of California, Division of Agriculture and Natural Resources, Oakland, CA, 327 pp.

Henn, T. and R. Weinzerl. 1990. *Beneficial Insects and Mites*. University of Illinois Extension, Circular 1298, Champaign, IL, 24 pp.

Koehler, C. S. and W. S. Moore. 1983. Resistance of several members of the Cupressaceae to the cypress tip miner, *Argyesthis cupressella*. *J. Environ. Hortic.* 1: 87–88.

Olkowski, W. 1974. A model ecosystem management program. *Proc. Tall Timbers Conf. Ecol. Anom. Control Habitat Manag.* 5: 103–117.

Orton, D. and T. Green. 1989. *Coincide: The Orton System of Pest Management*. Plantsmen's Publications, Flossmoor, IL.

Raupp, M. J., J. A. Davidson, C. S. Koehler, C. S. Sadof, and K. Reichelderfer. 1989. Economic and aesthetic injury levels and thresholds for insect pests of ornamental plants. Florida Entomologist.

Raupp, M. J., C. S. Koehler, and J. A. Davidson. 1992. Advances in implementing integrated pest management for woody landscape plants. *Annu. Rev. Entomol.* 37: 561–585.

Stairs, N. 1999. Riding the bio' wave. *Landsc. Manag.* 38(6): 20–21.

Stewart, C. D., S. K. Braman, B. L. Sparks, J. L. Wilimas-Woodward, G. L. Wade, J. G. Latimer. 2002. Comparing an IPM pilot program to a traditional cover spray program in commercial landscapes. *J. Econ. Entomol.* 95(4): 789–796.

University of California IPM Online. http://www.ipm.ucdavis.edu/index.html. Retrieved July 12, 2019.

U.S. EPA. Biopesticides. https://www.epa.gov/pesticides/biopesticides. Retrieved July 12, 2019.

Zilahi-Balogh, G. and D. G. Pfeiffer. Understanding degree-days and using them in pest management decision making http://www.virginiafruit.ento.vt.edu/Understanding_Degree_Days.html. Retrieved July 12, 2019.

13 Energy
Sources and Uses

OBJECTIVES

Upon completion of this chapter, the reader should be able to

- Identify all of the sources of energy used in the United States
- Discuss the differences between types of coal
- Discuss the differences between oil extraction methods
- Identify the major alternative energy sources
- Explain the role of government policies on energy usage
- Relate energy needs and solutions
- Explain how the energy distribution grid works
- Describe the role of alternative fuels for landscape equipment
- Discuss the role of alternative fuels for transportation

TERMS TO KNOW

Anthracite
Biodiesel
Biofuels
Biogenic
Biomass
Bituminous
Cellulosic ethanol
Enhanced oil recovery (EOR)
Ethanol
Fission
Geothermal
Grain-based ethanol
Heliostat
Hydroelectric
Light-emitting diode (LED)
Liquefied petroleum gas (LPG)
Lignite
Lumens
Photovoltaics (PV)
Primary recovery
Propane

Renewable resource
Secondary recovery
Tertiary recovery
Thermogenic

INTRODUCTION

In an economic impact study conducted by researchers of University of Illinois (Campbell, et al. 2001), 10 percent of non-labor nursery and turfgrass production expenses in that state went towards fuel, oil, propane, and other energy needs. Tools, equipment, and transportation account for most of the fuel used in the landscape industry. Indoor and outdoor lighting are other examples of energy usage, amounting to about 20–50 percent total energy use for most businesses.

Fuel prices are unpredictable. As a result, they are difficult to budget for. Most business cannot pass on rising fuel costs to their customers. Whereas companies can spend time and effort hedging fuel costs or locking in prices by purchasing in advance, there is still a limit to the amount of price control they can achieve. By looking at ways to reduce fuel usage, a landscape company can achieve a higher level of control over the costs of running the business. In addition to being more energy efficient, the savings that are realized can be pocketed as profit. Many companies are finding that energy efficiency is good not only for the environment, but also for the bottom line.

To evaluate energy use, landscape companies would do well to examine their fleet of vehicles, equipment, and power tools, and devise a replacement plan that allows them to purchase more energy-efficient models over time. A cost analysis that considers fuel and maintenance costs will likely prove favorable for newer energy-efficient models. For lighting, fixtures and lamp types should be considered. **LEDs** are more energy efficient than other lighting sources, such as fluorescent, halogen, and incandescent lamps. Solar-powered landscape lights have solar panels built into them and besides providing their own energy have the added advantage of not requiring wiring or the special installation skills needed for wired systems.

In addition to transportation, tools, and equipment, energy is also consumed in the cooling, heating, and lighting of offices and shops, and of lighting the outdoor landscape. All of these areas may be subject to better efficiency or energy alternatives. This chapter will focus on the energy that is required to operate tools and equipment, including transportation, such as pick-up trucks, and the energy that is required to generate electricity. Discussion includes the sources and landscape industry uses of traditional energy such as oil, coal, and natural gas as well as newer or "alternative" energy, such as biofuels and propane. Energy-efficient lighting alternatives will also be addressed.

In addition to considering alternative energy sources, businesses may be able to improve the energy efficiency of their buildings. The Green Building Advisor (http://www.greenbuildingadvisor.com) is an excellent resource. The Green Building Council (https://new.usgbc.org/) implements the Leadership in Energy-Efficient Design (LEED) program, as discussed in Chapter 1.

ENERGY SOURCES

Traditional energy sources for electricity-generating power plants are fossil fuels and hydroelectric power. Nuclear energy has been around for several decades, even though it is a controversial source of energy. Traditional energy sources for fuel are gasoline and diesel, both petroleum products. Alternative energy sources include such renewable resources as **biomass**, wind, solar, and **geothermal**. The latter one is not incorporated into electric generation at power plants, but can be quite useful in bringing about both energy efficiency and energy independence.

SCOPE OF THE PROBLEM

There are several problems related to energy consumption. The quantity and availability of energy sources such as fossil fuels is a concern, since they are a non-renewable resource. The price of energy is a concern, since availability, political situations, infrastructure for delivering energy, and other factors affect the price of energy. Emissions are an environmental and health concern caused by some sources of energy production. These are the primary concerns about energy consumption, but other factors may also play a role in the ongoing discussion about energy supply and usage.

Non-renewable Resource

Fossil fuels are a non-renewable resource, and one of the big questions concerning their use is how much longer can known underground reserves last? The answer to this question changes as new oil fields are discovered. Further, some oil is relatively easy to get because underground pressures force it out when a hole is drilled into layers of overlying rock, whereas other oil is more difficult to bring up from underground.

Expense

Another unknown factor is the question of how expensive fossil fuels, especially petroleum products, will get over the next 20–50 years. Fuel prices and dependence on foreign oil are cited as reasons by some individuals, landscape companies, and other entities, to switch to alternative energy, fuels, vehicles, and equipment. Figure 13.1 shows the retail prices for gasoline in the United States between 1995 and 2020. Table 13.1 shows the top countries that the United States imports oil from. In 2011, the United States was a net exporter of fossil fuel products.

Emissions

In addition to concerns about carbon dioxide and its contribution to global climate change, there are additional concerns about other emissions, such as nitrous oxides and sulfur dioxide. Regulations controlling emissions have been implemented at the federal, state and municipal levels. The California Air Resources Board (CARB) is a leader of such regulations, sometimes even exceeding federal standards. The CARB

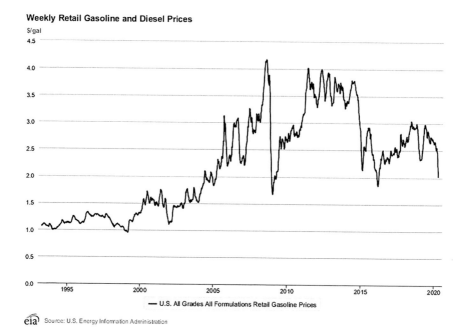

FIGURE 13.1 Gasoline prices in the United States.

TABLE 13.1
Crude Oil Imports to the United States: Top 15 Countries (1,000 bbl per day)

Country	2019	2018
Canada	3,810	3,707
Mexico	600	665
Saudi Arabia	500	870
Iraq	331	518
Colombia	318	295
Ecuador	199	176
Russia	132	73
Nigeria	186	175
Brazil	121	126
Brunei	32	15
Trinidad And Tobago	46	7
Kuwait	45	78
Libya	60	55
Cameroon	9	3
United Kingdom	64	57

Source: https://www.eia.gov/petroleum/imports/companylevel/.

Energy: Sources and Uses

is a part of the California Environmental Protection Agency, which reports directly to the Governor's office. Their stated mission is: "To promote and protect public health, welfare and ecological resources through the effective and efficient reduction of air pollutants while recognizing and considering the effects on the economy of the state". Even though no other state has such a commission, other states are following California's lead on clean air issues and regulations.

OTHER PRESSURES

Customer demand will drive the adoption of alternative fuels and use of more sustainable technologies in landscape equipment and tools. It remains to be seen what the extent of impact this has on the landscape industry, but the idea of sustainability is pervasive on local, national, and international levels. Furthermore, as businesses realize savings in their operating budgets, the incentive to continue to improve their bottom-line through fuel efficiency is increasing.

GOVERNMENT SUPPORT FOR RENEWABLE ENERGY

In 1978, the US Congress passed the National Energy Act, which, among other things, encouraged the use of renewable resources. The term "gasohol" was used to describe a blend of gasoline with at least 10 percent alcohol, which would be produced from renewable resources. In 1979, the Energy Tax Act provided subsidies for gasohol. The Volumetric Ethanol Excise Tax Credit (VEETC) enacted in 2004 set a 51 cent per gallon credit for any blender of ethanol into the petroleum gasoline stream (see Table 13.2 for other federal subsidies for alternative fuel). Currently, gas with 10 percent ethanol (E-10) is the norm at the gas pump. In addition to being a renewable resource, ethanol serves as an "enleanment" in gasoline, allowing it to burn more cleanly.

In 2005, the Renewable Fuel Standard program was created as a mandate of the Energy Policy Act. It required 7.5 billion gallons of renewable fuel to be blended into gasoline by 2012. In 2007, this act was expanded to include diesel; and increased the volume to be blended from 9 billion gallons in 2008 to 36 billion gallons by 2022,

TABLE 13.2
Select Federal legislation Providing subsidies for Alternative Energy

Legislation	Year	Subsidy
Alternative Motor Fuels Act	1988	Vehicle manufacturer incentives for producing motor vehicles that can operate on certain alternative fuels
Alternative Fuel Motor Vehicle Credit (Energy Policy Act of 2005)	2005	Tax credits for qualified vehicles that operate using hybrid technology, fuel cells, alternative fuels, and advanced lean-burn
Business Energy Investment Tax Credit (ITC)	2015	Corporate tax credit for investments in solar, wind, fuel cells, and geothermal technologies

Source: https://afdc.energy.gov/laws/key_legislation#amfa. Retrieved 7/7/2020.

among other things. The goal is to reduce emissions, reduce reliance on petroleum imports, and encourage expansion of renewable fuel production. By 2016, 45 percent of federal energy subsidies were associated with renewable energy.

In 2019, Michael Bloomberg launched a campaign called "Beyond Carbon", which aims to move the United States towards clean energy. His central goal is to close every coal-fired power plant in the United States by 2030. He has partnered with The Sierra Club in their "Beyond Coal" campaign, an initiative with a goal "to end America's reliance on dirty coal and shift the US Power sector to a cleaner future".

ENERGY FOR ELECTRICITY

For heating and air conditioning in offices, many companies rely on electricity distributed through the local power grid. Energy for electricity is generated by power plants throughout the United States. In 2017, there were 8,652 power plants in the United States. The largest percentage of electricity has traditionally been generated using fossil fuels, with some power provided by nuclear and hydroelectric sources. But newer sources have been developed, and other alternatives are currently being explored. Such sources and alternatives include wind turbines and solar cells, also known as **photovoltaics** (**PV**). Biomass energy is another alternative source for energy and, like solar and wind, is a **renewable resource**.

Approximately 25 percent of the energy used in commercial buildings is for lighting, and a similar amount is used for heating and cooling. By designing workspaces that rely on natural light (a practice sometimes referred to as "daylighting"), many companies are able to eliminate artificial lighting altogether. To reduce heating and cooling costs, businesses should conduct an energy audit on their buildings. Big energy savings may be realized with relatively minor improvements, such as adequate insulation. Turning lights off in unoccupied rooms, and other practices that waste electricity, should be avoided.

There are many changes nationally with regard to traditional sources for generating electricity. Some individual businesses have gone so far as to install their own photovoltaic systems or heat pumps. Heat pumps use the earth's constant temperature as the air exchange medium instead of relying on outside air. The steady temperature underground acts as an insulator that prevents unnecessary heat loss during water circulation through pipes. The water can be used to cool a structure in summer and to heat it in winter. Another innovation is the use of methane from landfills to heat water that can then be used to heat a building.

GENERATION OF ELECTRICITY

Electricity is produced at an electric power plant. A fuel source, such as natural gas, coal, crude oil, or nuclear fuel, is used to create heat. The heat is used to boil water and create steam, which then spins a turbine. As the turbine spins, electricity is generated using a system of magnets. With hydroelectric power, water that flows over a dam turns the turbines to generate the electricity.

Alternative sources of energy for power plants include geothermal, which uses steam extracted from hot underground rocks; photovoltaics; wind turbines; biomass,

TABLE 13.3
US Sources of Electricity (Quadrillion Btu)

Year	Coal	Natural Gas	Oil	Nuclear	Hydroelectric	Biomass	Geothermal	Wind	Solar
2000	22.73	22.31	12.37	7.87	2.79				
2003	22.07	22.07	12.02	7.94	2.81				
2006	23.78	21.37	10.79	8.24	2.84				
2009	21.79	24.22	11.23	8.29	2.64	3.96	0.37	0.66	0.07

Source: EIA August 2010. US Energy Background Information.

which can come from plant material, municipal solid waste, or landfill methane; waste heat from industrial processes; and solar thermal. Table 13.3 shows the US Sources of energy for recent years.

Non-renewable Fossil Fuel Energy

Fossil fuel is the umbrella term that includes petroleum, coal, and natural gas, the primary sources of energy today. The fossils from which they are derived are the partially decomposed remains of carbon-based organisms. Fossil fuels require millions of years to form and, when burned, give off carbon dioxide, methane, nitrous oxide, and hydrocarbons. Of these, carbon dioxide is by far the major greenhouse gas.

In 2018, 45 percent of global carbon emissions from fossil fuels came from petroleum, much of which was used for transport. Coal accounted for 26 percent of global carbon emissions that year, and the remaining 29 percent came from natural gas.

Natural Gas for Electricity

Natural gas is used in power plants to generate electricity. Natural gas is also used in homes for heat, cooking, water heaters, and gas-burning clothes dryers. It is a fossil fuel that is often found in association with other fossil fuels, notably oil and coal. It is composed mainly of methane, but can have impurities in it that must be removed prior to use. There are two main processes for generating natural gas, and they are known as **biogenic** and **thermogenic** processes. Biogenic gas is created in marshes, bogs, landfills, and shallow sediments, whereas thermogenic gas is created from organic material that is buried deep in the earth where temperatures and pressures are much greater than those closer to the surface.

A new development in the process for extracting natural gas is hydrofracturing, nicknamed "fracking". This process uses high-pressure injection of water along with sand and some other additives, to create cracks in underground rock formations to release reserves of natural gas or petroleum that is stored there. Immense amounts of water are used in this process. Fracking is controversial, and the practice has been banned in a number of countries. Some of the effects that are blamed on fracking include contaminated ground and surface water, release of methane, exploding water wells, and tremors, or small earthquakes caused by various activities associated with this practice.

Oil

Oil has traditionally been extracted from underground using natural forces, along with some artificial techniques, such as pumps. This relatively easy method of extraction is known as **primary recovery**. About 5–15 percent of the oil present can be recovered this way. After that, **secondary recovery** techniques are implemented, in which water or gas is injected into the areas where the oil is located, driving the oil to a central location where it can be recovered, such as a wellbore. Twenty to forty percent of the oil can be recovered using this technique.

Tertiary recovery is also known as **enhanced oil recovery** or simply **EOR**. One advantage of EOR is that as most oil fields have not produced half of the original oil present; infrastructure is still present for continued efforts; and a skilled oil-field labor force remains in place from the primary and secondary oil extraction efforts. The US Department of Energy estimates that 264 billion barrels of oil remain in existing US oil fields. However, *proved* reserves are those reserves that can be economically and feasibly recovered. Thus, approximately 30–40 billion barrels of oil are currently estimated to exist in the proved reserves in the United States.

There are three major processers included in this category: thermal recovery, gas injection, and chemical injection. Of these three, the first two each account for about 50 percent of the US EOR production. Thermal recovery involves the use of steam to reduce the viscosity of heavy oil, improving its ability to move through the system. Gas injection relies on the use of nitrogen, natural gas, or carbon dioxide, to drive the oil to a central location. Other oils are sometimes used in this technique, to reduce the viscosity of the oil and increase its flow to a central location, such as a wellbore. All of these techniques are relatively costly, and their effectiveness can be difficult to predict.

Coal

Coal is a sedimentary rock that usually occurs in layers or veins called coal beds or coal seams. There are different types of coal, including relatively hard **anthracite** coal, the softer **bituminous** coal, and the lower energy **lignite**.

Coal is partially decomposed plant matter that has accumulated at the bottom of water. During the Carboniferous period, atmospheric carbon was trapped in the ground in immense peat bogs that were eventually covered over and buried by sediments. Eventually, the plant material metamorphosed into coal. This ancient history puts coal into the category known as fossil fuels, which is largely responsible for energy in the current day. Because of its biological origins, coal is a carbon-based energy source that releases carbon dioxide when it is burned for its energy.

Anthracite has the highest carbon count and contains the fewest impurities of all coals, between 92 and 98 percent, although it has lower heating value. Anthracite ignites with difficulty and burns with a short, blue, smokeless flame. Because of its higher quality, anthracite is around two to three times the cost of regular coal. Anthracite was first experimentally burned as a residential heating fuel in the United States in 1808.

Bituminous coal (black coal) is a relatively soft coal that contains the tarlike bitumen. It is of poorer quality than anthracite coal, but higher quality than lignite coal. The carbon content of bituminous coal is around 60–80 percent. Lignite is

Energy: Sources and Uses

brownish-black in color and has a carbon content of around 25–35 percent. Bituminous and lignite coal are the primary types of coal used in electric power plants.

Coal is primarily used to produce electricity and heat through combustion. In 2008, approximately 49 percent of the electricity used in the United States came from coal. In 2018, that amount had been reduced to about 27 percent.

Nuclear

In 2009, there were 65 nuclear power plants in the United States. In 2019, this number had been reduced to 59. In all, they generated about 800 billion kilowatt-hours (kWh) of electricity or around 20 percent of the total US demand.

Nuclear energy relies on nuclear **fission** or splitting of atoms, which results in a release of energy. Uranium^{235} is a so-called heavy isotope of uranium that is the common fuel used in nuclear reactors. Uranium^{235} is a fissionable atom, meaning that it is subject to being struck by neutrons in the environment, and when that happens, it splits in two and also releases a neutron. This neutron then strikes one of the two atomic nuclei, causing it to split into two and also releasing more neutrons. This chain reaction continues until the uranium is spent, meaning it has used all its free neutrons. Each time the atom splits, enormous amounts of energy are released. To compare the energy released in a nuclear reaction, consider that a pound of highly enriched uranium such as what is used to power a nuclear submarine is equal to about a million gallons of gasoline.

Nuclear energy is considered clean energy because there are no carbon-based emissions. Even so, it is not embraced widely because spent fuel rods have to be disposed of, yet are hazardous. Furthermore, nuclear energy (nuclear power plants/uranium for the power plants) can be re-allocated to build nuclear bombs.

The history of nuclear energy is clouded by two events at nuclear power plants. In 1979, a partial core meltdown at Three Mile Island nuclear power plant in Pennsylvania released radioactive gases and resulted in widespread public opposition to nuclear power plants, especially in the United States. In 1986 in Chernobyl, Russia, the worst nuclear power plant accident in history occurred. As a result, a nuclear reactor vessel ruptured, releasing a plume of radioactive fallout into the air. Hundreds of thousands of people throughout Belarus, Russia, and Ukraine had to be evacuated and re-located. Twenty-eight people died due to acute radiation syndrome, with many others dying later on.

In 2011, another disaster, this time in Japan, occurred when an earthquake and tsunami damaged the Fukushima Daiichi nuclear power plant. The earthquake of magnitude 9.0 occurred off the eastern coast of Japan, followed by a tsunami on the same day. Electric power was provided by emergency diesel-powered generators to cool the reactor cores. Numerous incidents in the weeks and months to follow resulted in fires at the plant and release of radiation into the environment. Many workers as well as the local populace were exposed.

In addition to acute radiation poisoning, a major health concern associated with nuclear power is cancer caused by excessive radiation. Disposal of nuclear waste has not been satisfactorily addressed and is a concern due to the long half-life of radioactive materials. According to Socolow and Pacala, disposal of nuclear waste and prevention of accidents are the two major factors preventing scale-up of nuclear power production.

Renewable Energy

Renewable energy is the most sustainable form of energy available. However, the technology to use it has not been fully developed on a global scale. Nevertheless, companies and individuals have been able to implement various forms of renewable energy technology to meet much or all of their energy needs. The main sources of renewable energy are water, wind, and the sun. Plants can provide renewable energy in the form of biofuels and biomass.

Hydroelectric

Hydroelectric generation dates to ancient times. It is included as a traditional source of electricity, because it has been in use since 1881 in the United States. It is, in fact, the most widely used form of renewable energy. In hydroelectric generation, it is water that turns the turbines to generate the electricity. In 2015, hydroelectric plants produced 6.1 percent of all the power generated in the United States and 35 percent of renewable energy.

Other advantages of hydroelectric power are that it produces no waste and releases far fewer emissions than fossil-fuel burning plants. However, most hydroelectric power plants rely on dams and reservoirs, which undoubtedly interfere with nature. Fish migration, native plants and animals, and even water temperature are all affected.

The future of hydroelectric is not so bright because most of the good spots to build dams have been used already (Table 13.4).

Wind

Wind energy, in addition to being a renewable source of energy, reduces air pollution emissions, including sulfur dioxide, nitric oxide, nitrogen dioxide, and particulate matter. By 2050, wind energy could avoid the emission of 12.3 gigatonnes of greenhouse gases.

Wind energy is one of the fastest growing sectors of energy in the United States. Between 1999 and 2009, wind power capacity grew from 2,472 MW (megawatts) to 35,159 MW. In 2018 alone, the US Wind Industry added 7,588 MW of new wind capacity. This brought the total capacity to 97,223 MW. This is estimated to be a fraction

TABLE 13.4

Some of the Largest Hydroelectric Operations in the United States (The United States currently has over 2,000 Hydroelectric Power Plants which Supply 35 percent of Its Renewable Electricity)

Location	River	Year Completed	Generation Capacity (MW)
Grand Coulee Dam	Columbia	1942	6,809
Hoover Dam	Colorado	1926	1,345
Niagara Falls	Niagara	1961	2,515
Chief Joseph Dam	Columbia	1958	2,620
John Day Dam	Columbia	1949	2,160
The Dalles Dam	Columbia	1981	2,038

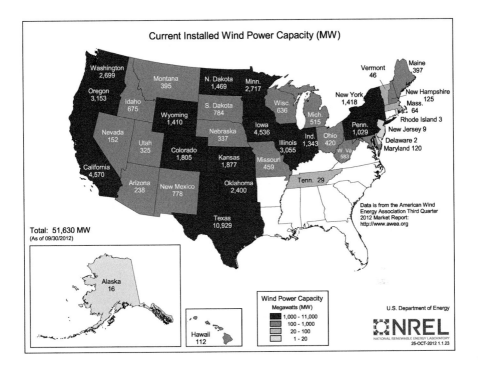

FIGURE 13.2 Wind power capacity.

of the total capacity that is possible, based on calculations by the National Renewable Energy Laboratory (Figure 13.2), which indicate a total of 10, 640, 080 MW.

Federal tax credits are provided to wind energy producers to help them invest in the technology and equipment. Many states provide additional incentives such as property tax exemptions and "green credits". The Energy Improvement and Extension Act of 2008 extends tax credits for producing wind energy. In 2015, the US Department of Energy released *Wind Vision: A New Era for Wind Power in the United States*, which examined and quantified the economic and environmental costs and benefits of a scenario in which wind would provide 10 percent of US electrical demand in 2020, 20 percent in 2030, and 35 percent in 2050. A series of topical meetings that took place throughout 2016–2017 resulted in a new report, *2016–2017 Status Assessment and Update on the Wind Vision Roadmap: Findings from Topical Working Sessions report*. This new report provides new actions and other changes and adjustments to the 2015 report.

Solar

Solar energy is another fast-growing renewable energy resource. Today, there are 89 petawatts (PW, 10^{15} watt) of potential solar energy production available on earth, making solar the world's most abundant available source of power. Solar energy does not produce pollutants during its production; however, manufacturing and disposing of retired solar panels present issues that must be addressed.

Solar energy converts sunlight into electricity using one of the two methods: concentrated solar power (CSP) which occurs on a large scale or photovoltaics (PV). Large-solar scale projects use parabolic mirrors to focus the sun's energy onto fluid-filled collector tubes. The fluid is heated and sent to a boiler, which sends live steam to a turbine to produce electricity. Other technologies to capture and focus the sun's thermal energy include power towers that use **heliostats** or tracking mirrors, Fresnel reflectors, and parabolic dishes.

Photovoltaic panels are composed of photovoltaic cells that are commonly made of silicon. Taking advantage of the properties of silicon as a semi-conductor, light energy in the form of photons is turned directly into electrical energy when an electron is knocked loose and travels through the conducting material. The result is direct current (DC) that can be used to power electric devices or stored in a battery for future use. Whereas individuals may install solar panels for their individual use, community systems have been installed in numerous places across the United States. Back-up batteries allow excess energy to be stored and used later, such as at night. Excess energy may also be banked with a power plant and drawn from during high-energy-demand times, such as in the summer when air conditioning is used.

Although initially it was quite costly to produce and inefficient in its energy conversion, the past decade has brought about many technological advances. Solar power has been adopted around the world, and costs for materials are declining. The efficiency of solar panels has also improved over the past several years, going from 17 to 18 percent in 2014 to 20 to 23 percent by 2019.

FUEL FOR TOOLS, EQUIPMENT, AND TRANSPORTATION

Fuel usage can be divided into two main areas: on-road use and off-road use. On-road use of fuels refers to fuel used for transportation. Off-road use includes that fuel used for heavy and light equipment and for power tools. The primary fuel used by the landscape industry is in the form of gasoline and diesel for power tools, landscape equipment, and transportation. Gasoline and diesel are both derived from petroleum and are collectively referred to as fossil fuels, due to their origin. Biofuels are a renewable resource as well as an alternative fuel.

In America, automobiles consumed about 143 billion gallons of gasoline in 2018. By contrast, mowers used about 580 million gallons of gas per year or about 4 tenths of 1 percent of the amount used by automobiles. Furthermore, a lawn mower emits over 10 times more pollution than a car. Landscape equipment is expected to receive increasing attention with respect to emissions controls because large gains have already been made with automobiles, but little has been made with mowers and other equipment.

Skid loaders, trenchers, mowers trucks, and many other pieces of equipment have relied on the use of diesel or gasoline for many years. Some alternative-fuel powered machinery is now available, with more becoming available in future years.

Non-renewable

Just as there are non-renewable sources of energy for use in electric power generation, there are also non-renewable sources that are used for fuel. **Propane** and natural gas are two alternatives to the traditional gasoline and diesel fuel.

Energy: Sources and Uses

Propane

Propane is a byproduct of natural gas processing and crude oil refinement. It is also referred to as **LPG (liquefied petroleum gas)**. It is the third most common engine fuel after gas and diesel.

Using propane can cut fuel costs in half compared to gasoline. It can be purchased everywhere, including large retail chain stores, and the owner of a propane-powered piece of equipment can realize a return on their investment in as little as 3 years. The machines are more costly than comparable gasoline-powered machines, but the fuel costs less.

Case Study: Propane Power

Pristine Green, Greenscape Services, and Merry Acres
  Reference: Chilcote, Lee. March 2011. Propane Power. Lawn & Landscape. http://www.lawnandlandscape.com/lawn-landscape-0311-propane-power-alternative-fuel.aspx. Viewed December 30, 2011.

  Shannon Wilson, owner of Greenscape Services in Sarasota Florida, Jon Dozier of Merry Acres Landscaping in Albany, Georgia, and Jesse Triick of Pristine Green in Byron Center, Michigan, have one thing in common. They have all switched over to propane-powered mowers. Citing environmental sustainability, economic benefits, and predictability of pricing as some of the reasons for the switch, they also cite higher up-front costs and special fueling requirements as barriers to change.

  The propane-powered machines produce about 60 percent less carbon monoxide and 25 percent less greenhouse gases than gas-powered mowers, while offering similar power as gas mowers. The propane comes mainly from US sources. Fuel savings were around 40 percent for Dozier's Merry Acres Landscaping Company.

  The convenience of filling up with propane relies on local availability of both tanks and fuel, since the gas is pressurized to maintain a liquid state. Dozier had a propane storage tank installed on his property. This adds a level of convenience for refilling the tanks that are mounted on the mowers. It also allows Dozier to lock-in prices a year at a time. In addition to fuel costs that were about 57 percent lower than those for gasoline, less frequent oil changes are needed.

  Triick echoes the need to negotiate availability and cost of fuel, preferably on an annual basis. He points out that propane prices are lower in summer, making it a good time to negotiate.

Natural Gas Vehicles

Natural gas is clean-burning and non-toxic, and evaporates if spilled. It can be used in both light- and heavy-duty vehicles, including buses. It produces about 30 percent less carbon dioxide for each unit of energy released, as compared to petroleum oil and 45 percent less than coal.

The disadvantages of natural gas are that it is pressurized, requiring tanks approximately 67 percent larger than a gas tank to run for the same amount of time before refueling is required; it provides less horsepower by 10–15 percent compared to gas-powered engines; fuel is available through a different distribution system than gasoline; and ease of refueling is not as good.

Renewable

Other alternatives for powering vehicles include **biofuels**, electric vehicles, and hydrogen-powered vehicles. Combinations of energy sources are also available, as in electric plug-in hybrid automobiles.

Biofuels

There are two categories of **biofuels**, and both are made from plant material. One is **biodiesel** and the other is **ethanol**. Ethanol is further divided into two types: that made from starches and that made from cellulose. The latter is referred to as **cellulosic ethanol**, whereas the former is **grain-based ethanol**. Ethanol is added to both gasoline and diesel, serving as an oxygenator in both cases. Gasoline has either 10 percent (E-10) or 85 percent (E-85) ethanol blended in (Figure 13.3). Table 13.5 shows US biodiesel production for recent years.

Ethanol was first used as an additive to gasoline in 1988. At the time, methyl tertiary butyl ether (MTBE) was the primary gasoline additive. After California began replacing MTBE with ethanol in 1990, other states followed. Likewise, some American cities and school districts have converted their entire fleets to ethanol-diesel.

Biodiesel is made from vegetable oils, including soybeans and used cooking oil, or animal fats, and are usually blended with petroleum diesel. Blends are usually at 50 percent (B50) or 20 percent (B20) biodiesel.

Grain-based ethanol is made from corn, wheat, or barley, and is blended with gasoline at 10 or 85 percent levels, although variations on these percentages do exist. (States are passing legislation on percentages required/allowed on a state-by-state basis.) Currently, access to 85 percent ethanol-gasoline (E85) is limited and thus presents a problem that usually results in on E10 being used.

Some caveats are to use only biodiesel that has been tested and certified. Water and other contaminants may be present in every batch, so testing must be performed according to ASTM standards. Landscape companies are advised to request certificates of analysis from their suppliers.

FIGURE 13.3 E-85 gasoline has a high content of ethanol and can only be used in vehicles that are specifically designed for it.

TABLE 13.5
US Biodiesel Annual Production Capacity

Year	Capacity (million gallons)
2001	62.3
2002	92.3
2003	153.3
2004	162.3
2005	240.3
2006	332.9
2007	672.3
2008	690
2009	516
2010	343
2011	967
2012	991
2013	1,359
2014	1,279
2015	1,263
2016	1,568
2017	1,596
2018	1,832

Source: Statista. https://www.statista.com/statistics/509875/production-volume-of-biodiesel-in-the-us/.

Electric Vehicles

The ability to power vehicles using electricity relies on two sources of energy: batteries and power plants. As discussed earlier in this chapter, electricity can be generated using renewable or non-renewable sources. In general, it is thought that electric-powered vehicles are less polluting than gas-powered vehicles. Batteries that are used in electric vehicles are nickel-metal-hydride (NiMH). Lead–acid is an older battery type that has been used, but future developments are now focusing on lithium-ion batteries. These batteries are rechargeable using regular or specialized electrical outlets. Hybrid vehicles that run on gas and electricity thus use energy that may have come from either domestic or foreign sources. Hybrid vehicles use both fuel and electricity. Kenworth Truck Company manufactures diesel-electric hybrid trucks as well as propane- and natural gas-powered trucks.

Hydrogen Power

The Gale Group reports that there were 200 hydrogen-powered public vehicles in use in October 2009, mostly in California. Whereas fuel cell efficiency on small scales required for landscape equipment has been lacking, breakthroughs are occurring and research is being conducted. Automobile introductions such as the Honda FCX and FCX Clarity, Hyundai's Blue^2 (Blue Square) represented early versions of hydrogen-powered cars, also sometimes called fuel cell vehicles. As of 2016, there were three models of hydrogen cars publicly available in select markets: the Toyota Mirai, the Hyundai Nexo, and the Honda Clarity. As of 2014, 95 percent of hydrogen was made from natural gas.

ENERGY-EFFICIENT LIGHTING

Energy Star is an energy efficiency rating system developed by the EPA and the US Department of Energy. Outdoor lighting that is rated as an Energy Star fixture is more energy efficient than others. Solar-powered landscape lighting is widely available. One advantage of solar-powered landscape lights are that there is no wiring installation required. Many attractive styles are available.

Light-emitting diodes (LEDs) have been developed into light fixtures that can provide a bright source of locally focused light, which is helpful for lighting walkways without creating excessive lighting beyond the area where it is needed. Both indoor and outdoor light bulbs and fixtures are available. LED bulbs are approximately ten times more efficient than a comparable incandescent bulb, and 2.5 times more efficient than fluorescent bulbs. The expected lifespan of an LED is ten times longer than that for compact fluorescent light bulbs (CFLs). Comparisons of the brightness of LEDs to other bulb types are made possible by measuring the perceived light output or **lumens**. In the United States, all light bulb packages are required to provide this information as of 2012. Typically, a 45-watt incandescent bulb is equivalent to a 13-watt LED. Both bulbs produce 450 to 650 lumens. LEDs have two other advantages over incandescent and fluorescent bulbs: they do not emit much heat, which contributes to their energy efficiency.

Motion-activated lights provide energy savings by only turning on when someone approaches or is the vicinity of the light fixture. When motion is no longer detected, the light will switch off. These are available in many different styles and fixture types.

SUMMARY

Energy used in the landscape industry comes from a limited number of sources and can be thought of in terms of that used for fueling tools, equipment, and transport vehicles and that used for offices and outdoor lighting. The latter is usually in the form of electricity, which may also be used to power some landscape equipment and tools.

Fossil fuels provide much of the energy used in power tools and landscape equipment. This includes petroleum, gasoline, diesel fuel, and, to a limited extent, natural gas. Gasohol describes a blend of gasoline with at least 10 percent alcohol (ethanol).

The continued use of fossil fuels in energy for buildings and for transportation and to power equipment faces significant challenges in the coming years. Easily extracted oil is expected to become available in decreasing amounts, causing a rise in the cost of oil in the future. Fracking is a technology that has been developed to retrieve oil that is less easily extracted by traditional drilling. Coal is readily available in large amounts, but there are many problems associated with its extraction, including serious environmental and health concerns. Nuclear power plants have been in existence for several decades, but the technology has suffered setbacks.

Energy: Sources and Uses

Government support for alternative energy has been in place for three to four decades, although it has fluctuated in terms of subsidies. The National Energy Act, the Energy Policy Act, and others have encouraged the use of renewable resources and development of alternative energy. The California Air Resources Board has been a leader in setting standards to control emissions, including carbon dioxide, a major greenhouse gas.

Alternative energy includes those renewable sources that supply power plants, such as hydroelectric power, wind power, and solar power, as well as those that can fuel vehicles, such as ethanol and biodiesel. Battery-powered electric vehicles, mowers, and other equipment, along with propane-powered mowers, are some examples of emerging alternatives in the field. Hydrogen power is still in the research stage, but breakthroughs are occurring. Geothermal systems can be added to an existing structure or part of a design when building a new facility. Solar lighting, LED lights, and motion-activated lights are examples of energy-efficient lighting.

REVIEW QUESTIONS

1. Name two traditional sources of energy in the United States and two alternative sources.
2. What is the name of the California agency that sets emissions regulations for that state?
3. What is "fracking" and why is it controversial?
4. Name two types of ethanol used as additives in fuel, and explain how they differ.
5. How much oil is there in US oil fields? How does this compare to 2000 figures?
6. What are the three major types of coal? Which are used in electric power plants?
7. Why is nuclear energy considered "clean"?
8. Name two disadvantages of natural gas.
9. What kind of batteries are used to in electric vehicles and what type will be used in the future?
10. Why is hydrogen power not used in landscape equipment?

ACTIVITIES

1. Using the wind power map at https://www.nrel.gov/gis/wind.html, write a summary of how wind power how changed in your state in the last 10 years. Research websites for wind energy in your state to discover plans for future wind energy development in your area.
2. Visit a landscape equipment retailer or contact an equipment sales rep in your area and interview them about new energy-efficient technology on popular equipment (mowers, chainsaws, etc.). Find out whether plans for continued energy efficiency are planned.
3. Conduct an energy efficiency audit on a landscape company in your local area. Make a list of energy-efficient changes they could make to their building and equipment.

REFERENCES

110th Congress Public Law. The Energy Independence and Security Act of 2007. https://www.govinfo.gov/content/pkg/PLAW-110publ140/html/PLAW-110publ140.htm. Retrieved July 12, 2019.

American Wind Energy Association. http://www.awea.org. Retrieved July 12, 2019.

California Air Resources Board. http://www.arb.ca.gov/homepage.htm. Retrieved July 12, 2019.

Campbell, G. E., R. J. Brazee, A. G. Endress, T. B. Voigt, D. F. Warnock, J. L. Hall. 2001. The Illinois Green Industry: Economic impact, structure, characteristics. Department of Natural Resources and Environmental Sciences. Report 2001-01, Urbana, IL, 106 pp.

Center for Sustainable Systems. Commercial Buildings factsheet. http://css.umich.edu/factsheets/commercial-buildings-factsheet. Retrieved July 12, 2019.

Coombs, J. Should ethanol subsidies be renewed? http://www.altenergystocks.com/archives/2010/06/should_ethanol_subsidies_be_renewed.html. Retrieved July 12, 2019.

Kammen, M. 2006. The rise of renewable energy. (Special Issue: Energy's Future Beyond Carbon). *Scientific American*. September, 84–93.

Goerold, W. T. 2002. *Sources of United States Oil Supply*. Lookout Mountain Analysis. Golden, CO.

Hoye, S. and S. Hargreaves. 2010. 'Fracking' yields fuel, fear in Northeast. CNN. September 3, 2010. http://www.cnn.com/2010/US/09/02/fracking/index.html. Retrieved July 12, 2019.

Kargbo, D. M. 2010. Biodiesel production from municipal sewage sludges. *Energy Fuels*. 24 (5): 2791–2794. EPA Region III, Office of Innovation, Environmental Assessment & Innovation Division, Philadelphia, Pennsylvania 19103. Publication Date (Web): April 13, 2010.

Kunzig, R. 2009. The Canadian oil boom: Scraping bottom. *National Geographic*. March. 42–59.

Mullins, E. 2007. The alternate route. *Lawn & Landscape*. October. 29–38.

National Biodiesel Board. U.S. biodiesel production. http://www.biodiesel.org/production/production-statistics. Retrieved July 12, 2019.

Natural Gas. Natural Gas Supply Association. http://www.naturalgas.org/. Retrieved July 12, 2019.

NREL. Wind maps. https://www.nrel.gov/gis/wind.html. Retrieved July 12, 2019.

Socolow, R. H. and S. W. Pacala. 2006. A plan to keep carbon in check. *Scientific American*. September 2006. 50–57.

U.S. D.O.E. 2008. 20% Wind Energy by 2030: Increasing Wind Energy's Contribution to U.S. Electricity Supply. https://www.energy.gov/eere/wind/20-wind-energy-2030-increasing-wind-energys-contribution-us-electricity-supply. Retrieved July 12, 2019.

U.S. D.O.E. Enhanced oil recovery/CO_2 injection. http://fossil.energy.gov/programs/oilgas/eor/. Retrieved July 12, 2019.

U.S. Energy Information Administration. How much petroleum does the United States import and export? https://www.eia.gov/tools/faqs/faq.php?id=727&t=6. Retrieved July 12, 2019.

U.S. EPA. Overview for Renewable Fuel Standard. https://www.epa.gov/renewable-fuel-standard-program/overview-renewable-fuel-standard. Retrieved July 12, 2019.

USGS. Water science school: Hydroelectric power water use. https://www.usgs.gov/special-topic/water-science-school/science/hydroelectric-power-water-use?qt-science_center_objects=0#qt-science_center_objects. Retrieved July 12, 2019.

14 Tools and Equipment

OBJECTIVES

Upon completion of this chapter, the reader should be able to

- Discuss the advantages and disadvantages of manual and power tools
- Explain the role of tools and equipment in landscape work
- Discuss landscaping tasks and the tools required to accomplish them
- Identify the power tools and equipment used by landscapers
- Explain the role of labor cost versus tool and equipment cost in deciding which to use
- Develop an understanding of equipment-related issues such as noise, pollution, and work efficiency
- Describe the role of the landscape industry in regulation implementation
- Discuss health and environmental issues of concern with regard to landscape tools and equipment
- Discuss regulations implemented in communities concerning bans on power tools and equipment

TERMS TO KNOW

California Air Resources Board (CARB)
Decibel (dB)
Four-stroke engine
Fugitive dust
Hydrocarbons
Particulate matter
Two-stroke engine

INTRODUCTION

Power tools and landscape equipment are typically powered by fuel, and the primary fuel used is gasoline. This chapter looks at current equipment usage and some of the issues involved that may drive changes in the way equipment is designed in the future. In addition to fuel costs, other environmental issues concerning the use of landscape equipment and power tools are noise and air pollution. Fuels and alternative fuels were discussed in Chapter 13; however, engines and equipment must be specifically designed or possibly retro-fitted for use of fuels other than the one they were originally designed to use. Both fuel efficiency and job productivity are important factors in equipment design. Manual tools are discussed in this chapter as replacements for power tools where appropriate. The use of manual tools instead of power tools constitutes a sustainable landscaping practice and should be used whenever it is feasible to do so. A good example of the use of a manual tool in place of a power tool is when there is a small amount of debris or leaves that are as easily and quickly removed using a broom or rake as a power blower.

POWER TOOLS USED IN THE LANDSCAPE

Landscaping uses a wide array of power tools and equipment, from handheld tools to riding mowers to tree stump grinders and back hoes. Many lawn and garden tools and equipment have small engines with power output that is less than 25 horsepower. These units are considered to be off-highway/utility engines. Whereas these engines have historically been designed for fuel-rich combustion and high output power, affordability, and durability, they are presently the subject of clean air regulations that have been in effect since 1990 and have become more stringent since then.

The Clean Air Act was passed by Congress in 1970 and has been amended since then several times. It defines the responsibilities of the Environmental Protection Agency (EPA) for protecting and improving the nation's air quality, including the stratospheric ozone layer. Total US emissions have risen by nearly 14 percent between 1990 and 2008, However, in 2009, emissions decreased by 6.1 percent from 2008. This was attributed to decreased spending across all sectors and replacement of natural gas for coal as an energy source at electric power plants. By 2014, total emissions in the United States were around 6,760 metric tons of carbon dioxide equivalents, a 9 percent decrease from 2005. Emissions in 2017 were even lower, at 6,456.7 metric tons of carbon dioxide equivalents, a drop of 4.6 percent in a 3-year period.

Approximately 35 million small engines are sold annually in the United States. They are used about 1/100th to 1/1000th the amount of time a car or light truck is used, but their emissions are 100–1000 times greater. For typical residential landscape maintenance, mowers are used to the greatest extent, with homeowners spending an average of 40 hours each year mowing and landscape companies acquiring around 39 percent of their annual sales from mowing. In the process, lawn mowers consume 580 million gallons of gas annually. Landscape companies get 39 percent of their annual sales from mowing, according to industry surveys. Automobile emissions have been steadily declining since the 1970s, whereas emissions from landscape tools and equipment began to come under increased scrutiny in the 1990s.

For residential purposes, both walk-behind and riding mowers are used, but for larger areas, landscape companies, states and municipalities, highway departments, grounds departments, and others favor riding mowers for the wider mowing decks resulting in greater productivity. Zero-turn mowers are popular for their maneuverability. Some of the smaller power tools that are used in landscape maintenance activities include hedge trimmers, leaf blowers, chainsaws, string trimmers, edgers, chipper/shredders, leaf mulchers, and snow blowers (Figure 14.1).

For landscape installation, both the compact utility loader and lawn and garden tractor can be used with a variety of implements. Yet many jobs may be performed with individual pieces such as the rototiller, augur, or plate compactor. There are

Tools and Equipment

FIGURE 14.1 Leaf blowers are regularly used in landscape maintenance.

many specialized pieces of equipment, such as tree spades, trenchers, and sod removers. In addition to the many tools employed on a regular basis, the landscape industry relies on trucks and trailers for transportation of workers and equipment to each job site.

Certain landscape tools have come under greater scrutiny and regulation than others. For example, leaf blowers have come under regulatory control in many municipalities due to their widespread use, high noise level, and the fact that they stir up **particulate matter** which can trigger respiratory problems and allergic reactions in many people. These problems and regulations to address them are discussed in the following sections.

TYPES OF ENGINES

Gas-powered landscape equipment and tools incorporate one of two engine types. They are the **two-stroke** (or two-cycle) and the **four-stroke** (or four-cycle) engines. The major distinction pertains to the number of strokes of the piston involved in a working cycle. The four strokes are intake, compression, combustion, and exhaust. This engine model was first installed in a car by German engineer Nikolaus Otto; thus, the four-stroke engine is commonly known as the Otto cycle. In a two-stroke engine, the intake and compression stroke are combined, as are the combustion and exhaust stroke. Dugald Clerk was a Scottish engineer who patented his design for the two-stroke cycle in 1881. Lightweight landscape tools, such as chainsaws and string trimmers, have two-stroke engines. Two-stroke engines have the advantage of remaining operable even when held at an angle. They can also produce as much as twice the power than four-stroke engines of the same size.

Two-stroke engines exhaust a high percentage of fuel as unburned **hydrocarbons**. The amounts of unburned fuel ranges from 25 to 50 percent, depending on the equipment, how well it is tuned, and whether they are run with or without air filters.

For the user, some major differences are that two-stroke engines require the use of a fuel–oil mixture, commonly mixed at anywhere from 30:1 to 50:1 ratio, depending on engine design, and they have higher emissions than four-stroke engines. On the other hand, they are louder than four-cycle engines. This is a major disadvantage in locations where noise regulations limit the time and use of some landscape equipment. Nevertheless, two-stroke engines are preferred for their portability, lightweight, and ability to operate in any position. This is expected to change in coming years, as manufacturers respond to consumer demand for less-polluting, quieter engines and as emissions regulations become more stringent.

Electric-powered landscape tools are often available and are generally quieter than gas-powered tools. They provide equal or lesser power, and do not require the regular maintenance that a gas-powered engine requires. Emissions are shifted to the power plant that generates the electricity, which relies largely on coal in many areas of the United States. Coal-fired power plant emissions are controlled, and cleaner energy alternatives, such as wind power, are being used and developed, as discussed in Chapter 13. Battery-powered tools are also quiet and may provide less power than gas-powered engines. Some considerations in choosing to use battery-powered equipment are amount of time it takes to charge the battery, length of time a charged battery will last, combined weight of the battery and tool, and battery replacement cost.

In 2011, Honda introduced a hybrid snow blower, the Honda HS1336i snow blower, which combined a gasoline engine that simultaneously drove the auger/fan apparatus and charged the battery, with the electric motors (two in parallel) controlling the track drive forward propulsion. By 2019, several companies had released hybrid and electric (battery-powered) models of snow blowers. In addition to fuel efficiency and lower emissions, they are quieter than conventional snow blowers, while retaining substantial power.

SUSTAINABILITY ISSUES CONCERNING LANDSCAPE TOOLS AND EQUIPMENT

Sustainability issues concerning landscape tools and equipment include emissions that contribute to air pollution and smog development as well as particulate matter that can cause health problems and noise. Under energy issues, the sustainable sites initiative (SITES) rating system suggests supporting manufacturers who implement sustainability measures. They specifically include objectives such as reduced emissions, reduced energy consumption, use of renewable energy, and reduced water consumption.

AIR POLLUTION FROM LANDSCAPE TOOLS AND EQUIPMENT

Landscape tools and equipment emit as much as 10–1,000 times as much air pollution as automobiles. The US EPA has shown that the typical leaf blower generates as much emission in 1 hour as an automobile does while traveling over 200 miles. One difference is that leaf blowers emit that amount of pollution in a smaller, more concentrated area than an automobile. In the Chicago metropolitan area in 2006, emissions from leaf blowers were estimated to be just under 2,000 tons per year.

This is equivalent to emissions from over 100,000 passenger cars. According to Riyaz Shipchandler (2008), 77 percent of this amount can be attributed to commercial sources, with the remaining 23 percent from residential sources. Kavanaugh observed similar results in a 2011 study. In the study, the author compared two leaf blowers: one with a two-stroke engine and the other with a 4-stroke engine, to a car and a pick-up truck. Overall, the two-stroke engine produced twice the emissions of the four-stroke engine. The four-stroke leaf blower produced 18 times the emissions of a 2012 Fiat 500 car and 13 times that of a 2011 Ford Raptor pick-up truck.

Major pollutants released from gas-powered engines are hydrocarbons, including benzene, 1,3-butadiene, acetaldehyde, and formaldehyde; and carbon monoxide. Ozone (O_3) is formed by a reaction between nitrogen dioxide and hydrocarbons emitted in the exhaust fumes of gas-powered engines, including leaf blowers. Ozone is an air pollutant, a smog precursor, and a known irritant that can cause damage in as little as one hour, with permanent lung damage and respiratory problems from repeated exposure.

Benzene is a known carcinogen that also depresses the central nervous system. Acetaldehyde, formaldehyde, and 1,3-butadiene may also be carcinogens and can lead to eye, skin, and respiratory tract irritation.

Carbon monoxide can be absorbed into the bloodstream from the lungs, reducing the oxygen concentration in the blood. This is unhealthy for anyone, but particularly risky for pregnant women, the elderly, infants, and people with respiratory problems or heart disease.

Air pollution has become a political issue with respect to landscape power tools and equipment. Gas-powered equipment is the biggest culprit, emitting around 5 percent of the country's ozone-harming pollutants by 1990. In that year, the **California Air Resources Board (CARB)** proposed regulations that would limit exhaust from non-road, gas-powered engines. These are known as the Tier I regulations. Also in that year, the federal EPA instituted its Phase I emission reductions at the national level. The EPA reductions were to take full effect by 1997 for most engines. The regulations reduced the combined limit of hydrocarbons (HCs) and nitrous oxides (NOX) by 25 percent. Such a reduction was not considered drastic or unreachable by engine manufacturers.

In 2000, the CARB implemented its Tier II standards reducing the combined emissions limit by an additional 70 percent. The EPA's Phase II standards allow for a scheduled reduction over a period of 5 years, starting in 2002, with an overall reduction of an additional 80 percent, making the EPA's emission levels more stringent than California's for the first time ever.

Tier III standards, which apply only for engines from 50 to 750 horsepower, were phased in beginning in 2006. Tier IV standards were phased in over the period of 2008–2015.

The Tier IV emission standards introduce substantial reductions of nitrous oxides and particulate matter, as well as more stringent HC limits. Due to the regulations put in place by the CARB, small engines are 40–80 percent cleaner than those were in 1990. Going forward, the CARB plans to reduce small engine emissions by 80 percent in 2031. A major shift to zero-emission equipment will be needed to meet this goal.

FUGITIVE DUST AND PARTICULATE MATTER

Leaf blowers were introduced into the landscaping industry in the 1970s. When leaf blowers are used, they stir up dust and other particulate matter. **Fugitive dust** is a term that refers to airborne sidewalk and roadway dust, including garden debris, leaves, and grass. Since leaf blowers move air at wind speeds similar to hurricane force, dust, fecal matter, fungi, pollen and other allergens, pet and animal dander, spores, and other harmful substances are picked up and distributed through the air, where they remain airborne for minutes to hours to days. Columns of dust and particulate matter blown into the air five or six ft high by leaf blowers are not unusual. With a typical leaf blower, approximately five pounds of particulate matter is blown into the air per hour.

Whereas particulate matter less than 10 μm in size can be inhaled and deposited on airway surfaces of the lungs, particles that are 2.5 μm or less can penetrate deep into lung tissue. Respiratory ailments and increased hospital admissions have been linked to particulate matter.

Steve Zien, owner of Living Resources Company, a landscape management service in California reported that the hurricane force wind generated by leaf blowers removes topsoil and mulch from landscaped areas. It also damages and stresses plant foliage or removes it altogether and can cause plant dehydration. It is further possible that plant disease organisms are spread to other plants and neighboring properties through the use of leaf blowers.

NOISE FROM LANDSCAPE TOOLS AND EQUIPMENT

A **decibel** is a measurement of sound. With total silence measuring a zero on the decibel scale, normal conversation is at 60 decibels, and 140 decibels is painful to the ear and causes immediate hearing loss (Table 14.1). Both the level of noise and the length of exposure to it determine its ability to cause hearing loss. Sounds louder than 80 decibels are considered potentially dangerous, and gradual hearing loss will occur at prolonged exposure to 85 decibels or more.

A chainsaw, at 110 decibels, may cause permanent hearing loss with 1-minute or more of exposure. The typical lawn mower emits around 90 decibels during use. The decibel scale is logarithmic rather than linear. Therefore, 70 decibels is 10 times louder than 60 decibels, 80 decibels is 100 times louder, and 90 decibels is 1000 times louder. Leaf blowers range in decibel level, depending on manufacturer and model, but at the operator's ear, they are often as high as 90–100 decibels or more. This is a factor of five to fifteen times louder than the maximum level which causes hearing loss. Leaf blowers also emit noise at higher frequencies, which are more damaging to the ear than lower frequency sounds. Both hair cells in the ears and hearing nerves can be permanently damaged by continuous or repeated exposure to noise. Leaf blowers are often targeted as being the greatest source of noise from landscape equipment due to their high-intensity and high-frequency (high pitch) sounds.

The city of Claremont, California, banned the use of blowers in 1990 and found that maintenance of city property required only 6 percent more labor using rakes and brooms. This did not include time spent on equipment maintenance and mixing the

TABLE 14.1
Decibel (dBA) Levels of Some Common Sounds and Some Common Landscape Tools and Equipment

Source	Decibel (dB) Level
Silence	0
Whisper	15
Normal conversation	60
Vacuum cleaner	75–90
Automobile (at 25 ft)	80
Motorcycle (30 ft)	88
Diesel truck (30 ft)	100
Jet plane (100 ft)	130
Chainsaw	60–75[a]
Lawnmower	90–110[a]

Each increase of 10 decibel units represents a ten-fold increase in loudness.

[a] Standard published decibel ratings for equipment are measured at 50 ft from the equipment.

fuel/oil mixture used to operate the blowers. Furthermore, when ground crews started using a sidewalk vacuum, they realized a savings in labor of close to 14 percent.

Other cities have followed Claremont's action, both in California and across the nation. In 2010, as many as 400 cities had reportedly banned or restricted the use of leaf blowers. Some municipalities have gone further by banning or restricting the use of other landscape tools. For example, a proposed ban in Coral Gables, Florida, would not only restrict the use of leaf blowers, but also lawn mowers, chain saws, lawn edgers, weed trimmers, and chippers.

Electric blowers generate less noise than gasoline-powered models having comparable power. They have the additional advantage of requiring less maintenance. However, they still stir up dust and particulate matter.

SOLUTIONS

A number of solutions can be implemented to address the problems associated with power tools and equipment. Analysis of the landscape and forethought during the design phase are good places to start a change. Reduction of maintenance activities and switching to less-polluting methods can be achieved in several ways. In addition to the actual equipment used, including trucks used to transport crews to job sites, there is also an organizational component that can help save on energy consumption. For example, scheduling job sites in the same proximity to follow one another may reduce the amount of travel required. Hand labor should be used when it is determined that hand labor either (1) provides improved quality of workmanship, such as hand-shearing hedges or (2) it is actually more efficient to use hand labor, such as when raking wet leaves.

Reducing Emissions

E10 unleaded fuel started to become widely available in the 1990s, when states replaced MTBE (methyl tertiary-butyl-ether) with 10 percent ethanol as an **enleanment additive** or oxygenator. Both ethanol and MBTE have been added to gasoline to help provide a cleaner burning fuel, thus reducing tailpipe emissions by permitting more complete combustion than fuel that does not have such an additive. There have been some disagreements about the effects of the ethanol additive on gaskets and other rubber parts in internal combustion engines, particularly small engines. According to equipment manufacturer Husqvarna,

> it (fuel with ethanol) is not recommended, especially in handheld products. Gas with ethanol separates while being stored in your gas tank. The 2 stroke oil remains bonded to the gasoline but not to ethanol. Thus, the ethanol and water part of the mixture contains no oil for engine lubrication. This leads to poor lubrication, performance issues and costly repairs over time.

Electric lawn mowers and manual push mowers with a reel mechanism for cutting are good choices for smaller areas of lawn. Both are quieter, of course, and the manual push mower uses no fuel, has no emissions, and requires little maintenance beyond keeping the blades sharp and oiled.

Electric lawn mowers are of two types: those that run on a battery and those that plug into an outlet during use. In either case, they are quieter than gas mowers. Battery-operated lawn mowers can run for about an hour on a charge, requiring multiple mowers to complete a job or to be available for multiple jobs in succession. Some disadvantages are that batteries drain more quickly on wet lawns, and they can take 30–90 minutes to recharge. Batteries make the mower heavier, and so they can be difficult to push uphill.

Plug-in electric mowers can be challenging to use over larger areas due to the power cord becoming twisted, tangled, caught on obstacles, and getting in the way of the mower operator. Another disadvantage of electric mowers, and manual push mowers, too, is their reduced power as compared to a gas-powered mower. Nevertheless, with a little planning and foresight, they can be an asset to the landscape company that has clients with small areas of flat lawn, and they have the additional benefits that employees are not exposed to toxic gasoline or its fumes, equipment runs much more quietly, there are no emissions, and there is no need to change oil or spark plugs.

Other Technological Advances

Exhaust-gas recirculation (EGR) is used in the Ryobi Pro-4-Mor four-stroke-engine string trimmer. The ¾ horsepower engine replaces the traditional two-stroke engine that has long been used to power string trimmers.

Re-design of combustion chambers and valve-gear for operation at higher temperatures is quite feasible, according to G.S. Brereton at Michigan State University. After-treatment or more sophisticated carburetion (sensor controlled) may prove to be effective solutions that are more cost-effective than re-design of the induction system.

Evatech is a "Green Design" company that has built a robotic hybrid mower called a GOAT that generates its own electricity. The GOAT can mow on slopes using a

Tools and Equipment

remote-control device. They have also developed a second model, the TREX, for which a snowplow blade is available. The company is beginning to integrate artificial intelligence and memory into their robots so that they will be totally autonomous. Husqvarna also has a robotic mower, called Automower, which can be operated and monitored through an app on your smart phone. There are several other manufacturers of robotic mowers, as well.

Reducing Noise

Manufacturers can reduce the noise factor of leaf blowers and other equipment by creating larger mufflers and using rubber-isolated engine mounts. Running leaf blowers on low throttle speeds as much as possible can reduce noise. Using a full-nozzle extension can also help by placing the air stream as close to the ground as possible. Mufflers, air intakes, and air filters should be checked regularly and replaced or repaired as necessary for optimum performance and minimum noise.

Sidewalk Vacuum

The City of Claremont, California, stopped using leaf blowers in the maintenance of city property, replacing it with the use of a sidewalk vacuum. In addition to saving an hour every day on each crew, there was also a reduction in dust and pollen. From 50 ft away, the decibel level was 69, bringing the noise level below that of gas blowers although slightly more than electric blowers.

Reducing use of Power Tools

In addition to the technological advances, another way to reduce emissions and noise is to reduce the use of power tools altogether. While this may appear at first to imply an increase in manual labor, that is not necessarily the case. Manual performance of certain tasks that have been performed using power tools may not decrease the overall efficiency of performing that task. In an oft-cited contest, Diane Wolfberg outperformed both gas and electric blowers in three different tasks involving leaf removal using a rake and broom. In addition to the quietness and lack of toxic emissions, she was able to move wet pine needles and small stems more effectively than either of the blowers. Ergonomic rakes and snow scoops may be useful for such tasks as well.

It is always a good idea for landscape companies to record and evaluate labor efficiency of each task. Only by keeping good records and evaluating the information can a landscape manager or company know for certain that the most efficient method is being used. Some of the less obvious labor costs of power tool usage include the time it takes to load, maintain, and service the equipment.

Transportation Efficiencies

All landscape companies use transport vehicles, whether it is to meet with clients, conduct a site analysis, install, or maintain a landscape. Landscape companies should evaluate their transportation fleet and consider purchasing fuel-efficient, low emissions vehicles over a phase-in period. Another consideration that can be made by landscape companies is the efficiency of job-site planning. In larger metropolitan areas, planning jobs for each crew within a close area can eliminate unnecessary travel back to the shop site and traveling long distances during the day to move from one job site to another. Proper planning can also ensure that a work crew has the necessary tools and equipment for all jobs throughout the day rather than requiring travel back to the workshop to pick up tools or equipment during the day.

Case Study: Fuel-efficient business practices.

Lawn & Landscape. http://www.lawnandlandscape.com/lawn-landscape-0611-pump-pressure-gas-prices.aspx. Viewed Dec. 30, 2011.

Rising gas prices are difficult for landscape companies to contend with due to the difficulty or reluctance of passing on increased prices to the client. For Dowco Enterprises of St. Louis, Missouri, streamlining routes is a key factor in helping keep fuel surcharges at a minimum. They keep routes within their "domination zone" to minimize travel time to work sites. The company also uses GOS software and monitors driver behavior. Gas cans are kept in a locked room with camera monitoring. For sales personnel, fuel-efficient Pontiac vibes are now used.

In Orlando, Florida, Bel Air X Lawn & Landscapes also uses tight routing to maintain fuel efficiency in the company. One tactic that has worked is offering discounts for customers who sign up their neighbors for mowing services.

The Brickman Group has tried a somewhat different approach to fuel efficiency – they train drivers not to idle the engine when they are loading up their vehicles. In addition, they switched from Ford F-350 trucks for their managers to the smaller, four-cylinder Ford Ranger and began using Toyota Prius hybrids. Finally, they use hedging to lock in fuel prices, bringing a predictability into fuel budgeting and a predictable price for customers.

In all areas of landscaping, from design to installation to management, improvements or changes may be made that can help to reduce the use of power tools and equipment. Many ideas for plant usage in the design phase of landscaping are presented in earlier chapters. Some ideas related to tools and equipment in all areas of landscaping are provided in the following section.

LANDSCAPE DESIGN

The design phase occurs before the landscape maintenance crew is on the job. This is the crucial time when it is important to avoid design flaws that require high levels of maintenance with power tools. Some ideas for designing for reduced use of gasoline-powered engines are given as follows:

- Heated sidewalks reduce the need for snow and ice removal. Hot water circulates beneath the pavement to provide the heat.

Tools and Equipment

- Replace grass with lower-maintenance plants.
- Avoid tiny areas where it is difficult to grow plants, as these often are susceptible to weed problems. These areas often creep into a design unnoticed. They may occur where a sidewalk meets a lawn or flower bed at an acute angle, for example.
- Avoid using fast-growing, weak trees that suffer regular breakage requiring pruning or chipping.

LANDSCAPE MAINTENANCE

Other practices can be applied during the maintenance phase. Some ideas for reducing maintenance include the following:

- Manually prune when other considerations (time) are less important or make less of a difference (manual hedging may not take that much longer, depending on the situation).
- Rake leaves and debris out of small spaces, wet leaves, etc., when blowers are inefficient.
- Use push reel or electric mowers for trim areas (the clients and their neighbors will love you for the noise reduction).
- Mulch around trees to reduce the need for weed-eaters close to the trunks; this also reduces turf area and eliminates trunk damage.
- Apply mulch in all beds at a depth of 2–4 in. to keep weeds down.
- Control weeds when they are small.
- Blowers that are used to remove lawn clippings from the street may not be necessary if a mulching mower or sidewalk vacuum is used.

SUMMARY

Power tools and equipment used in the landscape are typically powered by fuel, mainly gasoline, and to a lesser extent, diesel. Environmental issues, cost, and noise are some of the factors contributing to a re-evaluation of power tools and equipment use in the landscape. Many lawn and garden tools and equipment have engines that operate at 25 horsepower or less, putting them in the category of off-highway/utility engines.

Gas-powered engines are either four-stroke or two-stroke engines. Two-stroke engines require the use of a fuel/oil mixture. Widely used two-stroke engines in landscaping include chain saws, string trimmers, edgers, and leaf blowers. They release more HCs due to the large percentage of unburned fuel and are louder than four-stroke engines.

Electric-powered tools and equipment are generally quieter than those that are gas-powered. They often provide less power for the engine size, but on-site emissions are eliminated. Emissions are released by landscape tools and equipment in a smaller, more concentrated area than those emitted by vehicles. Major pollutants released from gas-powered equipment include HCs and carbon monoxide. Noise from leaf blowers has come under opposition since they were first used in the 1970s, with an increasing number of municipalities regulating their use each year. In 2010, at least 400 cities had banned or restricted the use of leaf blowers. Electric blowers generate less noise than comparable gas-powered blowers, although they still stir up dust, another objection to the use of leaf blowers in general.

Some solutions to the problems that have been encountered with landscape power tools and equipment are the use of E10 fuel (10 percent ethanol-gas blend), which permits more complete combustion of fuels, electric or battery-powered tools, and manual push mowers. Other technological advances and more complex engine designs may contribute to better fuel efficiency in the future. Such solutions have disadvantages, such as less power or more expensive equipment. However, the costs of continuing to pollute at the current levels may prove to be economically unsustainable.

Reducing the use of power tools is an important aspect of increasing sustainability in landscaping. This requires planning during the design phase, as well as analysis of labor efficiency. Studies suggest that it is as efficient to rake or sweep leaves in some situations as it is to use a leaf blower.

Landscape companies need to evaluate their practices in light of environmental and human health concerns, and develop standard operating procedures accordingly.

REVIEW QUESTIONS

1. What is the typical power output (in horsepower) for many lawn and garden tools and equipment?
2. Compare the use and emissions of small engines to cars or light trucks in the United States.
3. What Act of Congress was passed in 1990 that regulates air quality? What agency is responsible for its implementation?
4. Which landscape tool has come under regulatory scrutiny due to its noise?
5. Name the two gas-powered engine types used in many landscape tools.
6. List two advantages and two disadvantages of two-stroke engines used in landscape tools.
7. What fuel has replaced MTBE as an enleanment additive or oxygenator in landscape tools, equipment, and transport vehicles.
8. How can landscape companies reduce transportation energy usage other than switching to alternative-energy vehicles?
9. Name two design strategies that may be used to help reduce energy consumption in maintenance of the landscape.
10. List three ideas for reducing energy consumption during landscape maintenance.

ACTIVITIES

1. Interview a local landscape company, and determine the amount of time that is spent servicing each type of power tool. For each tool that has an electric version available, compare the purchase price, costs for spark plugs, fuel and oil, and number of hours of expected use between the

gasoline-powered version. Adding that to the service costs, calculate which piece of equipment is more economical to own and operate.
2. Conduct a survey of battery-powered landscape tools that are available. Compare the types of batteries used, battery charging time, and battery usage time. Develop a list of options for common landscape tools.

FURTHER READING

American Green Zone Alliance. https://www.agza.net/ojai-agza-green-zone. Retrieved July 29, 2019.
American Speech-Language-Hearing Association. Noise and hearing loss. http://www.asha.org/public/hearing/disorders/noise.htm. Retrieved July 29, 2019.
Anon. 2001. Gas-powered equipment ban passes in Texas. *Amer. Nurserym.* 193(6): 14.
Banks, J. L. and R. McConnell. National emissions from lawn and garden equipment. https://www.epa.gov/sites/production/files/2015-09/documents/banks.pdf. Retrieved July 29, 2019.
Citizens for a quieter Sacramento. Leaf blower facts. http://www.nonoise.org/quietnet/cqs/leafblow.htm#conseq. Retrieved July 29, 2019.
City of Long Beach. 2017. Qualitative risk assessment on leaf blowers. http://www.longbeach.gov/globalassets/city-manager/media-library/documents/memos-to-the-mayor-tabbed-file-list-folders/2017/january-10--2017---report-and-recommendations-on-leaf-blowers. Retrieved July 29, 2019.
Harrison, M. Which is better, a 2 stroke or 4 stroke engine? http://www.deepscience.com/articles/engines.html. Retrieved July 29, 2019.
Husqvarna. What is ethanol fuel and why is it bad for your small engine outdoor power equipment? https://www.husqvarna.com/us/forest/basics/ethanol-free-fuel/. Retrieved July 29, 2019.
Kavanaugh, J. 2011. Emissions test: car vs. truck vs. leaf blower. https://www.edmunds.com/car-reviews/features/emissions-test-car-vs-truck-vs-leaf-blower.html. Retrieved July 29, 2019.
Mollenkamp, B. 2003. Overblown? *Grounds Maint.* 38(7): 17.
Mullins, E. 2007. The alternate route. *Lawn & Landscape.* October. 29–38.
Pew Center on Climate Change. http://www.pewclimate.org/. Retrieved July 29, 2019.
Quiet Communities. https://www.quietcommunities.org/who-we-are-qc/. Retrieved July 29, 2019.
ROBOT GOAT. Evatech http://www.evatech.net/. Retrieved July 29, 2019.
Shipchandler, R. 2008. VOC emissions from gas powered leaf blowers in the Chicago Metropolitan Region. *Waste Manage. Res. Cent.* https://www.lincolntown.org/DocumentCenter/View/6453/Shipchandler_VOC-Emissions_Waste-Mgmt-Res-Center_Illinois?bidId. Retrieved July 29, 2019.
Stahl, J. 2001. Texas bans morning commercial equipment use. *Landsc. Mgmt.* 40(1): 16–19.
U.S.E.P.A. April 2010. Trends in greenhouse gas emissions. U.S. EPA # 430-R-10-006. https://www.epa.gov/ghgemissions/inventory-us-greenhouse-gas-emissions-and-sinks. Retrieved July 29, 2019.
Valdemoro, T. 2010. Noise builds over Gables' effort to outlaw leaf blowers. *Miami Herald*, October 12. http://www.miamiherald.com/2010/10/12/1870420/noise-builds-over-gables-effort.html#ixzz15fsJfWY7. Retrieved November 18, 2010.
Weber, M. 2006. Cutting edge. *Grounds Maintenance*, May. 13–24.

15 Sustainable Landscape Materials and Products

OBJECTIVES

Upon completion of this chapter, the reader should be able to

- Describe traditional materials used in landscape construction
- Identify sources of recycled materials in landscape products
- Discuss renewable sources of landscape product materials

KEY TERMS

Chain of custody
Crumb rubber
Life cycle assessment
Plastic lumber
Sustainably harvested lumber
Urban wood
Waste stream

INTRODUCTION

The phrase "reduce, re-use, recycle" has been associated with environmental concerns for a long time. The EPA has added the term "rebuy", meaning "buy recycled products", as there are now many fine landscape products made from recycled tires, plastics, twine, glass, and many other items that could otherwise clog up the waste stream and end up occupying space in landfills. Benefits of recycling are given as follows:

- Protects and expands US manufacturing jobs and increases US competitiveness.
- Reduces the need for landfilling and incineration.
- Prevents pollution caused by the manufacturing of products from virgin materials.
- Saves energy.
- Decreases emissions of greenhouse gases that contribute to global climate change.
- Conserves natural resources such as timber, water, and minerals.
- Helps sustain the environment for future generations.

Table 15.1 shows the kinds of materials that have been recovered from the municipal solid waste stream.

TABLE 15.1
Generation and Recovery of Materials in Municipal Solid Wastes, 2009 (in Millions of Tons and Percent of Generation of Each Material)

Material	Weight Generated	Weight Recovered	Recovery as a Percent of Generation (%)
Paper and paperboard	68.43	42.50	62.1
Glass	11.78	3.00	25.5
Metals			
Steel	15.62	5.23	33.5
Aluminum	3.40	0.69	20.3
Other nonferrous metals[a]	1.89	1.30	68.8
Total metals	*20.91*	*7.22*	*34.5*
Plastics	29.83	2.12	7.1
Rubber and leather	7.49	1.07	14.3
Textiles	12.73	1.90	14.9
Wood	15.84	2.23	14.1
Other materials	4.64	1.23	26.5
Total Materials in Products	*171.65*	*61.27*	*35.7*
Other wastes			
Food, other[b]	34.29	0.85	2.5
Yard trimmings	33.20	19.90	59.9
Miscellaneous inorganic wastes	3.82	Neg.	Neg.
Total other wastes	*71.31*	*20.75*	*29.1*
Total municipal solid waste	242.96	82.02	33.8

Source: US EPA.
Includes waste from residential, commercial, and institutional sources.
Neg., less than 5,000 tons or 0.05 percent.
[a] Includes lead from lead–acid batteries.
[b] Includes recovery of other municipal solid waste (MSW) organics for composting. Details may not add to totals due to rounding.

LANDSCAPE CONSTRUCTION MATERIALS

Landscape construction materials are used for paved surfaces, decks, fences and gates, walls, and items such as statuary, gazebos, pergolas, and trellises. In addition to the landscape feature being constructed, there may be additional materials required to support, underlay, or otherwise augment the item. For example, a brick or concrete paver patio area requires an underlayment, which may be gravel topped by a layer of sand, or it may be asphalt or concrete. Sand is also frequently used as filler between pavers or bricks. Some common materials used in various landscape projects are

- Asphalt
- Brick
- Concrete blocks

- Concrete pavers
- Gravel and crushed stone
- Landscape timbers
- Lumber
- Metal
- Mortar
- Plastic
- Poured concrete
- Sand
- Stone of several types

Mulch for walking paths or landscaped beds is also a landscape material, despite not being used strictly as a construction material.

Both softwoods and hardwoods are used for construction of landscape products. Trees that are resistant to decay are often favored due to their longevity in the outdoor environment where they are constantly exposed to environmental degradation due to ultraviolet radiation, water, fluctuating temperatures, and so on. Tree species that are resistant to decay include black cherry, chestnut, black locust, oak, sassafras, black walnut, old growth bald cypress, cedars, Arizona cypress, redwood, and Pacific yew. Table 15.2 lists tree species that are used in outdoor products.

The SITES rating system addresses construction materials in Section 5: Site Design – Materials Selection. Prerequisite 5.1 requires eliminating the use of wood from threatened tree species. Suggested alternatives include using wood products from sustainably managed forests, and using recycled plastic or composite lumber instead of wood. Plastic lumber may contain recycled plastic, which is within the goals and criterion for SITES accreditation. They also propose the use of salvaged

TABLE 15.2
Tree Species Used in Outdoor Products

Hardwoods	Uses	Native Areas
Red oak	Fence posts	Pennsylvania west to Minnesota and Iowa
White oak	Fence posts	Maine to Florida and west to Minnesota and Texas
Softwoods		
Incense cedar	Fencing	CA, OR, NV
Western larch	Poles	MT, ID, OR, WA
Lodge pine	Posts, poles	Rocky Mountains and Pacific coast to Arkansas
Ponderosa pine	Posts, poles	AZ to NM and SD and west to Pacific coast mountains
Red pine	Posts, poles	NE states, NY, PA, and the Great Lake states
Western red cedar	Posts, poles	Pacific coast north to Alaska; ID, MT
Redwood	Outdoor furniture, fencing	Sierra Nevada of CA
Engelmann spruce	Poles	High elevations of Rocky Mountains
Tamarack	Fencing, poles	From ME to MN

materials and appropriate plants in order to conserve resources and avoid landfilling useful materials. All of these objectives are discussed in more detail in this chapter.

RECYCLED MATERIALS FOR LANDSCAPE PRODUCTS

Depending on the level of constructed features in the landscape and the actual materials used, some landscapes can be energy intensive, represent significant carbon emissions, and make significant demands on limited resources. Many landscape companies are beginning to implement alternative practices and use alternative materials. Recycled materials are one such alternative. Some of the recycled landscape products available include

- Plastic lumber for outdoor structures, made of recycled bottles and bags
- Rubberized asphalt made from recycled tires
- Concrete that contains fly ash and/or other recycled materials
- Stepping stones, tiles, and pavers from recycled glass
- Trash containers from composite materials
- Irrigation dripline from recycled plastic
- Pavers from rubber and plastic

In addition to recycled materials, other materials that come under the "sustainability" umbrella are salvaged materials, re-manufactured materials, renewable materials, sustainably harvested lumber, and locally sourced materials. All of these terms allude to the fact that there is a finite supply of certain resources or raw materials, and alternatives will need to be found to replace them as they become scarce or difficult to obtain. In some cases, though, such as rubber, wood, glass, and plastic, the material can often be re-used, either in its current form or in a re-manufactured form. For example, plastic bottles are re-manufactured into plastic lumber, and rubber tires are shredded into crumb rubber.

Salvaged materials are taken from a site where they are no longer desired and simply placed into the new site or situation. Stones, pavers, and bricks are examples of salvaged materials.

Using local materials is a component of sustainable landscaping. It involves considering the cost in terms of fuel consumption and the carbon footprint of shipping across long distances in terms of emissions. For example, shipping one ton of material one mile typically uses between 2,000 and 6,000 BTUs. In addition to saving in shipping costs, another benefit of using locally available material is the addition of local flavor to the landscape design.

LIFE CYCLE ASSESSMENT

The term **life cycle assessment** (LCA) is used to describe and measure the environmental costs of a product. It is applied across disciplines, from meat production, to household goods, and includes construction materials. According to Gail Hansen at the University of Florida, the six stages of the "life cycle" of landscape construction materials and products are

Sustainable Landscape Materials & Products

1. Acquisition
2. Fabrication
3. Transportation
4. Installation
5. In-place performance
6. Demolition and disposal

Acquisition practices for sustainability include using local materials or lumber that was certified as sustainably harvested. This subject is discussed in more detail later in this chapter. Fabrication practices include minimally processed materials or naturally occurring materials. Transportation impacts can be reduced by using materials found on the site, such as boulders and stones, or materials that were obtained from a local source. Sustainable installation practices include using materials efficiently to avoid waste and in-place performance means using techniques to protect the materials from the elements so they will last longer. Demolition and disposal practices for construction materials are receiving increasing attention due to the large amount of materials that go to landfills. In 2015, 48 million tons of construction and demolition debris were generated in the United States, more than twice the amount of generated municipal solid waste. Demolition represented more than 90 percent of total construction and demolition debris generation, while construction represented less than 10 percent.

The environmental impacts that are taken into consideration during a LCA include energy requirements, human health issues, air pollution and carbon emissions, landfill space, impacts on recycling, landscape disruption, and local ecosystems and climate impact.

LCAs are complex calculations that require consideration of economic and environmental data from diverse sources. However, some basic principles can aid the landscape professional in choosing materials and implementing sustainable practices during the design, construction, installation, and management phases. Each phase has one or more components unique to it, but there is also overlap in some practices. For example, specifications for materials are made in the design phase and purchased during the construction and installation phase. Decisions could be made during either of these phases as to the exact materials used, based on local availability, whether salvaged materials could be substituted or other factors.

Software has been developed to aid in LCAs such as Athena Institute's Ecocalculator or Impact Estimator and the National Institute of Standards and Technology's Building for Environmental and Economic Sustainability (BEES).

WASTE MANAGEMENT

Household and construction waste are taken to municipal solid waste facilities, commonly called landfills. All municipal solid waste landfills are required to comply with certain state and federal laws that are designed to prevent contamination of soil, groundwater, and wetlands; reduce odor; control rodents; and protect public health.

RECYCLED MATERIALS

Many materials in the **waste stream** can be re-used in some form or another. Sometimes, the material has to be treated or broken down into a more basic material, and then, there may be further processing before it can be shaped or formed or otherwise transformed into a new product. Some of the major types of products that are recycled are certain kinds of plastic, rubber from used tires, and glass. Sawdust and particles of wood can be incorporated into composite materials.

Plastic Lumber

Plastic lumber is widely accepted, and does have standards set for testing and performance. It is typically made from recycled milk jugs, plastic wrap, and other sources of high-density polyethylene (HDPE), but can also be made from other types of plastic, such as polyethylene, polypropylene, and polyvinylchloride.

Standards for plastic lumber were developed and implemented in 1977 by the American Society for Testing and Materials (ASTM). However, testing is expensive, and only a few companies have completed the testing process and now meet the new standards. Therefore, structural characteristics vary among manufacturers, and the buyer must research products before committing to a final purchase.

Plastic lumber is commonly used in decks and railing, parks and playgrounds, picnic tables, and docks. Benefits include lack of splintering, no need for toxic preservatives, attractive appearance, choice of colors, durability, reduction of landfill waste, and saving trees.

Rubber

There are around 300 million tires scrapped every year in the United States, amounting to 1.1 tires for every person (Figure 15.1). Tires do not biodegrade and take up considerable space in landfills. Tires are often left sitting around rather than being disposed of, which creates good conditions for mosquitoes to breed, due to their tendency to hold water and provide mosquito breeding areas. One rubber recycler in California obtains several products from used tires, including

- Playground cover
- Athletic surfaces
- Rubber asphalt concrete
- Nylon fibers for concrete mixes, carpet underlays, and carpet blends
- Molded products
- Tire derived fuels

Azek Pavers makes pavers from recycled rubber and plastic. The pavers are 30 percent lighter than concrete pavers and easier and faster to install. The pavers themselves sit into a grid system that lays over a traditional base of gravel and sand. They can be used for traditional paved surfaces such as parking lots, patios, driveways, plazas, and sidewalks. They can also be installed on roofs to provide a walking surface

Sustainable Landscape Materials & Products

FIGURE 15.1 Used tires create a surplus of rubber.

that is much lighter than other paving materials. Since they are made from recycled materials, they require 94 percent less energy to produce with only 11 percent of the carbon emissions. They are also permeable, providing an infiltration rate of 480 in. per hour due to the 12 percent open area accommodated in their design.

KBI Flexi-Pave is a poured material comprised of rubber from recycled tires and aggregate, plus a special urethane that remains flexible after pouring. It can be used as an overlay on existing asphalt, concrete, wood, or other surfaces, or it can be installed as a new surface. It is resistant to ultraviolet rays, damage from snowplows, salt water, transmission fluid, and other hazardous materials. It is best used where pervious pavement is desired due to its porosity. It can be used as a perimeter surface to save on costs, as it aids in water infiltration and reduces the need for a retention pond in large paved areas. Other uses are

- Around trees in paved areas
- As a base for playground areas
- Golf cart paths
- Trailways and walkways
- Sidewalks
- Driveways

Crumb Rubber

Made from used tires, **crumb rubber** is primarily used for paving or surfacing, but can also be recycled into mulch and used for trail surfacing. It is commonly used on athletic fields and golf courses having artificial turf, to provide a softer landing during sports. However, some athletes do not like the crumb rubber surface because it tends to adhere to shoes, clothing, and skin. Soccer goalies often experience abrasions during a game, and the rubber crumbs get into their wounds. More serious concerns have arisen since the use of crumb rubber first began. For example, artificial turf fields heat up 10 to 15 degrees hotter than the ambient temperature, and crumb rubber fields may emit gases from polycyclic aromatic compounds, and other toxic fumes, that are inhaled. Mercury, lead, benzene, polycyclic aromatic hydrocarbons, and arsenic, heavy metals, and carcinogens have been found in tires. On any given field, thousands of different brands of tires may be present, so studies on what exactly is present in the crumb rubber are difficult to do. Research so far has been inconclusive, but health concerns do exist.

CONCRETE AND ASPHALT

Cement is the primary ingredient in concrete, the second most consumed material on the planet. The manufacture of cement is energy-intensive and uses a large amount of limestone as a raw ingredient. As a whole, the cement industry produces about 5 percent of the anthropogenic or manmade carbon dioxide emissions. The cement industry has formed a consortium of businesses and has generated the Cement Sustainability Initiative as a response to concerns about global climate change. They would like to reduce carbon dioxide emissions, and some of the strategies they have proposed include

- Innovation in improving the energy efficiency of processes and equipment.
- Switch to lower carbon fuels.
- Use alternative raw materials to reduce limestone use.
- Develop carbon dioxide capture and sequestration techniques.
- Take advantage of market mechanisms such as emission trading and voluntary initiatives

In addition to changes at the production end, there is also the possibility of re-using concrete from demolition sites. Crushed concrete can be used as fill, to provide drainage in a swale or French drain, and is sometimes used to protect banksides. Concrete can also be recycled into concrete aggregates, but embedded reinforcing rods and other items must be removed first. The quality of recycled aggregates depends on the quality of the recycled material used. In addition to steel, it may also be contaminated with asphalt, sand, clay, chlorides, glass, gypsum board, sealants, plaster, wood, and roofing materials.

Asphalt can be re-used on location or removed and stockpiled for use in other areas, such as driveways. Hanover Architectural Products produces asphalt blocks that can be used as pavers that contain varying amounts of post-industrial recycled asphalt.

GLASS

Numerous landscape products are made using recycled glass, including stepping stones, pavers (http://www.tiletechpavers.com/recycled-glass-pavers/), tiles (https://www.mineraltiles.com/recycled-glass-tile/, https://www.lightstreamsglasstile.com/), and planter boxes (http://www.terragreenceramics.com/).

RENEWABLE RESOURCES

Some products are made from renewable resources that can be continually replaced, whereas other products come from finite source materials. Lumber belongs to the first group. However, historically, trees have not always been harvested in a sustainable way. Before 1900, forest growth could not keep pace with the volume of timber cut. Lumber barons and lumber mills went hand-in-hand with clear cutting and deforestation in the early history of the United States. There were no reforestation programs, and often only the best part of the tree was used, leaving large volumes of wood left behind on forest floors. In addition, timberland was cleared to provide large tracts of land for agricultural use.

In 1897, Congress passed the Forest Management Act explicitly stating the purpose of Forest Reserves (later National Forests) as resources for lumbering, mining, and grazing. In the early 1900s, Gifford Pinchot, who eventually founded the National Forestry Service, was instrumental in establishing and promoting scientific forest management. Aldo Leopold continued in the same vein, expanding upon the ideas of Gifford. Leopold promoted species diversity in wilderness environments, which directly opposed prevailing thoughts at the time of human dominance over nature. He also saw the environment as an interconnected whole rather than as a place to simply harvest products of discrete value. Thus, for Leopold, a forest was a wildlife habitat, as opposed to merely a stand of timber to be removed. He developed a philosophy that he called "The Land Ethic", describing it as follows: "The land ethic simply enlarges the boundaries of the community to include soils, waters, plants, and animals, or collectively: the land".

SUSTAINABLY HARVESTED LUMBER

The Forest Stewardship Council, created in 1993, intended "to change the dialogue about and the practice of sustainable forestry worldwide". To that end, the Council has developed standards to encourage sustainable harvesting of the world's lumber, known as **sustainably harvested lumber**. They set forth criteria that address economic and social concerns in conjunction with environmental concerns. Their criteria impact LEED (Leadership in Energy and Environmental Design) certification, and they work to ensure chain of custody certification as an integral part of the LEED certification process. Sustainably harvested wood that is FSC-certified has been identified as the most specified green-building product in a database of 60,000 project specifications collected annually. Worldwide, over 371 million acres of forest had been certified by the end of 2011. In the United States, around 34 million acres had been FSC-certified.

Forest Management and Chain of Custody are the two types of certification the FSC provides. In both types of certification, independent FSC-accredited certification bodies verify that all FSC-certified forests conform to the requirements contained within an FSC forest management standard. The following North American certification bodies are accredited by the FSC to certify forest products:

1. Advanced Certification Solutions (http://advancedcertificationsolutions.com)
2. BM Trada (https://www.bmtrada.com)

3. Bureau Veritas Certification (http://www.us.bureauveritas.com)
4. Control Union Certifications (https://www.petersoncontrolunion.com/en)
5. DNV GL Business Assurance (https://www.dnvgl.us/assurance)
6. PricewaterhouseCoopers (https://www.pwc.com/us/en.html)
7. QMI-SAI Global (http://www.qmi.com/registration/forestry/fsc/)
8. NEPCon (formerly Rainforest Alliance) (https://www.rainforest-alliance.org/business/solutions/certification/forestry/)
9. SCS Global Services (https://www.scsglobalservices.com/services/forestry-and-chain-of-custody)
10. SGS Systems & Services Certification USA (https://www.sgs.com/en/certification)
11. Soil Association – Woodmark (https://www.soilassociation.org/certification/forestry/)

To be considered "sustainably harvested", forest products must be part of a certification process. This is a voluntary program that has certain criteria that must be met, as well as a documentation procedure. The ultimate goal is to ensure that long-term productivity is sustained for future generations while desired outputs are produced.

The five steps to certification are given as follows:

1. The interested company will contact an FSC-accredited certification body. The company will need to provide information and will also be provided with information about certification.
2. The company will decide on a certification body and sign an agreement.
3. The company will then undergo a certification audit to assess their qualifications for certification.
4. Data collected during the audit will be compiled into a report which then serves as a decision-making tool.
5. If approved, a certificate is provided. Otherwise, suggested changes must be implemented and further audits are conducted until certification is achieved.

Some of the criteria that have been established to meet the certification requirements are that the practices to harvest the wood improve long-term ecological processes and productivity. The certification process requires that actions be analyzed at the landscape level, and not the stand level, as is common in traditional forest management. In addition, smaller cuts of wood are used, and organic matter is left for decomposition and recomposition. Clear cutting a forest is not a sustainable practice. The long-term productivity of a forest is recognized as important, and forest resources are recognized as being equally important to economic factors.

Sustainability goals have been adopted in the recent past by foresters throughout the world through this voluntary program. In the program, a forest owner requests a forest inspection to determine if predefined management standards are being achieved. Products can then be eco-labeled signifying their lands are sustainably managed and their practices are environmentally acceptable. This is the start of the **chain of custody** process in which products are monitored to ensure customers they are receiving the true certified material. There are currently over a 1,000 chain-of-custody wood suppliers in the United States.

The American Forest and Paper Association (AF&PA) has its own Sustainable Forestry Initiative in which participants follow a set of mandatory guidelines. They are required to file annual reports documenting their compliance.

Case Study: Sustainably Harvested Lumber

Arcata, California.
  Reference: Suutari, Amanda. http://www.ecotippingpoints.org. Viewed Dec. 27, 2011.
  The city of Arcata, California, owns 622 acres of redwood forest. The Arcata Community Forest was the nation's first forest to be certified by the Forest Stewardship Council as eligible for the designation of "Smartwood" or sustainably harvested wood.
  The Community Forest has many species of trees in addition to California Redwood, including Douglas Fir, Grand Fir, Sitka Spruce, and Western Hemlock. Trees are harvested every other year in general, bringing in revenues of over half a million dollars. The money then goes towards purchasing additional forest land or to grant conservation easements that allow land owners to sustainably harvest a given parcel of land.
  In addition to sustainable timber harvesting, the forest provides water to the community, recreational and educational opportunities, and wildlife habitat.

SALVAGED MATERIALS

Salvaged materials include anything that has been used on a site but is no longer useful at that site. Bricks, stones, lava rock, deck boards, and other materials that are replaced following re-design of a site still have usefulness and may be "re-purposed" at another site. The Sustainable Sites Initiative provides credits for diverting construction and demolition materials from disposal.

URBAN WOOD

Urban wood is a product made from reclaimed materials such as lumber and lumber products, and mulch. According to the Solid Waste Association of North America, wood waste accounts for about 17 percent of the total waste received at municipal solid waste landfills in the United States. A number of communities, businesses, and governmental and non-governmental organizations are spearheading efforts to recycle wood waste into furniture, mulch, and boiler fuel (Figure 15.2).

LOCAL MATERIALS

Local materials vary in availability, depending on geographical location, materials desired, and quantity and quality. Rocks and stones having local origin add to a sense of place that is appropriate for a given locale. The Sustainable Sites Initiative provides

FIGURE 15.2 These used pallets will be shredded and sold as landscape mulch. (Photo by author)

credit under the construction component for use of excess vegetation, rocks, and soil generated during construction. Although it is not always appropriate or possible, milling wood from trees that have to be removed from a landscape or construction site can provide a source of locally available material, while exploiting an otherwise wasted resource. Non-conventional woods, such as that harvested from hackberries (*Celtis* spp.), trees of heaven (*Ailanthus altissima*), and others, can provide wood for trellises, arbors, tables, or shredded for mulch.

SUMMARY

Reduce, re-use, recycle, and re-buy are four words associated with sustainable landscape materials. Landscape products made from recycled materials allows such materials to be removed from the waste stream. There are other benefits, including protecting jobs and increasing competitiveness, reducing the need for landfilling and incineration, and preventing pollution caused by the manufacturing, among others.

Traditional landscape construction materials include asphalt, brick, wood, concrete and concrete products, gravel, sand metal, and stone. Some recycled landscape products include plastic lumber, rubberized asphalt, recycled concrete, and stepping stones from recycled glass, and many landscape products made from composite materials.

In addition to recycled materials, other materials that come under the "sustainability" umbrella are salvaged materials, re-manufactured materials, renewable materials, sustainably harvested lumber, and locally sourced materials.

Materials standards were developed and implemented in 1977 by the ASTM. There are many positive attributes to products made from recycled materials, For example, plastic lumber does not splinter, does not require toxic preservatives, has

Sustainable Landscape Materials & Products

an attractive appearance, is available in a choice of colors, and is durable. Rubber can be recycled into mulch and used for trail surfacing, playground cover, or to topdress athletic fields to provide cushioning and reduce injuries.

Sustainably harvested lumber refers to a program designed by The Forest Stewardship Council. Sustainable forest management and wood harvesting practices are incorporated into the program. Forest resources are recognized as being equally important along with economic factors.

The American Forest and Paper Association (AF&PA) has its own Sustainable Forestry Initiative in which participants follow a set of mandatory guidelines.

Salvaged materials include anything that has been used on a site but is no longer useful for some reason. Various types of salvaged landscape materials are urban wood, bricks and pavers, and decorative stones. Local materials having local origin add to a sense of place that is appropriate for a given locale.

REVIEW QUESTIONS

1. What are the four "R's" of sustainable materials and products in landscaping?
2. Name three benefits of recycling.
3. Name four recycled materials that are used to make new landscape products.
4. List four benefits of plastic lumber.
5. What is the primary source of recycled rubber?
6. Who establishes standard for testing the strength and other properties of plastic lumber?
7. What organization developed the sustainably harvested lumber program?
8. What are the criteria set forth for sustainably harvested lumber?
9. What is "urban wood"?
10. Wood waste accounts for what percentage of total waste at municipal landfills in the United States?

ACTIVITIES

1. Using an online LCA tool, research what steps are involved in conducting an LCA. Conduct an LCA on several landscape features, such as a deck, a hardscaped patio, and a gazebo.
2. Conduct a LCA for a landscape product that is commonly used in your area. Use the environmental impacts stated in this chapter, and evaluate each of the six stages listed.
3. Make a list of commonly used landscape materials in your area. For each, indicate whether it could be considered part of a sustainable landscape. If not, suggest replacements. Justify your choices.
4. Visit the local waste management facility, both the landfill and the recycling center. Interview the manager(s) to identify the largest contributors to the waste stream and what materials are being recycled. For each category, research the possibility of using waste materials in recycled products for the landscape.

REFERENCES

Andrew, R. M., 2018. Global CO2 emissions from cement production, 1928–2017. *Earth Syst. Sci. Data.* 10: 2213–2239. https://www.earth-syst-sci-data.net/10/2213/2018/essd-10-2213-2018.pdf. Retrieved July 31, 2019.

Azek Pavers. https://www.azek.com. Retrieved July 29, 2019.

Bear Creek Lumber. https://www.bearcreeklumber.com/. Retrieved July 29, 2019.

Calkins, M. 2009. *Materials for Sustainable Sites: A Complete Guide to the Evaluation, Selection, and Use of Sustainable Construction Materials.* John Wiley & Sons, Inc., Hoboken, NJ.

Carpenter, J. D. 1976. *Handbook of Landscape Architectural Construction.* The Landscape Architecture Foundation. McLean, VA. 772 pp.

Forest Stewardship Council. https://us.fsc.org/en-us. Retrieved July 29, 2019.

Goetzl, A., P. Ellefson, P. Guillery, G. Dodge, and S. Berg. 2008. Assessment of Lawful Harvesting & Sustainability of US Hardwood Exports. http://www.ahec.org/publications/AHEC%20publications/AHEC_RISK_ASSESSMENT.pdf. Viewed January 1, 2013.

Hanover Architectural Products. https://www.hanoverpavers.com/component/content/article/26-products/155-chesapeake-collection-asphalt-block?Itemid=101. Retrieved July 31, 2019.

Hansen, G. Sustainable landscape construction: materials and products—life cycle assessments. http://edis.ifas.ufl.edu/ep402. Viewed February 2, 2012.

Hornbostel, C. 1991. *Construction Materials: Types, Uses and Applications.* 2nd ed. John Wiley & Sons, Inc., New York, NY, 1023 pp.

Landscape for Life. https://landscapeforlife.org. Retrieved July 29, 2019.

Leopold, A. 1968. *A Sand County Almanac.* Oxford University Press, Oxford, 240 pp.

Marinelli, J. 2009. How green is your garden? *Nat. Wildl.* 47(3): 46–50.

Paloma Pottery. https://www.palomapottery.com/. Retrieved July 29, 2019.

Plastic Lumber Trade Association. http://plasticlumber.org. Retrieved July 29, 2019.

Porous Pave Incorporated. http://www.porouspaveinc.com. Retrieved July 29, 2019.

Rerubber. http://www.rerubber.com. Retrieved July 29, 2019.

Rubber Mulch. https://www.rubbermulch.com/. Retrieved July 29, 2019.

Smith, C., A. Clayden, and N. Dunnett. 2008. *Residential Landscape Sustainability: A Checklist Tool.* Blackwell Publishing Ltd, Oxford, UK.

Standards Worldwide. https://www.astm.org. Retrieved July 29, 2019.

Thompson, J. W., and K. Sorvig. 2000. Sustainable landscape construction. Island Press, Washington, DC, 348 pp.

United States Environmental Protection Agency. Wastes. https://www.epa.gov/environmental-topics/land-waste-and-cleanup-topics. Retrieved July 29, 2019.

Appendix A
Sustainability Audit

OBJECTIVES

Learn about the basic components of a landscape
Observe how landscapes meet sustainability criteria
Assess a landscape for sustainable practices
Develop an appreciation for sustainable practices
Incorporate design aspects and management issues using sustainability goals

TERMS TO KNOW

Sustainability audit

SUSTAINABILITY AUDIT

This chapter introduces the concept of a **sustainability audit**. Now that the material has been covered in detail in the preceding chapters, it may be compiled and organized into the audit. This will serve as a concise reference for incorporating sustainable practices into the landscape.

The audit serves as a detailed gathering of information that resembles the more traditional site analysis that is conducted prior to designing a new landscape project. One purpose of the audit is to aid the landscape professional in identifying features of the landscape that present problems, and to help guide his or her thinking about possible solutions. Some problems and solutions are inherent in the site itself, whereas others have been introduced during previous landscaping efforts.

HOW TO IMPLEMENT THE SUSTAINABILITY AUDIT

An audit consists of a walk through the site accompanied with note-taking on the problems that are identified throughout the process. Following the audit, a thorough review of the problems and potential solutions should be conducted and documented. Two pieces of documentation can be used during the audit walk-through: a checklist and a diagram of the property. Table A.1 is a checklist that can be used as is or modified to allow space for writing comments. The diagram of the property can be as simple as a sketch drawn approximately to scale, all the way to a photocopy of original blueprints of the house or other structures and the entire property

TABLE A.1
Sustainability Audit Checklist

Plants	Comments
Invasive plants	
Turfgrass and other groundcovers	
High-maintenance plants	
Soil properties	
Condition	
pH	
Essential plant nutrients	
Soil texture	
Terrain	
Organic matter (%)	
Sun exposure	
House orientation to the sun	
Heat gain in summer	
Eaves, overhangs, and awnings	
Tree canopy	
Latitude	
Shady areas in the landscape	
Wind	
Prevailing direction	
Proximity of house to other shelters or plants	
Plants for energy efficiency	
Air conditioning unit exposure	
Heat loss	
Water	
Average annual rainfall and other precipitation	
Wet or dry areas in the landscape	
Drainage patterns around the house	
Downspout placement	
Signs of erosion	
Utilities	
Lighting	
Pavement permeability	
Driveway	
Sidewalks	
Patio	
Paths	
Steps	

showing boundary lines. The diagram will be used in a manner similar to a functional drawing used in landscape design projects. Notes will be made directly on the diagram, indicating undesirable plants, hot spots in the landscape, and other problem areas.

Case Study Sustainability Audit: Loehrlein-Green residence

Reference: Author's property, Macomb, IL.

The Loehrlein-Green residential property is located in Macomb, Illinois, situated in the west-central part of the state. What began as a typical backyard approximately 60 ft wide and 60 ft deep, grew into nearly an acre of land when two adjoining properties were purchased. The adjoining properties had small rental houses on them which faced the opposite direction of the main house, and were actually located on a parallel street.

Mowing the entire property usually required two hours on a rider mower. The intention was to remove as much turfgrass as possible, maintaining turf areas for circulation throughout the property. Some turf would be left for front and backyard aesthetics and for seating and outdoor entertainment. The remaining area would be turned into woodland, flower gardens, vegetable plots, and fruit growing. Materials would be retained and reused whenever possible.

A brick sidewalk that led from a back door of one of the rental properties to the end of the backyard was excavated, and all the bricks were reused for pathways and edging throughout the property. Stepping stones were obtained from another property when the owners no longer needed them and wanted to dispose of them. The stepping stones were used at the bottom of some landscape steps that had been installed. Many of the plants had been obtained from other people as divisions or cuttings. Every year, mulched hardwoods are provided by the city forester as part of the city policy, allowing homeowners to request mulch during regular tree maintenance activities. A rain garden was established adjacent to one of the rental units, using plants that were already growing in other gardens on the property. Drainage materials under the rain garden were mostly rubble from a decommissioned chimney from one of the rental houses. Fencing on one property line was obtained from discarded materials from another project.

Initially, a 1:2 slope approximately 100 ft long wrapped around the backyard area. It was completely covered in turf, making mowing somewhat hazardous. Two sets of steps were installed. The steps on the south side of the backyard provided easier access between the back door of the house and the driveway. The steps on the east side of the backyard provided easier access to the vegetable garden and rental properties. Groundcover was planted between the two sets of steps, leaving only enough turf for mower access to the area immediately surrounding the back patio area. Mowing was reduced from 1 hour to less about 15 minutes for this area.

Between the residence and the rental homes, approximately 13,200 ft^2 of turf was removed to install a woodland containing native maples, red twig dogwood, and herbaceous groundcover. This woodland added to the riparian habitat along the creek on the south side of the property. On another 8,400 ft^2 of turf area in the backyard of a rental property, approximately 6,000 ft^2, was planted to fruiting shrubs, vines, and trees, including grapes, raspberries, and apples. The vegetable garden was expanded as well.

ASSESSMENTS INCLUDED IN THE SUSTAINABILITY AUDIT

The assessments included in the sustainability audit follow the topics provided in the chapters of this book. The main topic areas are the plantscape, sun, wind, water issues, soil health and organic matter, pesticides, tools and equipment, energy usage, and materials.

The Plantscape

Look at the overall plantscape: trees, shrubs, borders and beds, and turf areas. Note issues with weeds, slopes, and proper plant placement, and identify areas that remain wet or dry during the growing season. Check for "right plant, right place" with respect to moisture, sun, wind, and mature plant size.

High-Maintenance Vegetation

Look for maintenance issues: overgrown trees and shrubs; poorly structured woody plants; neglect suffered from failure to prune regularly; and decline due to stresses in the landscape (pests, drought, sun, wind, shade).

Identify invasive plants that may be present, including weed problems.

Make an inventory of problematic plants and planting situations and indicate them on the diagram. Indicate the invasive species and high maintenance plants.

Make an inventory of problematic plants and planting situations, and indicate them on the diagram.

Turfgrass and Other Groundcovers

Assess whether turf areas could be converted into shrub borders or into a habitat area such as a woodland, prairie planting, meadow, wetland, or other similar planting.

Identify weed issues. Note whether shaded areas have healthy turf growing in them.

On the diagram, indicate difficult spots for growing turf, such as under shade trees or in the shade of buildings or fences, narrow strips between pavement, and slopes that are too steep to safely mow.

Sun Exposure

A house's exposure to the sun can be influenced by the direction the house faces as well as the amount of shade provided by trees or nearby structures.

On the diagram, indicate the four directions, and note which side of the house faces west. This will be the hottest place in the warmer months, unless shade is provided.

Indicate other hot-spots. These would include areas that are not shaded by trees or structures at all throughout the day. It would also include paved areas that are not shaded, and windows that receive direct sun during the day.

Evaluate the amount of shade received at windows facing south, east, and west.

Evaluate the shade for larger paved areas, noting whether the surfaces are dark or light.

Make note of walls or other surfaces than may absorb a lot of heat in the summer.

Note shady areas in the landscape.

Appendix A: Sustainability Audit

WIND

Take note of exposed north- and west-facing sides of the house.
  Indicate whether there is enough space for a wind screen, if appropriate.
  Identify the location of the air conditioning unit.
  Look for plants that provide dead air space near the house.

WATER ISSUES

Identify the drainage pattern on the property. Note whether there could be issues with fertilizer or pesticide runoff.
  If you are in a low-rainfall region, identify plants that require excessive amounts of water.
  If an irrigation system exists, check for leaks and that the system functions properly.
  Indicate the location of downspouts.
  Calculate the roof area and estimate the amount of rain that could be harvested.

Paved Areas

Note locations of paving, both permeable and impermeable.
  Calculate areas with impermeable paving, and estimate the amount of rain that could be harvested.

SOIL HEALTH

Take note of any areas in danger of erosion or where erosion is already occurring.
  Conduct soil tests for pH, nutrient levels, organic matter, and texture.
  Provide fertilizer recommendations based on the type of plant grown: turf, tree, or ornamental beds.
  Indicate whether proper mulching practices are being observed and whether more or less mulch is required.

PESTICIDES AND PEST MANAGEMENT

Examine the pesticide storage area.
  Interview client about pesticide usage practices.

EQUIPMENT

Inventory the landscape equipment and vehicles, and identify sustainable replacement options in both equipment and fuels.
  Indicate if hand-labor may be better suited in various areas of the landscape, such as leaf and debris removal, and pruning.

ENERGY CONSUMPTION

Evaluate building energy usage.

Check to see if there is unnecessary electrical usage, such as for lights.
Look for energy-efficient lighting.
Evaluate whether computers are used in power-saving modes during breaks, nights, and weekends.
Assess whether job vehicles are used in an efficient manner.

SUSTAINABLE LANDSCAPE MATERIALS

Indicate whether materials used are appropriate for the site.
Indicate if replacement materials are required and suggest sustainable materials.

COMPLETED CHECKLIST AND DRAWING

When all areas have been evaluated, the checklist should be complete (Table A.2 shows a completed checklist). The final step is to prepare an evaluation to accompany the list. This evaluation will provide observational notes, suggestions for improvement, and other ideas. The diagram of the property should be included, with indications of changes or trouble spots in the landscape. The landscape manager should then meet with the client and review the audit and decide upon a course of action. During this process, both the landscape manager and the client will gain a full appreciation for the strengths and weaknesses of the landscape, as well as the opportunities for improvement.

TABLE A.2
Sustainability Audit Checklist Completed

Plants	Comments
Invasive plants	Poison ivy, English ivy
Turfgrass and other groundcovers	10,784 ft^2
High-maintenance plants	No
Soil properties	
Condition	Compaction in backyard due to construction activities
pH	6.8
Essential plant nutrients	High soil fertility
Soil texture	Clay loam
Terrain	Steep slope in backyard
Organic matter (%)	2.5
Sun exposure	
House orientation to the sun	House faces west; large windows on west side
Heat gain in summer	Yes, in the afternoon
Eaves, overhangs, and awnings	Front porch shades west windows somewhat
Tree canopy	Shade trees on east and west sides
Latitude	38°
Shady areas in the landscape	Various

(Continued)

Appendix A: Sustainability Audit

TABLE A.2 (*Continued*)
Sustainability Audit Checklist Completed

Plants	Comments
Wind	
Prevailing direction	North
Proximity of house to other shelters or plants	Neighboring house on north side
Plants for energy efficiency	Shade trees in summer
Air conditioning unit exposure	Protected on two sides
Heat loss	Minimal
Water	
Average annual rainfall and other precipitation	35–40 inches
Wet or dry areas in the landscape	Under eaves – dry
Drainage patterns around the house	Good
Downspout placement	Good
Signs of erosion	Along creek
Utilities	
Lighting	None
Pavement permeability	
Driveway	Concrete – impermeable
Sidewalks	Concrete – impermeable
Patio	Concrete – impermeable
Paths	None
Steps	Concrete – impermeable

THE NEXT STEP: FINDING SOLUTIONS

For each of the problem areas discussed in this chapter, there are suggestions for improvement. Individual chapters provide more details in this area. Undesirable plants may require removal, areas of high heat gain may require shading, and windy areas that suffer heat loss in winter may require a wind screen. Potential conflicts may arise in achieving one's goals. For example, turf is the best groundcover for certain uses, such as entertainment, yet it uses non-renewable resources and contributes to greenhouse gas emissions. Such conflicts should be evaluated and discussed with the client in order to prioritize the options that are available. There will undoubtedly be compromises during the process of achieving a sustainable landscape. However, the best choices can be made when all possible options are examined and ultimately one is chosen. As technology and ideas continue to develop and emerge, better, more viable options may become available.

REVIEW QUESTION

1. Discuss ways in which the sustainability audit resembles and differs from a typical site analysis.

PROJECT IDEA

1. Conduct a sustainability audit on a residence. If possible, meet with the owners and discuss management practices and options for improvement.

FURTHER READING

Smith, C., N. Dunnett, and A. Clayden. 2007. *Residential Landscape Sustainability: A Checklist Tool.* Wiley-Blackwell, Hoboken, NJ, 208 pp.

United States Environmental Protection Agency. Greenscapes. http://www.epa.gov/greenscapes/. Viewed July 18, 2008.

University of Minnesota. Sustainable Urban Landscaping Information Service. http://www.sustland.umn.edu/. Viewed July 18, 2008.

van Mansvelt, J. D., and M. J. van der Lubbe. 1999. *Checklist for Sustainable Landscape Management.* Elsevier Science, Amsterdam, 202 pp.

Appendix B
Important Websites Used as Resources in this Book

Encyclopedia of Earth. https://www.iucn.org/content/explore-encyclopedia-earth
Lawrence Berkeley Laboratory. https://www.lbl.gov/
NASA Earth Observatory. https://earthobservatory.nasa.gov/
National Atmospheric and Oceanic Administration. https://www.noaa.gov/
National Center for Appropriate Technology. https://www.ncat.org/
Sustainable Landscaping. Colorado State University. https://extension.colostate.edu/topic-areas/yard-garden/sustainable-landscaping-7-243/
Sustainable Sites Initiative. http://www.sustainablesites.org/
USDA Forest Service. https://www.fs.usda.gov/
USDA Natural Resources Conservation Service. https://www.nrcs.usda.gov/wps/portal/nrcs/site/national/home/
US Department of Energy, Energy Information Administration. https://www.eia.gov/
US Environmental Protection Agency. https://www.epa.gov/
US Geological Survey National Water Quality Assessment Program. https://www.usgs.gov/mission-areas/water-resources/science/national-water-quality-assessment-nawqa?qt-science_center_objects=0#qt-science_center_objects
US Geological Survey Publications. https://pubs.er.usgs.gov/

Index

above-ground water storage 112–113
acetylcholinesterase inhibitors 211
acquisition practices, for sustainability 279
action thresholds, IPM 236, 241
active ingredient, in pesticide 208
acute effects, of pesticide 214
adjuvant 219
aeration 21
aesthetic injury level (AIL) 237
air pollution 8, 14
  from landscape tools and equipment 264–265
air quality 8–9
Akbari, H. 63
albedo 56
Allen, George 103
aluminum sulfate 174
amendments 192
American Forest and Paper Association (AF&PA) 285, 287
American Rainwater Catchment Systems Association (ARCSA) 113
American Society for Testing and Materials (ASTM) 280, 286
American Society of Landscape Architects (ASLA) 2
animal manure 179–182
anthracite coal 250
anthropogenic climate change 1
anthropogenic heat 52
AquaBasin™ water reservoir 114
aquascape 114
aquifers 88
Arizona, three-tier system 122, 126
asphalt 282
Audubon Cooperative Sanctuary Program 101
Audubon Society 45
Augustus F. Hawkins Natural Park 153
Aurora Municipal Xeriscape Garden 118
avoidance, IPM 227
Azek Pavers 280

Bacillus thuringiensis 237–238
Backyard Habitat Program 45
bactericides 205
Ballard Public Library Seattle Washington 142–143
battery-operated lawn mowers 268
battery-powered tools 264
below-ground water storage 113–114
benzene 265
biodiesel 256–257

biofuels 256–257
biogenic gas 249
biological controls, IPM 227, 237–238
  botanicals 239, 240
  insect growth regulators 240–241
  non-toxic pesticides 239–240
biomass 245
biomes 47
bioremediation 98, 99, 106
bioretention 136
birds and mammals 39–45
bituminous coal 250–251
black water 121
Blickle, Alayne 182
Bloomberg, Michael 248
"blue baby syndrome" 90
bone meal 179, 185, 196
Bonita Bay Club East 46–47
Bordeaux mixture 239
botanicals 237, 239, 240
branch beating 234
Brazilian pepper trees 25–26
Brittin, C. L. 193
brownfields 152–153, 167

calcium 174
calcium magnesium acetate (CMA) 160
California Air Resources Board (CARB) 245–246, 265
California Environmental Protection Agency 247
Camelot 239
Carballo, T. 184
carbamates 209
carbaryl 95
carbon 198
carbon cycle 18, 19
carbon dioxide 18
carbon emissions 8, 14
carbon footprint 8
carbon monoxide 265
carbon sequestration 17, 18–19
carcinogen 211, 214, 265
Carson, Rachel 97, 209, 210
cation exchange capacity (CEC) 158–161
cations 158, 159
caution 172, 239
CEC *see* cation exchange capacity (CEC)
cellulosic ethanol 256
cement 282
Cement Sustainability Initiative 282
chain of custody process 284

299

Chalker-Scott, L. 120, 179
chelates 180
chlorinated hydrocarbons 208–209
cholinesterase inhibitors 216–218
chronic effects, of pesticide 214
clay soils 158
Clean Air Act 9, 262
Clean Water Act 134, 183
climate change
    anthropogenic 1
    global 1, 6–8, 14
CMA *see* calcium magnesium acetate (CMA)
coal 250–251
coal-fired power plant emissions 264
Coincide (Orton and Green) 236
cold air 83
    trapping on slope 75
combined sewer overflow (CSO) 133
compaction 151, 157–158, 165
    cultural practices 229
compost
    carbon and nitrogen 198
    compost bin system 197
    ingredients 197
    moisture 199
    proper conditions 198
    turning 199
compost bin system 197
composted materials 166
compost extract 184
composting 166
compost solutions 184–185, 199
compost tea 184–185
concentrated solar power (CSP) 254
concrete 282
constructed wetlands 98–99, 106
    built components of 104–105
    design of 101–103
    natural components of 103–105
    uses for 100–101
contaminated water, prevention and treatment 97
convection 59
Cook, Tom 185
cooling effect
    of plants 19–20, 62
        shade 62–64
        transpiration 62
    of wind 75–77
cooling paved surfaces 65–67
Cooperative Sanctuary Program 46
core aeration 157, 165
cowsmo brand compost 182
Crane, D.E. 19
Crassulacean Acid Metabolism (CAM) 116
crumb rubber 281–282
CSO *see* combined sewer overflow (CSO)

cultural practices
    irrigation/watering 228–229
    sanitation 228
    soil health and compaction 229
cypress mulch 194

Daniels® Plant Food 185
dead air space 75, 76, 83
decibel (dB) 266
decomposers 162
deep tillage 165
degree days, IPM 234–236, 241
deicers 159–160
demolition practices 279
desert ecosystems 34–35, 37
diatomaceous earth 239
diazinon 95
dichloro-diphenyl-trichloroethane (DDT) 210–211
dieldrin 95
diesel-electric hybrid trucks 257
diflubenzuron (Dimilin) 241
direct current (DC) 254
disposal practices 279
drainage 136, 145
drip line 120, 195
drought 110–111
drought-tolerant plants 116, 126
drought-tolerant turf 20–21

earthworms 161–162
ecological landscaping 37–38
Ecology and Design (Johnson and Hill) 211
economic thresholds, IPM 236–237
ecosystem services 5, 47
ectomycorrhizae 162, 163
effluent irrigation 123
effluent system 88, 92
electric blowers 267, 272
electricity 248
    fuel for tools, equipment, and transportation 254–257
    generation of 248, 249
        non-renewable fossil fuel energy 249–251
        renewable energy 252–254
    US sources of 249
electricity-generating power plants 245
electric lawn mowers 268
electric-powered landscape tools 264
electric-powered tools and equipment 271
electric vehicles 257
electromagnetic spectrum 52
electronic temperature monitoring devices 235–236
El Nino 7
emergent species 104
emissions, energy 245, 247
emissivity 56

Index

emittance 56
endocrine disruptors 93, 211, 215–216
endocrine systems 215
endomycorrhizae 162, 163
energy 244, 258
    for electricity 248
        fuel for tools, equipment, and transportation 254–257
        generation of 248–254
    energy-efficient lighting 258
    government support for renewable energy 247–248
    scope of problem 245, 247
        emissions 245, 247
        expense 245, 246
        non-renewable resource 245
    sources of 245
    usage in winter 78–79
energy consumption 293–294
energy-efficient lighting 258
Energy Policy Act 259
Energy Star 258
Energy Tax Act 247
engines, types of 263–264
English ivy 25
enhanced oil recovery (EOR) 250
enleanment additive 268
environmental issues
    air quality 8–9
    carbon emissions 8
    global climate change 6–8
    pesticides 218–219, 222
        usage and toxicity 9–10
    waste stream 10
    water issues 9
environment aspects, of plants 18
    carbon sequestration 18–19
    cooling effect 19–20
    oxygen release 19
    structural effects 20
EOR *see* enhanced oil recovery (EOR)
EPA *see* US Environmental Protection Agency (EPA)
epidemiology 212–214
erythrocytes 211
essential nutrient 158
ethanol 256
eutrophication 90
evapotranspiration (ET) 17, 47, 60, 62, 112, 125–126
Evatech 268–269
excessive nutrients, in polluted water 90–92, 105
excess water management 131–133, 145–146
    drainage 136
    green roofs 141–144
    green walls 144
    landscape swales 136

    permeable pavement materials 140–141
    rain gardens 136–139
    rainwater collection 139–140
    stormwater run-off 133–135
    urban water cycle 135
exhaust-gas recirculation (EGR) 268
expense, energy 245, 246
extensive green roofs 143

fabrication practices 279
Federal food, Drug, and Cosmetics Act (FFDCA) 209–211
Federal Insecticide, Fungicide, and Rodenticide Act (FIFRA) 209, 216, 221
fenoxycarb (Precision) 241
fertilizer analysis 172
fertilizer contamination 178, 186
fertilizers 171–173, 186
    forms of 175, 176
    sources of 175–176
        compost solutions 184–185
        green manure and inter-planting 185–186
        mineral fertilizers 176–179
        organic fertilizers 179–183, 185
fire-wise landscaping 45
fission 251
Food Quality Protection Act (FQPA) 216, 221
Forest Management Act 283
Forest Management and Chain of Custody 283
forests and woodlands landscapes 27–32
Forest Stewardship Council (FSC) 283, 287
formulation 219
fossil fuels 245, 249, 250, 258
four-stroke engines 263–265
fracking 249, 258
French drains 136
fuel-efficient business practices 270
fuel prices 244
fugitive dust 266
fulvic acid 191, 192, 200
fungicides 9, 205

Garry, Vincent 214
gasohol 247, 258
gasoline-powered engines 8, 14
gas-powered engines 265, 271
Gearheart, Robert 103
genetically improved plants 229–231
geothermal systems 245, 259
Givoni, B. 63
glass 282
global climate change 1, 6–8, 14
GOAT 268–269
Golf Course Habitat 45–47
grain-based ethanol 256
granule fertilizers 172
grass clippings 199

gray water 121–122, 126
    timing of 122–125
Greater Earth Organics 184
greenhouse effect 7
greenhouse gases 1, 2, 7
Green Infrastructure Development program 136–137
Green Infrastructure Research Program 134
green manure 185–186
green materials 194
green roofs 65, 141–142, 146–147
    Ballard Public Library Seattle Washington 142–143
    design 143–144
    and media depth 143
    plants for 144, 145
green spaces 78
Green, Tom 236
green walls 144
Griffin Industries LLC 185
groundcovers 292
groundwater 88, 105
guard cells 111

habitat development programs 45–47
halofenzamide (MACH-2) 241
Hansen, Gail 278
hardwoods 277
heat capacity 57
heat gain 52, 56, 68
    in summer 67
    in winter 68
heat island effect 52
heat loss 59
heat transfer 59
Heilman, J. L. 193
Heisler, Gordon 63
heliostats 254
herbicides 205
    polluted water 96–97
high-maintenance vegetation 292
Hill, K. 211
hormones 215
hot pepper wax 239
human health hazards 211–213
    acute and chronic effects 214
    carcinogens 214
    endocrine disrupters 215–216
    epidemiology 212, 214
    teratogenic effects 215
Humboldt Bay, Arcata, California 102–103
humic acid 191
humic substances 192, 200
humin 191, 200
humus 29, 161, 191, 200
hydrocarbon poisoning 218
hydrocarbons 263

hydroelectric power 245, 252
hydrogen power 257–258
hydrologic cycle *see* water cycle
hydrology 100, 103
hyphae 162, 163
hypoxia 90

IGRs *see* insect growth regulators (IGRs)
illicit discharge 135
infiltration 132, 157
infrared 52, 53
inorganic fertilizers, effect on soil health 178
insect growth regulators (IGRs) 240–241
insecticides 205
    polluted water 95–96
integrated pest management (IPM) 98, 209, 226–227, 241
    action thresholds 236
    aesthetic injury level 237
    alternative pest controls 237, 238
    avoidance 227
    biological controls 237–238
        botanicals 239, 240
        insect growth regulators 240–241
        non-toxic pesticides 239–240
    cultural practices 227–228
        irrigation/watering 228–229
        sanitation 228
        soil health and compaction 229
    economic thresholds 236–237
    genetically improved plants 229–231
    phenology and degree days 234–236
    pilot program *vs.* traditional spray program 227
    treatment 227, 230–231
        pest/disease presence determination 231–234
intensive green roofs 143
inter-planting 185–186
invasive plants 23–26
iron 175
iron chlorosis 175
Irrigation Association 115
irrigation system 114–115, 126, 228–229

jar test 155
Johnson, B. R. 211

Kavanaugh, J. 265
KBI Flexi-Pave 281
Kelling, K. A. 171–172
kinoprene (Enstar II) 241
Koehler, C. S. 237
Kuiper High Plains Garden 118

labels 210
"The Land Ethic" 283

Index

landscape equipment 261
landscape fabrics 119, 120
landscape irrigation, gray water for 122
landscape products, recycled materials for 278, 286
landscape swales 136
landscape tools and equipment
   air pollution from 264–265
   fugitive dust and particulate matter 266
   noise from 266–267
   solutions 267
      design phase 270–271
      emissions reduction 268
      fuel-efficient business practices 270
      maintenance 271
      noise reduction 269
      sidewalk vacuum 269
      technological advances 268–269
      transportation efficiencies 269–270
      usage reduction 269
   sustainability issues concerning 264–267
landscaping 5–10, 14
   construction materials 276–278
   and cooling 64
   ecological 37–38
   excess water management in 131–133, 145–146
   fire-wise 45
   organic matter in 190
      types of 192–196
   pesticides
      reduction in 97–98
      use in 205–208, 221
   power tools used in 262–263
   soil health for 164–165
      amending with organic matter 166
      composting 166
      correcting compaction 165
      preserving and replacing topsoil 165
      reducing subsoil at surface 165
      soil quantity for root growth 166
      using mulch 166–167
   sun effects on 52–54
landscaping industry, role in sustainability 11–12
La Nina 7
large-solar scale projects 254
lawn aeration 21
Lawrence Berkeley National Laboratory 53
LCA *see* life cycle assessment (LCA)
LD_{50} 211, 221
leaching 178–179
Leadership in Energy and Environmental Design (LEED) certification 2, 283
leaf scald 74
leeward 79
Leopold, Aldo 283
Lerman, S.B. 38

life cycle assessment (LCA) 278–279
light-emitting diodes (LEDs) 244, 258
light, reflecting and absorbing 56–58
lignite 250–251
liquefied petroleum gas (LPG) 255
living organisms, in soil
   decomposers 162
   earthworms 161–162
   nematodes 162
   symbionts 162–164
local materials 285–287
Loehrlein-Green residence 291
lumens 258

macronutrients 164, 173, 174, 186
Madder, D. J. 219
magnesium 174
Mahomet aquifer 89
manual push mowers 268
manual tools 261–262
Manufacturer's Safety Data Sheets (MSDS) 210
McEwen, F. L. 219
McPherson, E.G. 63
meadows 30, 32
methyl tertiary butyl ether (MTBE) 256, 268
microclimates 81
micronutrients 164, 173, 180, 183
microsurfacing 66
Milorganite 183
mineral fertilizers 176, 177
   contamination of environment 178
   inorganic fertilizers effect on soil health 178
   nitrogen 176
   nutrient run-off and leaching 178–179
   phosphorus 176, 178
   potassium 178
mineral soils 154
modern green roof technology 141–142
monitored, landscape plants 231
Moore, W. S. 237
motion-activated lights 258
mowing 21–22
MSDS *see* Manufacturer's Safety Data Sheets (MSDS)
mulch 119–120, 126, 192–193, 200, 277
   applying of 195–196
   on crape myrtle 193–194
   materials of 194–195
   problems with 196
   soil health using 166–167
Municipal Separate Storm Sewer System (MS4) 135
municipal solid waste 182–183
   landfills in United States 286
   stream, generation and recovery of 275, 276
municipal solid-waste compost (MSWC) 184–185
municipal water supplies 105, 125

mutagens 211
mutualists 161
mycelia 162
mycorrhizae 162–163

National Energy Act 247, 259
National Environmental Policy Act (NEPA) 209
National Forestry Service 283
National Pollutant Discharge Elimination System (NPDES) 135
National Wildlife Federation (NWF) 45, 47, 48
native plant selection 37–38
native pollinators 38–41
natural gas 249
natural gas vehicles 255–256
naturally occurring pesticides 209
neem extract 239
nematodes 162
NEPA *see* National Environmental Policy Act (NEPA)
Nira Rock 44–45
Niyogi, Dev 59
nitrogen 164, 173, 174, 176, 178, 198
nitrogen fixation 163, 172
nitrogen-fixing bacteria 163–164
nitrogen-fixing legumes 185
noise
    from landscape tools and equipment 266–267
    reduction of 269
non-Hodgkins lymphoma (NHL) 214
non-native plant selection 37–38
non-point-source pollution 90
nonpoint sources, urban runoff 134
non-renewable fossil fuel energy 249
    coal 250–251
    fuel for tools, equipment, and transportation 254–255
    natural gas 249
    nuclear 251
    oil 250
non-renewable resources 1, 2, 10, 11, 14, 245
non-synthetic pesticides 209
non-toxic pesticides 239–240
Nowack, D.J. 19
NPDES *see* National Pollutant Discharge Elimination System (NPDES)
nuclear energy 245, 251
nuclear fission 251
nuclear power plants 251, 258
nutrient cycling 158, 166
nutrient run-off 178–179
nutrients 171, 186, 190

off-road use of fuels 254
Ogallala aquifer 89
oil 250
Oke, T.R. 82

on-road use of fuels 254
on-site stormwater management 9
open-grid paving 66
organic amendments 192, 193, 200
organic fertilizers 179, 180, 186
    animal manure 179–182
    municipal solid waste 182–183
organic matter 166, 189–190, 199–200
    fate of 191
    grass clippings 199
    in landscape 190
        types of 192–196
    organic soil amendments 196
        compost 197–199
        compost solutions 199
        peat moss 197
    and soil health 191–192
organic mulches 167
organic phosphates 209
organic soil amendment 181, 196
    compost 197–199
    compost solutions 199
    peat moss 197
organic soils 154
organochlorides 208–209
organochlorine insecticides 92, 97
organocide controls 239
organophosphates 209
ornamental plants 163–164, 229
    in landscape setting 237
Orton, Don 236
oxygenator 268
oxygen release 19

Pacific Landscape Company 11–12
PAHs *see* polycyclic aromatic hydrocarbons (PAHs)
Parker, J. H. 64
particulate matter 263, 266
Paspalum grass *(Paspalum vaginatum)* 21
paved areas 293
peat moss 197
Percival, G. C. 120
Percolation test 101, 157
permeable pavement materials 140–141, 146
personal protective equipment (PPE) 220, 222
pest/disease presence determination 231–232
    branch beating 234
    pheromone traps 233
    sticky traps 232–233
pest emergence patterns 231
pesticide applicators 220
pesticides 9–10, 14, 204, 221–222, 293
    cholinesterase inhibitors 216–218
    environmental hazards 218–219
    health effects 218
    human health hazards 211–213

Index

acute and chronic effects 214
carcinogens 214
endocrine disrupters 215–216
epidemiology 212, 214
teratogenic effects 215
pesticide regulation 209
FFDCA 210–211
FIFRA 210
restricted use pesticides 211
in polluted water 92–93
human and environmental effects of 93–95
reduction in landscape 97–98
safety issues 219–220
seasonal patterns of 93
storage and disposal 220
types of 208–209
use in landscape 205–208
use, overuse, and mis-use of 222
pest management 293
Peterson, A. E. 171–172
PGRs *see* plant growth regulators (PGRs)
phenology, IPM 234–236
pheromone traps 233
pheromones 233
phosphate 174
phosphorus 173–174, 176, 178
photosynthesis 19
photovoltaic panels 254
photovoltaics (PV) 248
physiological time 234
phytoremediation 98, 99, 106
plant 125–126
essential nutrients 164
fertilizer requirements 173
calcium and magnesium 174
iron 175
nitrogen 173, 174
phosphorus and potassium 173–174
sulfur 174
for green roofs 144, 145
pests of 226
for rain gardens 138–139
safety concern, gray water for 123–125
soil quantity for root growth 166
water requirements of 111–112
in wetland 104
for windbreak 80–81
plant-derived oils 239
plant growth regulators (PGRs) 205
planting, for insulative properties 75, 76
plantscape 17–18, 292
attracting wildlife 38
birds and mammals 39–45
native pollinators 38–41
ecological landscaping 37–38
ecology and plants 26–27
ecoregions of United States 27–37

environment aspects of plants 18
carbon sequestration 18–19
cooling effect 19–20
oxygen release 19
structural effects 20
fire-wise landscaping 45
invasive plants 23–26
programs for habitat development 45–47
turfgrass 20–22
woody plants 22, 23
plastic-coated fertilizers 179
plastic lumber 277, 280
plastic mulches 120
plug-in electric mowers 268
point-source pollution 90
point sources, urban runoff 134
pollinators 47
polluted water 90
contributions from urban areas 93, 95, 96
excessive nutrients in 90–92
herbicides 96–97
human and environmental effects of pesticides in 93–95
insecticides 95–96
pesticides in 92–93
polycyclic aromatic hydrocarbons 97
polycyclic aromatic hydrocarbons (PAHs) 97, 133
Pomerantz, M. 60
Portland, Oregon 44
potable water 90, 105
potassium 173–174, 178
power tools 261, 271
usage reduction 269, 272
used in landscape 262–263
PPE *see* personal protective equipment (PPE)
prairie ecosystem 29–30, 33–35
precipitation 110
prestrike (S-Methoprene) 241
primary decomposers 162
primary recovery technique 250
propane 254, 255
pruning 22
Puoyat, R.V. 18
pyrethrum 239

quick-draining soils 137

radiative properties 56
rain barrels 112
rainfall 110
amounts calculation 114
rain gardens 136–137
area calculation for 137–138
building of 138
plants for 138–139
siting of 137

rainwater collection systems 112, 139–140
    above-ground water storage 112–113
    below-ground water storage 113–114
rainwater harvesting systems 145
Raupp, M. J. 237
recycled materials 280
    concrete and asphalt 282
    glass 282
    plastic lumber 280
    rubber 280–282
recycling 275
renewable energy 252
    fuel for tools, equipment, and transportation 256–257
    government support for 247–248
    hydroelectric 252
    solar 253–254
    wind 252–253
Renewable Fuel Standard program 247
renewable resources 1, 2, 10, 11, 14, 248, 283
    sustainably harvested lumber 283–285
restricted use pesticides (RUPs) 211
ribbon test 155, 156
riparian zone habitat 32, 34, 36
Robinette, G.O. 82
Rowntree, R.A. 63
rubber 280–282
runoff 90
Ruppert Landscape Company 12

Safe Drinking Water Act (SDWA) 216, 221
salinity 159–161
salt-tolerant turf 21
salvaged materials 285, 287
Sandifer, S. 63
sanitation 228
Schaffer, Matt 173
Scott, K. 63
secondary recovery technique 250
sediment runoff rates 151
sewage sludge 183
sewage treatment, in United States 182
shade 62–64
shade-loving plants 228
shading air conditioners 65, 69
shelterbelt *see* windbreak
Shipchandler, Riyaz 265
sidewalk vacuum 269
Silent Spring (Carson) 209, 210
Site Assessment and Environmental Planning Form 46
SITES rating system 277
slope, cold air trapping on 75
slow-release nitrogen fertilizers 179
sodium chloride 160
softwoods 277
soil chemistry, problems with 160–161

soil disturbance 157
soil-dwelling nematodes 162
soil health 149–151, 293
    brownfields 152–153
    cultural practices 229
    essential plant nutrients 164
    inorganic fertilizers effect on 178
    for landscaping 164–165
        amending with organic matter 166
        composting 166
        correcting compaction 165
        preserving and replacing topsoil 165
        reducing subsoil at surface 165
        soil quantity for root growth 166
        using mulch 166–167
    organic matter and 191–192
    problems with soil chemistry 160–161
    soil organic matter 161–164
    soils and construction activities 151–152
    sustainable fertilization 167
    testing 153–160
soil organic matter 161
    living organisms in
        decomposers 162
        earthworms 161–162
        nematodes 162
        symbionts 162–164
soil pH 158, 159
soil pollution, pesticides 218
soil profile 150
soil structure 156–158
soil testing 153–154, 173
    chemical properties of
        cation exchange capacity 158–160
        salinity and deicers 159–160
        soil pH 158, 159
    physical properties of
        soil structure 156–158
        soil texture 154–156
soil texture 154–156
solar energy 253–254
    electromagnetic spectrum 52
    heat capacity 57
    heat loss and heat transfer 59
    infrared 52, 53
    intensities of 52, 53
    reflecting and absorbing light 56–58
    solar heat gain 56
    Solar Reflectance Index 58, 59
    thermal emissivity 56–57
solar heat gain 56
solar-powered landscape lighting 258
solar radiation 52, 68
solar reflectance 56
Solar Reflectance Index (SRI) 58, 59
Solid Waste Association of North America 286
solid-waste material 183

Index

Spokane Valley-Rathdrum Prairie aquifer 89
spreader 219
Stapleton Xeriscape garden 118
Steingraber, Sandra 211
sticky traps 232–233
Stinson, J. M. 120
stomata 111
stormwater management 134
stormwater runoff 133–135
structural effects, of plants 20
subsoil, reduction of 165
substrates, in wetland 104
sulfur 174
sun
    effects on landscape 52–54
    exposure of 67
    heat gain
        increasing in winter 68
        reduction in summer 67
    solar energy
        electromagnetic spectrum 52
        heat capacity 57
        heat loss and heat transfer 59
        infrared 52, 53
        intensities of 52, 53
        reflecting and absorbing light 56–58
        solar heat gain 56
        Solar Reflectance Index 58, 59
        thermal emissivity 56–57
    solar radiance and heat 52
    structure orientation 67
    urban heat island 59–60
        cooling paved surfaces 65–67
        infrared waves 60–61
        landscaping practices to mitigate 61–65
sun exposure 292
surface reflectivity 56
surface water 88, 105
sustainability 2, 13
    landscaping industry role in 11–12
sustainability audit 13, 289
    assessments in 292
        energy consumption 293–294
        equipment 293
        pesticides and pest management 293
        plantscape 292
        soil health 293
        sun exposure 292
        sustainable landscape materials 294
        water issues 293
        wind 293
    checklist and drawing 294, 295
    finding solutions 295
    implementation of 289–290
    Loehrlein-Green residence 291
sustainability goals 284

sustainable fertilization 167, 171–172, 186
    fertilizers 172–173
        forms of 175, 176
        sources of 175–186
    plant fertilizer requirements 173–175
sustainable landscape materials 294
sustainable landscaping 1–2, 286–287
    construction materials 276–278
    environmental issues and landscaping 5–6
        air quality 8–9
        carbon emissions 8
        global climate change 6–8
        pesticide use and toxicity 9–10
        waste stream 10
        water issues 9
    history and background 2–4
    life cycle assessment 278–279
    local materials 285–286
    non-renewable and renewable resources 10, 11
    recycled materials 280
        concrete and asphalt 282
        glass 282
        plastic lumber 280
        rubber 280–282
    recycled materials for landscape products 278
    renewable resources 283
        sustainably harvested lumber 283–285
    salvaged materials 285
    sustainability audit 13
    Sustainable Sites Initiative 2, 4–5
    waste management 279
sustainable landscaping practices 204
Sustainable Sites Initiative (SITES™)
    excess water management, in landscape 135
    local materials 285–286
    pesticides 204
    soil health 151, 167
    sun 52
    sustainable landscaping 2, 4–5, 14
    water conservation 110, 125
    wind and energy conservation 74
sustainable sites initiative (SITES) rating system 264
sustainably harvested lumber 283–285, 287
Suutari, Amanda 103
swale 136
Swamp Lands Act (1849) 99
symbionts 162–164
symbiotic nitrogen-fixing bacteria 163–164
synthetic pesticides 208–209, 221

tebufenozide (Confirm) 241
temperate coniferous forests 27–30
temperate deciduous forests 29, 31–32
teratogenic effects, of pesticide 211, 215
tertiary recovery technique 250

thermal conductivity 59
thermal emissivity 56–57
thermal emittance 56
thermal imaging 54
thermal properties 56
thermogenic gas 249
thermography 54
three-tier system 122, 126
Tier IV emission standards 265
topsoil 150, 151
    preserving and replacing 165
total soluble salts (TSS) 181
toxic fertilizers 172–173, 186
toxicity 9–10
Toxic Substances Control Act 216
traditional landscape construction materials 286
transpiration 62, 74, 83, 111
trap 232
    branch beating 234
    pheromone 233
    sticky 232–233
treatment, IPM 230–231
    pest/disease presence determination 231–232
    branch beating 234
    pheromone traps 233
    sticky traps 232–233
tree funnels 75, 76, 83
Tres Rios project, Phoenix Arizona 102
TREX 269
TSS *see* total soluble salts (TSS)
Turf & Ornamental Greenbook 204
turfgrass 20–22, 47, 173, 186, 292
turgor 111
two-stroke engines 263–265

understory plants 27–29, 31–32
United States 47
    ecoregions of 27, 28
        desert ecosystems 34–35, 37
        forests and woodlands 27–32
        meadows 30, 32
        prairies 29–30, 33–35
        riparian zone habitat 32, 34, 36
    hydroelectric power in 252
    municipal solid waste landfills in 286
    peat moss in 197
    pesticides use in 205
    sewage treatment in 182
    wind energy in 252–253
United States Geological Survey (USGS) 91
Uranium[235] 251
urban areas, polluted water contributions from 93, 95, 96
urban boundary layers 81
urban canopy layers 81
urban canyons 74, 81

urban heat island 52, 53–54, 59–60, 68
    cooling paved surfaces 65–67
    infrared waves 60–61
    landscaping practices to mitigate 61–62
        cooling effect of plants 62–64
        green roofs 65
        shading air conditioners 65
urbanization 109, 125
urban landscape, wind in 81–82
urban pollution 82
urban runoff 134
urban soils 18
urban streams 105
urban water cycle 135, 145
Urban Wilds Initiative 44–45
urban wood 285, 286
US Environmental Protection Agency (EPA) 5, 95, 132, 134–136, 209, 221, 262
    and DDT 210–211
US Green Building Council (USGBC) 2, 5, 52

vegetation 53–54
vermicompost 192
Volumetric Ethanol Excise Tax Credit (VEETC) 247

Wasowski, A. 152
Wasowski, S. 152
waste management 279
waste products 162
waste stream 10, 14, 280
water conservation 9, 14, 109–110, 125–126
    drought and water shortage 110–111
    drought-tolerant plants 116
    gray-water use 121–122
        timing of 122–125
    irrigation and water-use efficiency 114–115
    mulch 119–120
    plant water requirements 111–112
    precipitation 110
    rainfall amounts calculation 114
    rainwater collection systems 112
        above-ground water storage 112–113
        below-ground water storage 113–114
    water-wise gardening 116–119
Water Conservation Garden (WCG) 119
water cycle 88, 105, 125
watering 228–229
water issues 9, 14, 88, 293
    bioremediation and phytoremediation 98, 99
    contaminated water prevention and treatment 97
    pesticides reduction in landscape 97–98
    polluted water 90
        contributions from urban areas 93, 95, 96
        excessive nutrients in 90–92

Index

herbicides 96–97
human and environmental effects of pesticides in 93–95
insecticides 95–96
pesticides in 92–93
polycyclic aromatic hydrocarbons 97
potable water 90
sources of 88–90
water cycle 88
wetlands and constructed wetlands 98–105
water management 9, 14
water pollution 14, 105
    pesticides 218
water quality 9, 14, 229
watershed 88
water shortage 110–111
water table 89
water-use efficiency 114–115
water-wise gardening 35, 116–119, 126
websites 297
weeds in turf 21
wetlands 98–99, 106
    components of 100
    nature, function, and value of 100
    plants in 104
    substrates in 104
    water in 103
whitetopping 66
Whitmore, W. H. 219
wildlife 38
    birds and mammals 39–45
    native pollinators 38–41
wind 293
wind and energy conservation 73–74, 83
    cold air trapping on slope 75
    cooling effects of 75–77
    planting for insulative properties 75, 76
    in urban landscape 81–82
    windbreaks 77–81
wind and energy efficiency 77–78
windbreak 77–78, 83
    designing 79
    energy usage in winter 78–79
    height of 79–80
    plants for 80–81
    shape and size 80
wind energy 252–253
windward 79
woody plants 22, 23, 47

Xerces Society for Invertebrate Conservation 39
xeric 116
xeriscaping 35, 116–118, 126
xerophytes 34

Zajicek, J. M. 193
zero-turn mowers 262
Zien, Steve 266

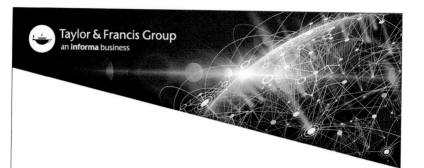